Birds of Lincolnshire

Lincolnshire Bird Club

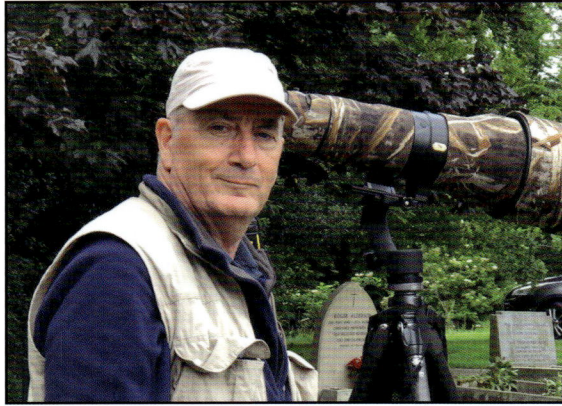

This book is dedicated to the memory of

Steve Keightley (1959-2021)

Founder member of the Lincolnshire Bird Club.
Lifelong Lincolnshire birder, former County Recorder, photographer and friend.

It is also dedicated to the birders of Lincolnshire, past and present, whose commitment to the recording and documentation of the birdlife of our "most brute and beastly shire" (Henry VIII, 1536) has made this enterprise possible, and also to the birders of the future who, we trust, will find this volume instructive and inspirational.

The bird records in this book have been verified by Lincolnshire bird authorities, BBRC, LBRC and the Lincolnshire County Recorder when preparing annual bird reports, as appropriate. Their accuracy has been checked as far as possible. If you are aware of any discrepancies or records that are not mentioned, please submit details to the Lincolnshire County Recorder. Contact details can be found at www.lincsbirdclub.co.uk or in the latest Lincolnshire Bird Report.

Published by Lincolnshire Bird Club, 28 Grimsby Road, Louth, Lincolnshire, LN11 0DY. More details about the Lincolnshire Bird Club can be found at www.lincsbirdclub.co.uk.

First edition published: 2021

ISBN 978-0-9538257-2-1

Printed by Swallowtail Print, Norwich

Recommended Citation

Casey, C., Clarkson, J.R., Espin, P. & Hyde, P.A. 2021. The Birds of Lincolnshire. Lincolnshire Bird Club. Louth.

Front cover: Pallas's Sandgrouse by Nik Borrow

ISBN 978-0-9538257-2-1

9 780953 825721

Contents

Introduction

On behalf of the Lincolnshire Bird Club (LBC) we are proud to present our update concerning the birds of Lincolnshire. The purpose of this book is to document the changes of the county's avifauna and to update our developing knowledge since the publication of the last complete treatment by Stephen Lorand and Keith Atkin (1989) *The Birds of Lincolnshire & South Humberside*. This update covers all records up to 2019 and a few significant records in 2020.

Throughout this book *The Birds of Lincolnshire & South Humberside* will be referred to as Lorand and Atkin (1989).

Lorand and Atkin (1989) is essential reading for anyone who has an interest in the county's birds and their history. This book carries on from where Lorand and Atkin (1989) left off, covering in the main from 1989 to 2019. After a huge amount of work validating all the records so that a reference is present, all accepted sightings details are listed in tables for rare species and many scarce species.

This book should be used in conjunction with the earlier work and, as that work is so complete and well researched, it has been decided not to revisit any of the early history. The main exception to this is where previously accepted records have, post Lorand and Atkin (1989), been declared no longer acceptable by the British Birds Rarities Committee (BBRC) which is the adjudication body for Britain's rarest bird species.

The records contained in this book cover the area of the historical county of Lincolnshire and throughout this book when we refer to Lincolnshire, we are referring to what the UK Government currently refer to as Greater Lincolnshire.

Greater Lincolnshire is made up of ten local authorities, the two unitary authorities of North Lincolnshire and North East Lincolnshire, together with Lincolnshire County Council which is made up of seven district councils (Boston, East Lindsey, North Kesteven, South Holland, South Kesteven, West Lindsey and the City of Lincoln).

The Local Government Act 1972, changed some addresses in the north of the county, and in response on Jul 1st 1974 the northern part of the county became the postal county of South Humberside.

In 1996 South Humberside became a "former postal county" and in that same year Humberside was abolished for local government purposes and the area south of the Humber became two unitary authority areas: North Lincolnshire and North East Lincolnshire.

The Greater Lincolnshire Nature Partnership (GLNP) was formed as a portal for the county to assist government departments, business and researchers with information on the wildlife of the county. The GLNP covers the area that LBC refers to as Lincolnshire. Building on the success of their predecessor organisation, the Lincolnshire Biodiversity Partnership, the GLNP was Government accredited in 2012 and is one of 48 Local Nature partnerships in England. The GLNP is primarily funded through a combination of Service Level Agreements, data searches and project funding. As an unincorporated partnership with no legal status one of the partners, Lincolnshire Wildlife Trust (LWT), provides the hosting body for the team known as the Lincolnshire Records Centre at Banovallum House, Horncastle.

LBC is one of the GLNP's main partners and under a data sharing agreement shares data on a regular basis for its use.

Since being computerised in 1998 it is estimated that LBC has dealt with over six million sightings records, many of which were duplicates. Once duplicates were removed, just over 2,625,000 sightings records were held in the LBC database up to and including 2019.

For completeness, when available, sightings records prior to 1998 have been added to the LBC database and at the time of print just over 20% of those sighting records relate to the earlier years. Many of these records are due to the excellent recording carried out at the Gibraltar Point Bird Observatory whose records date back to 1949.

The earliest record we have in the database is from 1783 when a Rose-coloured Starling was shot near Grantham.

The keeping of the county sightings records is a huge task and our thanks must go to the small army of observers and organisations who submit and share their records with LBC.

This book came about as a result of the hard work of a handful of people. After producing the Lincolnshire Bird Atlas from a box of 30 year-old parts, Colin Casey decided that during lockdown for Covid 19 he would see if it was possible to do an update to Lorand and Atkin (1989). After the lockdown had ended for most, a further 16 week enforced lockdown due to a health issue pushed him on and he persuaded the team who had worked on the *LBC Atlas* with him, along with a few others to comment and help.

Phil Espin, Chair of the LBC and compiler of the annual Rare Breeding Birds Report would write text and species accounts, Phil Hyde,

Lincolnshire County Recorder would validate the records and write and check the species write-ups and John Clarkson, respected photographer and former County Recorder, would find the images and proof-read the text.

Colin, LBC's Database Manager would produce all the information, make charts, export the sightings into tables, make the maps, decide which photographs would go where and assemble everything while asking the others what they thought; over a few short weeks a plan was put together.

This book was produced using Affinity Publisher, Wildlife Recorder and printed by Swallowtail Print, Norwich. The two Phils and John worked tirelessly to get things prepared and over to Colin so he could work on the book during his steroid induced sleepless nights. It was almost like a production line.

Graham Catley wrote a number of species accounts while Neil Smith and Steve Keightley helped by commenting on early page layouts and ideas.

Norfolk artist Nik Borrow kindly agreed to paint the stunning cover illustration for the Birds of Lincolnshire at short notice.

Others involved are mentioned in the Acknowledgements section.

In some instances, Colin's treatment stopped him from concentrating so something had to give. Readers will note some images are not tagged to credit the photographer's name. This was Colin's decision alone and he apologises for it, but the additional detailed work required to enter it would have been a step too far. However, all photographers are credited in the photographers section.

The amount of work put in by the team of four was staggering; this book was produced in record time, and the Lincolnshire Bird Club is very proud of it and would very much like to thank them all.

The records produced in this book are split into three main categories:

1. Rare species records confirmed by the British Birds Rarities Committee (BBRC). These species are listed where possible with all accepted records along with the date, site and other information where of interest.

2. Rare and Scarce records confirmed and accepted by the Lincolnshire Birds Records Committee (LBRC). These species have in some cases a list of all records along with the dates and location. There is also some other information.

3. Regular and Common species. These list status details and other information.

Recording sightings in the County

The number of recorded sightings since the publication of Lorand and Atkin (1989) has increased enormously due to easier transport links, more leisure time, the advent of computers, the internet, social media, bird information services and the consequent explosion of citizen science.

Almost everyone who goes to see birds now has a digital camera and good optical equipment, but some lack the knowledge of identification. Record gathering systems like the BTO BirdTrack or Cornell University's eBird are now open to anyone and the number of records has gone up. LBC is grateful to these bodies for sharing their data with the club. The data is generally collected for purposes different to county recording and the quality of those records can in some cases be less helpful. Over 40% of all the BirdTrack and eBird records do not have any sort of count whatsoever so the records are of less use towards the county records. Thousands of records are wasted because observers use site names such as "My Garden" or "Gravel Pit" which give no clue to their location.

From 2018 onwards even more problems were caused by the BirdTrack and eBird smart device applications, which both automatically select a site name from a device mapping system that was designed for use in car satellite navigation. This can give rise to site names such as "Unknown Road" or "Boston, PE21" with map references that can at times be over 20 miles away. These site names are too inaccurate to allow inclusion in LBC's database. Users of these devices should change the site name to something definitive or choose one from a list. LBC volunteers spend at least 60% of their time trying to sort out poorly entered site names and map references, the problem being that it's not possible to tell if the site name is correct and the map reference is not, or vice versa. As an example, in 2019 many records entered into one of these smart applications were found to be for Frampton Marsh, yet the app gave the site name as "Unknown Road, Boston, PE21" and the map reference was that of somewhere completely different, many miles away and we suspect the user may well have entered his records in a service station on the A1 on the way home.

The following chart shows the number of accepted records entered in the LBC Database for the 22-year period from computerisation in 1998 to 2019. About 40% of all records LBC receives are duplicates and in the main these duplicate sightings have been removed. A single rare bird present for a few days can generate several hundred sighting records and great care is taken to record the first and last date of rare and scarce sightings to reduce the number of records to a single entry.

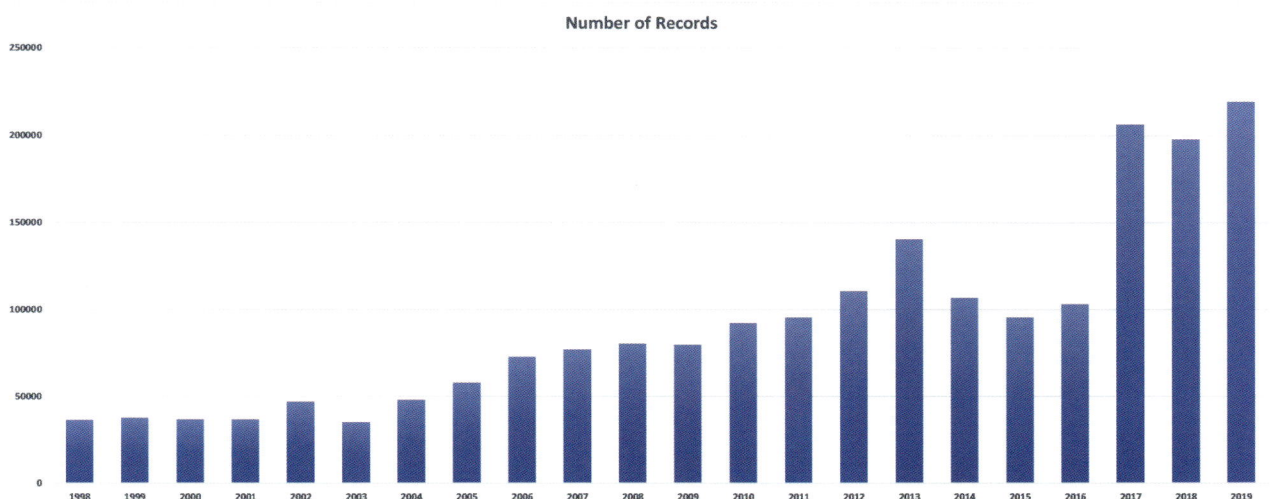

Number of Records

On an annual basis, once the records have been collated, they are entered on to a spreadsheet template to start the rather involved process of making them usable. The site names used by the observer are converted to the 890 'primary' site names the LBC uses. Errors are corrected and as far as is possible the records are then validated before being imported into the county database. The LBC has used the software program Wildlife Recorder to store its records since 1998 and have formed a close relationship with its writer, Jack Levene; currently LBC uses version 4.

Once these records have been imported into the database the information is exported on Excel spreadsheets and sent to what is a very small group of people that write up each species for inclusion in the annual Lincolnshire Bird Report.

Other than an 11-year period from 1997-2007, the LBC has produced a report for the county every year since its inception in 1979. The latest published report is for 2018. From 1997-2007 three summary reports covering 1997-99, 2000-02 and 2003-07 were produced. These reports provide valuable source material for this book.

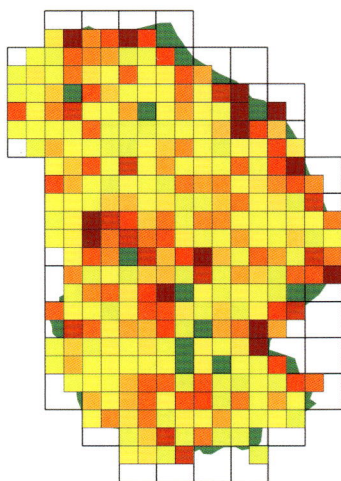

The map above shows the number of records sent in to the Lincolnshire Bird Club in each 5km square of the county, the darker the colour the more records were sent in for that 5km area.

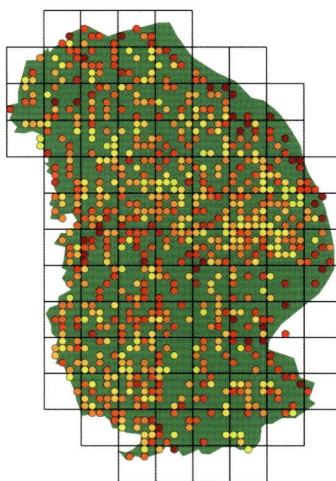

The map above shows the number of species seen in each 2km square (Tetrad), the darker the colour the more species were recorded in the 2km square.

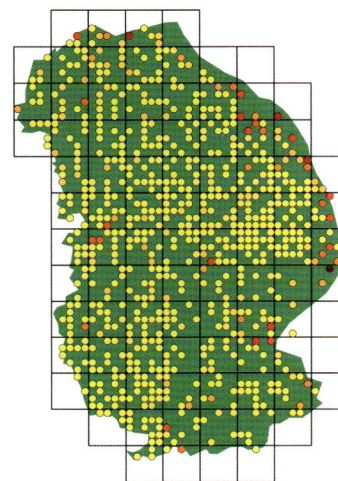

The map above shows the number of rare and scarce species recorded in each 2km square, the darker the colour the more rare species have been seen at that site,

Birds of Lincolnshire

It is fair to say that Lincolnshire is blessed with some of the finest farmland in Britain and is one of Britain's most agricultural counties. The hexagram map of predominant land use emphasises this point. It is reckoned that 86% of the county is given over to farming, much of it (74%) intensively farmed arable. Some of Lincolnshire's most characteristic birds like Grey Partridge, Turtle Dove, Lapwing, Skylark, Yellow Wagtail, Tree Sparrow, Linnet, Yellowhammer, Reed Bunting and Corn Bunting are all, of necessity, farmland specialists and all are Birds of Conservation Concern (BoCC) red data birds in Great Britain. They provide the bellwether for the health of the county bird populations.

While grassland makes up some 12.2% of the county, most of it is "improved" with the application of fertiliser and is just as intensively farmed as the arable; very little is Semi-natural Grassland/Marsh. The Wolds, an extensive area of rolling hills, was formerly a massive tract of open grassland which was only 25% cultivated up until the early 19th century, when it still supported droves of Great Bustards. Very little chalk downland remains. Our hay meadows all around the county are mostly gone and with them Whinchat and Corncrake. There are only 200ha of *Phragmites* reed bed in the county and extensive reed beds are mostly confined to the Humber bank area from Alkborough Flats to Killingholme Haven. However, the southern Fens are seeing a resurgence of reed swamp and wet Grassland at sites like Baston Fen, Frampton Marsh and Willow Tree Fen; though, as yet, nothing like the richness of the Fens, which had mostly been drained 200 years ago. The Lincolnshire Marsh still has areas of grazing land but they are generally not as wet as they need to be. Hopefully, the recently designated Lincolnshire Grazing Marshes Country Park between Chapel St Leonards and Sutton on Sea will address this over the coming years. Some DEFRA Higher Level Stewardship schemes such as Middlemarsh Farm near Skegness have already had an impact for species like Lapwing.

Lincolnshire is not famous for its Woodlands though some of them are unique to the county like the Lincolnshire Limewoods NR in the Bardney area. The county is 4.1% Woodland compared to 10% for England as a whole. Some of Lincolnshire's birds that are struggling the most like Lesser Spotted Woodpecker, Spotted Flycatcher, Willow Tit, Nightingale and Tree Pipit are Woodland species, as are two of our most recent losses as breeding birds, Hawfinch and Redstart. Even as far back as the 1870s, Cordeaux (1872) bemoaned the lack of ancient oak and beech Woodland in Lincolnshire that limited the range of Wood Warbler, which was lost as a breeding bird from its last location in the county in the north-west in the early 1950s. Even Song Thrush does relatively poorly in Lincolnshire compared to the rest of England according to the BBS and this is probably partly due to our relative lack of Woodland. Woodcock is increasingly confined to the west of the county. Crossbill is an irregular breeder and Siskin is rare. Having said that, there are a few surprises. Snipe Dales to the east of Horncastle is a fine Wolds valley Woodland that is managed to provide a mosaic of Woodland types and ages with glades. It is the last location where Willow Tit and Lesser Redpoll breed in the Wolds and indeed any part of south-east Lincolnshire.

Our Coastal habitats remain county jewels and cover 3.8% of the surface land area of the county. The estuary areas of The Wash (shared with Norfolk) and the Humber (shared with Yorkshire) are Britain's most important and sixth most important sites respectively for non-breeding waterbirds according to the annual report of the Wetland Bird Survey; Waterbirds in the UK 2018-19 published by BTO/RSPB/JNCC. The Wash holds some 380,000 waterbirds based on the five-year mean, twice as many as the next most important site, Morecambe Bay. The Humber Estuary holds around 140,000. Both sites are of international importance for an impressive range of migratory species. Sadly, the open sandy North Sea coast is too popular with holiday-makers and dog walkers as well as birders, which makes them much less valuable to breeding birds like Little Tern and Ringed Plover. Our coastal salt marshes, especially at the mouth of the Humber and The Wash provide valuable wintering habitat for a wide range of birds particularly Hen Harrier, Merlin, Short-eared Owl, Twite and Shore Lark as well as a host of waders and wildfowl. They comprise 6,393ha. Up until the 1980s they were threatened with drainage for agriculture and massive areas were lost over the previous hundreds of years. They are now protected and the previous "reclamations" are being rolled back under a policy of "managed retreat" which has created great coastal wetlands for birds at Alkborough Flats and Donna Nook on the Humber and Freiston Shore on The Wash.

Our Urban areas cover 3.5% of the county. In 2012, Lincolnshire had just over one million residents (1,046,900; 2012 Mid-year Population estimate) and had grown by 8.2% over the previous decade, this being slightly higher than the national rate of 7.7% over the same period. The county, however, is sparsely populated, and in 2012 it had an average of 150 people per sq.km compared with 411 per sq.km across England.

The population is characterised by a polycentric urban structure, with a small number of large towns, around 30 market towns, several coastal resorts, and then a fairly even distribution of villages across the county. The largest centres of population are the city of Lincoln (115,000), Grimsby with Cleethorpes (134,000), Scunthorpe (80,000), Grantham (44,000), and Boston (41,000). The city, towns and villages hold useful populations of Starling, House Sparrow and Swift. Increasingly, Herring and Lesser Black-backed Gulls are nesting on buildings in coastal towns such as Grimsby, Skegness and Boston and have started to spread inland to towns like Louth in the last five years.

Our Water bodies, comprising 2.4% of the county surface area, are primarily made up of rivers, drains, canals, flooded mineral workings like clay, gravel and sand pits and reservoirs, both hard and soft. These are sufficient to provide both breeding and wintering habitat for most British residents and visiting water birds which are generally well represented in Lincolnshire.

Our Sparsely Vegetated areas, including inland rock habitats and sandy dunes, are extremely limited, confined mainly to stone, sand and gravel quarries, they cover up to 0.08% of the county. Stone quarries provide a habitat that has been increasingly exploited by breeding Peregrines over the last 15 years. Our natural open sand dunes cover only 608ha, primarily along the north-east coast.

Heathland and Bogs are massively diminished; together with Scrubland they form just 0.03% of the county surface vegetation. Extensive parts of the north-west part of the county on the east side of the Trent between Scunthorpe, Brigg and Gainsborough, south to the west of Lincoln were formerly open sandy heaths with wet bog areas. These were lost to afforestation and industrial workings and, of course, agriculture from the latter part of the 19th century onwards. With them went Black Grouse and Stone-curlew. Places like Scotton Common, Manton Warren and Twigmoor Woods are remnants of what was once a major landscape in the county and survived intact as late as the 1920s. The Woodhall Spa area was similar but smaller. LWT Kirkby Moor NR provides a glimpse of what has been lost.

Scrubland is another habitat driven out by the intensification of agriculture and with its loss, birds like Willow Tit, Nightingale and Grasshopper Warbler get closer to the edge in the county. Some of our best Scrubland is on coastal dunes and provides some of the best birding sites in the county at migration hotspots such as, (from north to south) Donna Nook, Saltfleetby-Theddlethorpe and Gibraltar Point.

Redhill, The Wolds

8

Natural Regions

Throughout this book we refer to areas or natural regions which come from a system based on geographical and surface deposits proposed by J. W. Blackwood (1972) for general use in ecological surveys and this book broadly follows this system. As a result there are nine well-defined continuous geographical regions in Lincolnshire all with well-defined boundaries. The percentages mentioned below refer to the area of the surface of the county each natural region covers. This is depicted in a schematic map by tetrad, based on page six of the *Lincolnshire Bird Atlas 1980 to 1999: An Historical Perspective* (2020). The natural regions are described as follows:

Coast: the region between the Marsh, the Fens, the Humber, the North Sea and The Wash makes up 8% of the county. It is defined differently for geographical and habitat purposes hence the difference with the coastal habitat of 3.8% quoted above.

Isle of Axholme: the smallest region covering just 3% in the extreme north-west which lies to the west of the River Trent.

Trent Valley: this is a mainly lowland region of clays and gravel that extends along the western side of the county and covers 14%.

North Limestone Cliff: runs north from Lincoln to the Humber and consists of a narrow range of limestone hills which has a steep escarpment rising to over 80m to the west and a gentle eastern slope which covers about 6%.

South Limestone Cliff: runs to the south of Lincoln to the south-west corner of the county covering 15%, this limestone heath broadens towards Grantham where it is covered with boulder clay.

Central Vale: the area in the northern half of the county where a central value lies between the Cliff and the Wolds covering 10%.

The Fens: this region covers the whole of the south-east quarter of the county and covers the largest area at 25%.

The Wolds: a chalk upland region of rolling hills and river valleys rising to 168m, covering 11% of the county stretching from the Humber to the north edge of the Fens.

Marsh District: lies between the Wolds, Humber and North Sea coast and comprises 8% of the county from Barton-upon-Humber in the north to Skegness in the south.

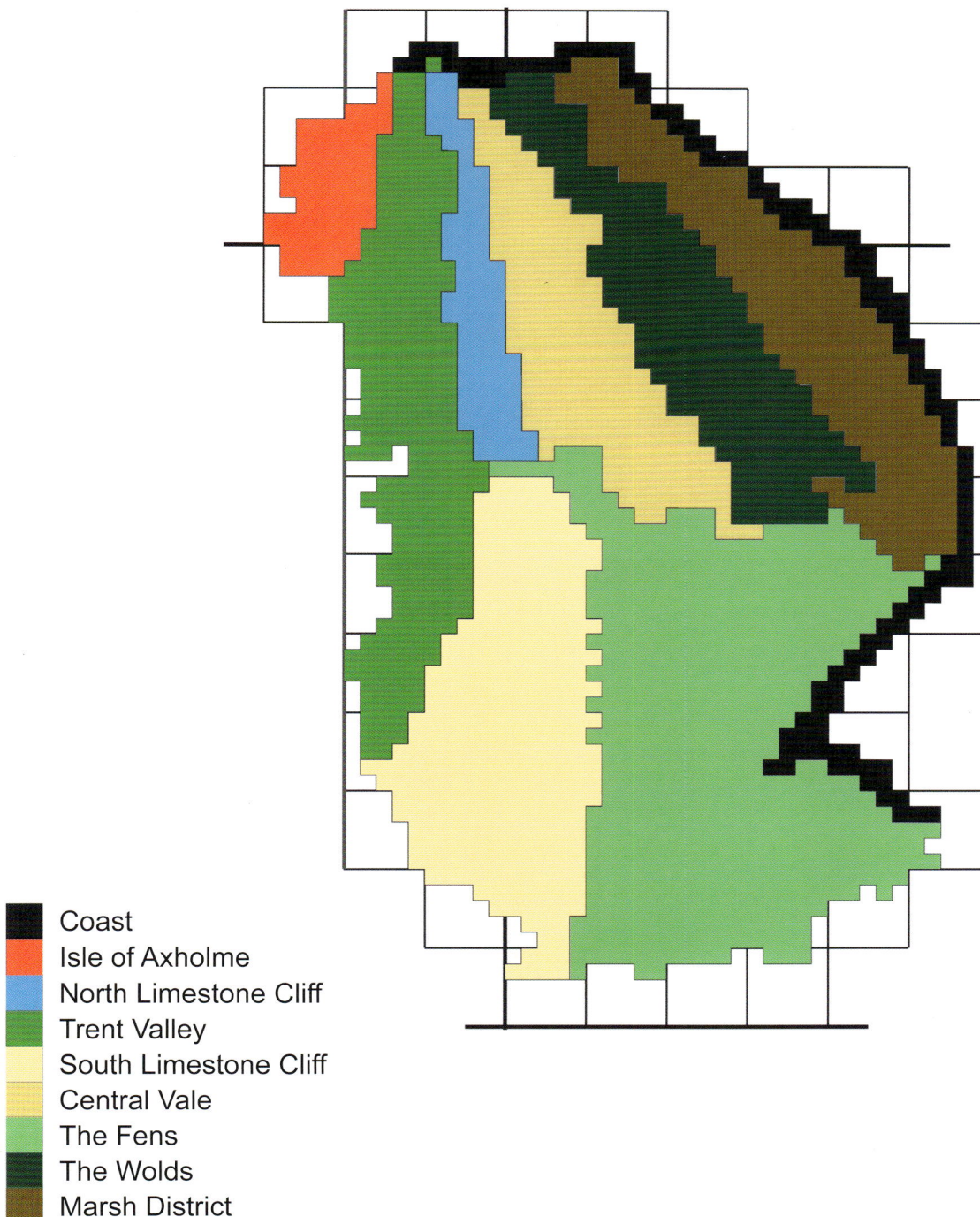

- Coast
- Isle of Axholme
- North Limestone Cliff
- Trent Valley
- South Limestone Cliff
- Central Vale
- The Fens
- The Wolds
- Marsh District

The Breeding Birds of Lincolnshire

The breeding birds of Lincolnshire have changed dramatically over the last 300 years. The drainage of the Fens, enclosures, the intensification of agriculture, quarries and mineral workings, urbanisation and the changing use of our coastline have all had a dramatic impact on the species that breed in Lincolnshire. We have had losses and new arrivals and Lorand and Atkin (1989) summarised the status of breeding county birds up to 1989.

Their book made the point that up until the publication of *The Breeding Birds of Britain and Ireland*, Sharrock (1976) (*BTO Atlas 1968-72*) when observers visited all 10 km squares in the county during 1968-72, there was no detailed knowledge of the distribution of breeding birds in the county. Until Lorand and Atkin (1989) the results of the *BTO Atlas 1968-72* were not summarised for Lincolnshire in a systematic way. Neither had the population of birds breeding in the county been rigorously estimated for all species. Lorand and Atkin pointed out that the *Lincolnshire Bird Atlas* project (a detailed investigation of the distribution of the county's birds based on tetrads, that ran from 1980-89) was intended to address that gap in our knowledge of Lincolnshire breeding birds. The publication of the *Lincolnshire Bird Atlas 1980 to 1999 An Historical Perspective*, (*LBC Atlas*) Lincolnshire Bird Club (2020) provided population estimates for the late 1980s based on the Atlas project itself, and additional information in the *New Atlas of Breeding Birds in Britain and Ireland:1988-1991* (*BTO Atlas 1988-91*), Gibbons, Reid and Chapman (1993).

The purpose of this section is to update our knowledge of breeding birds in Lincolnshire using information available up to Dec 31st 2019 and, in one or two exceptional cases, Sep 30th 2020. An analysis has been done of all species confirmed or suspected to have bred in Lincolnshire since records became available. These go back to Lincolnshire's first birder with a notebook, St Guthlac, a monk who resided at Crowland Abbey in the late 7th century. See Lorand and Atkin for a detailed history of Lincolnshire bird records. To date 160 BOU category A and C species have been confirmed as breeding in Lincolnshire and four others have been recorded showing signs of breeding short of confirmation of successfully rearing young. The latter four are Black-winged Stilt, Glossy Ibis, Marsh Warbler and Firecrest.

The analysis takes data for each breeding species from the first *BTO Atlas 1968-72*, summarised in Lorand and Atkin, the *LBC Atlas* covering 1980-89 and the *BTO Atlas 2007-11* and summarises their distribution and these categories have been assigned colour codes for inclusion in the species accounts as follows:

Very widespread species:		breeding in over 90% of Lincolnshire 10 km squares
Fairly widespread species:		breeding in 50-89% of Lincolnshire 10 km squares
Fairly local species:		breeding in 10-49% of Lincolnshire 10 km squares
Very local species		breeding in less than 10% of Lincolnshire 10 km squares
Irregular breeding species		species breeding irregularly in the last 50 years
Not recorded		species not recorded in the period
Former breeding species		species that have not bred in the county for over 50 years.

In order to bring the status of each species up to date the status of each species over the period from 2016 to 2019 has been analysed on the same basis using information available from BTO BBS, RBBP and reports to LBC published in Lincolnshire Bird Reports and held pending publication. A metric of change over the last decade has also been included as outlined below.

The BTO kindly provides LBC with a summary of BBS data each year from all BBS squares surveyed in the historic county of Lincolnshire. In order to look at the recent change in status of breeding birds the average incidence (the percentage of BBS squares visited in Lincolnshire in which each species was recorded) of each species was computed for the period of the *BTO Atlas 2007-11*. The same figure was computed for each species for the years 2016 -19, the latest year for which published data is available. The number of BBS squares covered in each year of the *BTO Atlas 2007-11* period was 63, 64, 64 and 61 respectively and in the recent period 82, 86, 77 and 74 respectively.

The advantage of this method is that BBS squares are randomly selected and incidence generally gives a good approximation of the distribution of each species throughout the county, though the method is of less use for scarcer wildfowl, waders, gulls and nocturnal species. These limitations have been addressed by using information from the other sources mentioned above and judgements have been made on the basis of them. The BBS incidence figures quoted run from 0 to 100%. The incidence change varies from -100% (Ruddy Duck extirpated) to +534% (the successful colonisation of Cetti's Warbler).

For each breeding species a summary strip has been produced and is illustrated by this example for Sparrowhawk.

Status Lorand and Atkin (1989) 1968-1972	Status LBC Atlas 1980-1989	Status BTO Atlas 2008-2011	Current Status 2016-2019	BBS Incidence 2016-2019	BBS Incidence change 2007-2011 to 2015-2019
				8%	-15%

In the *BTO Atlas 1968-72* period Sparrowhawk was still very local in Lincolnshire, its population having been knocked right back by agricultural pesticides in the late 50s early 60s. The population has recovered and the species has increased from fairly widespread to very widespread over the last 30 years. Sparrowhawk, though widespread, exists at fairly low density occurring in 8% of BBS squares visited between 2015-19. The fall in average incidence of -15% is suggestive of a recent decline.

The individual summary strip for each species is not always interpreted in the text account. For those species where BBS figures are not available, the final two columns have been omitted from the species accounts.

It should be understood that these percentage changes are purely Lincolnshire focused and are differently based to the much more accurate and statistically significant figures published in the annual BTO BBS report of annual and long-term population changes for British breeding birds. Where a species occurs in more than 30 squares in a region the BTO is able to calculate a regional index of the population changes for such species.

The LBC is grateful to the BTO for annually providing such index computations for the historic county of Lincolnshire for 29 of Lincolnshire's most widespread species.

Where such figures are held and have statistical significance, they are quoted in the individual species accounts. Graphs of the changes for selected species are published each year in the Lincolnshire Bird Report and for that reason are not reproduced systematically in this book.

Each year the *Lincolnshire Bird Report* also summarises the annual population changes of all red data Birds of Conservation Concern (BoCC) breeding in Lincolnshire on the inside back cover.

In summary the number of species breeding in Lincolnshire in each of the five categories from irregular to very widespread in each period examined was as follows:

Distribution Category	Status Lorand and Atkin (1989) 1968-1972	Status LBC Atlas 1980-1989	Status BTO Atlas 2007-2011	Current Status 2016-2019
Very widespread	41	51	55	54
Fairly widespread	26	23	19	19
Fairly local	30	29	29	31
Very local	20	12	23	20
Regular: sub-total	117	115	126	124
Irregular	18	6	10	17
Total	135	121	136	141

Over the last 50 years the table reflects an increase from 117 to 124 in the number of species regularly breeding in Lincolnshire. The overall total when including irregular breeders is currently 141 species.

Looking at the 124 species currently regularly breeding in Lincolnshire the breeding distribution pattern of 58 species shows no change, 48 species have increased their range since 1989 and 18 have decreased. Sixteen species breed irregularly in Lincolnshire and four have disappeared as breeders since 1989, (Whinchat, Hawfinch, Ruddy Duck and Redstart). A further 19 species that have bred previously in Lincolnshire no longer do so, as listed in the table below.

Species Name	Former Status	Last bred
Crane	Regular	1560s*
Spoonbill	Regular	1650s
Great Bustard	Regular	1810
Black Tern	Regular	1845
Merlin	Irregular	1850s
Hen Harrier	Regular	1872
Dunlin	Irregular	1890s
Stone-curlew	Regular	1904
Wryneck	Regular	1920s
Black Grouse	Regular	1930s
Scaup	Irregular	1944
Sandwich Tern	Irregular	1946
Wood Warbler	Regular	1951
Pied Flycatcher	Regular	1950s
Black-tailed Godwit (European)	Regular	1974
Red-backed Shrike	Regular	1978
Kentish Plover	Irregular	1979
Common Sandpiper	Irregular	1979
Arctic Tern	Irregular	1980s

(* See information below)

On the positive side, of the 48 species that have increased in range, 14 are "new" breeders. These 14 are listed below in order of increasing success. Seven are re-colonists having formerly bred: Black-necked Grebe, Red-crested Pochard, Raven, Red Kite, Marsh Harrier, Avocet and Buzzard. We have also had seven successful colonists: Cormorant, Mediterranean Gull, Herring Gull, Lesser Black-backed Gull, Peregrine, Cetti's Warbler and Little Egret.

Species Name	Status Lorand and Atkin (1989) 1968-1972	Status LBC Atlas 1980-1989	Status BTO Atlas 2007-2011	Current Status 2016-2019
Red-crested Pochard	[black]	[black]	[orange]	[orange]
Black-necked Grebe	[black]	[black]	[orange]	[orange]
Avocet	[black]	[white]	[yellow]	[yellow]
Mediterranean Gull	[white]	[white]	[red]	[orange]
Herring Gull	[red]	[red]	[orange]	[yellow]
Cormorant	[white]	[white]	[yellow]	[yellow]
Marsh Harrier	[red]	[yellow]	[green]	[green]
Lesser Black-backed Gull	[white]	[white]	[orange]	[yellow]
Little Egret	[white]	[white]	[yellow]	[yellow]
Red Kite	[black]	[white]	[yellow]	[yellow]
Peregrine	[white]	[white]	[yellow]	[yellow]
Raven	[black]	[white]	[orange]	[yellow]
Cetti's Warbler	[white]	[white]	[green]	[yellow]
Buzzard	[black]	[white]	[green]	[green]

One notable event in 2020, which is not included in the analysis above, deserves to be mentioned. Common Crane, a species that has not verifiably bred in Lincolnshire since the 16th century and has shown no signs of doing so, nested at the LWT reserve Willow Tree Fen. A pair successfully raised a single chick under the watchful eyes of a team of volunteer wardens. The birds had perhaps been tempted over from neighbouring Cambridgeshire or Norfolk and encouraged to breed by the peace and quiet engendered from the reserve being closed to public access due to the Corona virus pandemic. This is a location which 10 years previously had been arable fields, purposely rewilded to recreate fenland. It gives hope that other species long extinct as breeding birds in Lincolnshire could be encouraged to return. The fenland bird community formerly included Spotted Crake, Spoonbill, Black-tailed Godwit, Ruff and Black Tern. All these species might realistically return with suitable habitat creation along with Corncrake which is already on the cusp of returning.

On the negative side 18 species have decreased in range and these are all listed below in order of least to worst reductions. The most spectacular has been Lesser Redpoll which plunged from very widespread to very local in less than 20 years. Some species teeter on the edge of extinction as Lincolnshire breeders, notably Nightingale and Tree Pipit, while Montagu's Harrier may have gone already, never having had a real stronghold here. Other formerly widespread species like Turtle Dove, Willow Tit and Snipe have suffered alarming declines but might still be saved while the prospects for Corn Bunting, Cuckoo and Pochard also look bleak and action is needed if we are to avoid losing them.

Species Name	Status Lorand and Atkin (1989) 1968-1972	Status LBC Atlas 1980-1989	Status BTO Atlas 2007-2011	Current Status 2016-2019
Pochard	yellow	orange	orange	orange
Nightjar	yellow	orange	orange	orange
Cuckoo	dark green	green	green	green
Montagu's Harrier	orange	orange	orange	red
Willow Tit	green	green	yellow	yellow
Sand Martin	green	yellow	yellow	yellow
Grasshopper Warbler	green	green	yellow	yellow
Garden Warbler	green	green	green	green
Nightingale	green	green	orange	orange
Tree Pipit	green	green	orange	green
Bullfinch	dark green	green	yellow	green
Corn Bunting	dark green	green	green	green
Turtle Dove	green	green	yellow	green
Snipe	green	green	orange	orange
Short-eared Owl	yellow	yellow	orange	red
Spotted Flycatcher	dark green	green	green	green
Black Redstart	orange	yellow	red	red
Lesser Redpoll	green	green	orange	yellow

Population estimates of Lincolnshire Breeding Birds

Population estimates of breeding species in Lincolnshire were made in the *LBC Atlas* for the late 1980s. These were retrospective estimates made in the late 1990s based on the data available at the time. No other author has attempted to estimate Lincolnshire's breeding bird populations since. The latest *BTO Atlas 2007-11* did not make UK bird population estimates. This task is undertaken by the Avian Population Estimates Panel (APEP). Their latest estimates, (APEP4) were published by Woodward et al 2020 *Population estimates of birds in Great Britain and the United Kingdom in British Birds* 2020 113, pp. 69-104. They estimated breeding and wintering populations of 251 UK bird species for the year 2016.

Theoretically it should be possible to approximate Lincolnshire's share of the UK population of each breeding species by making adjustments and applying a ratio based on BBS counts of the total number of birds counted in Lincolnshire in 2016 as provided to LBC by BTO, divided by the total number of birds counted in the UK in 2016 as published on BTO's website.

An alternative method would be to take the population estimates made in the *LBC Atlas* and adjust them using the change in Lincolnshire bird numbers recorded by the BBS since 1994. One difficulty with this approach is the data is only statistically significant for 29 species. Also, there is a gap between the late 1980s and 1994. What change occurred in that five-year period?

Both approaches have been tried and it is difficult to get accurate converging estimates using these two different methods. The BBS while doing an excellent job only samples around 1% of Lincolnshire's close to 7,000 1km squares. The *LBC Atlas* visited all tetrads, 1,835 of them, providing a sounder basis for a much more accurate assessment of the population of each species.

Taking the example of Blackbird; Woodward *et al* estimated the UK population in 2016 at 5.05 million pairs. BBS data suggests that in that year 2.1% of UK Blackbirds were counted in Lincolnshire. Multiplying the two together derives a Lincolnshire population of 105,000 pairs.

However, using the data available for the late 1980s, the Lincolnshire Blackbird population was estimated in the *LBC Atlas* at 200,000 pairs. One might think at first sight that this is evidence that the population has fallen by half. But the annual BBS survey indicates that between 1994 and 2017 the Lincolnshire Blackbird population that stood at 200,000 in the late 80s increased by some 10%. That suggests the Lincolnshire population should be around 220,000. Squaring that with an estimate of 105,000 is not easy. Especially when one considers the APEP methodology is based on scaling forward from the *BTO Atlas 1988-91*, which is what the *LBC Atlas* estimate was partly based on.

A gut feel would suggest the 220,000 figure is much closer to the truth than 105,000, but gut feel does not seem like the best way to proceed. As a result, current Lincolnshire population estimates have not been included in the species accounts in the Systematic List where they seem to be out of line with other data. Estimates for 2016 made using APEP4 are mentioned for most species. The *LBC Atlas* population estimate is mentioned where available and an indication of the scale of change is given.

LBC makes annual reports to the Rare Breeding Bird Panel (RBBP) for species covered by that body. Reports are based on data collected annually. Although these are not always as complete as one might hope, they provide the best local estimate of population available for RBBP species. In the species account of RBBP species the average figure of breeding pairs over the five years to 2017 is cited, the latest year for which published figures were available at the time of writing.

Important sites in the County

Part of understanding the county avifauna is seeing where birds occur and why. The earlier section (pp. 6-9) covered the habitats and natural areas of the county. Here is where we get to the nitty-gritty of what sites have the most birds. Lincolnshire is England's second largest county and splitting it into areas is not a straightforward task. In the end we have gone for simplicity; splitting the county into six regions based on the OS grid system as shown on the map below. These have been imaginatively named North-West (NW), North-East (NE), West (W), East (E), South-West (SW) and South-East (SE). The LBC bird database has been described in the section named recording sightings in the county and one advantage of Wildlife Recorder is the ease with which a list of the birds for each site can be produced. Each region sub-section starts with a table of all sites in that region for which we hold records of more than 90 or more species. The database holds records for 890 sites of which 184 have recorded 90 species or more.

Each table shows a grid reference for the site, how many records are held in the database for the site, the total number of bird species that have been recorded at that site and the total number of rare and scarce species that have been recorded at that site. All of the 184 sites have been ranked in the order of the number of species recorded and the number of rare and scarce species recorded. They have been ranked in order from the largest number of species to the lowest. The rank appearing in the table is the overall rank not the regional rank.

It should be noted that we only have the records that have been submitted to us, shared with us or collected from public sources. If you think we are light on records for a particular site we are always happy to receive more and update our database.

The table is supported by a map showing the location of each site in the table and a text account of the key sites in the region. It should be noted that this section is not intended to be a detailed site guide. It is primarily to assist the reader in understanding where the locations mentioned in the text are, and how to find them. This selective approach means that in the narrative we have only mentioned sites where access is straightforward for the general public.

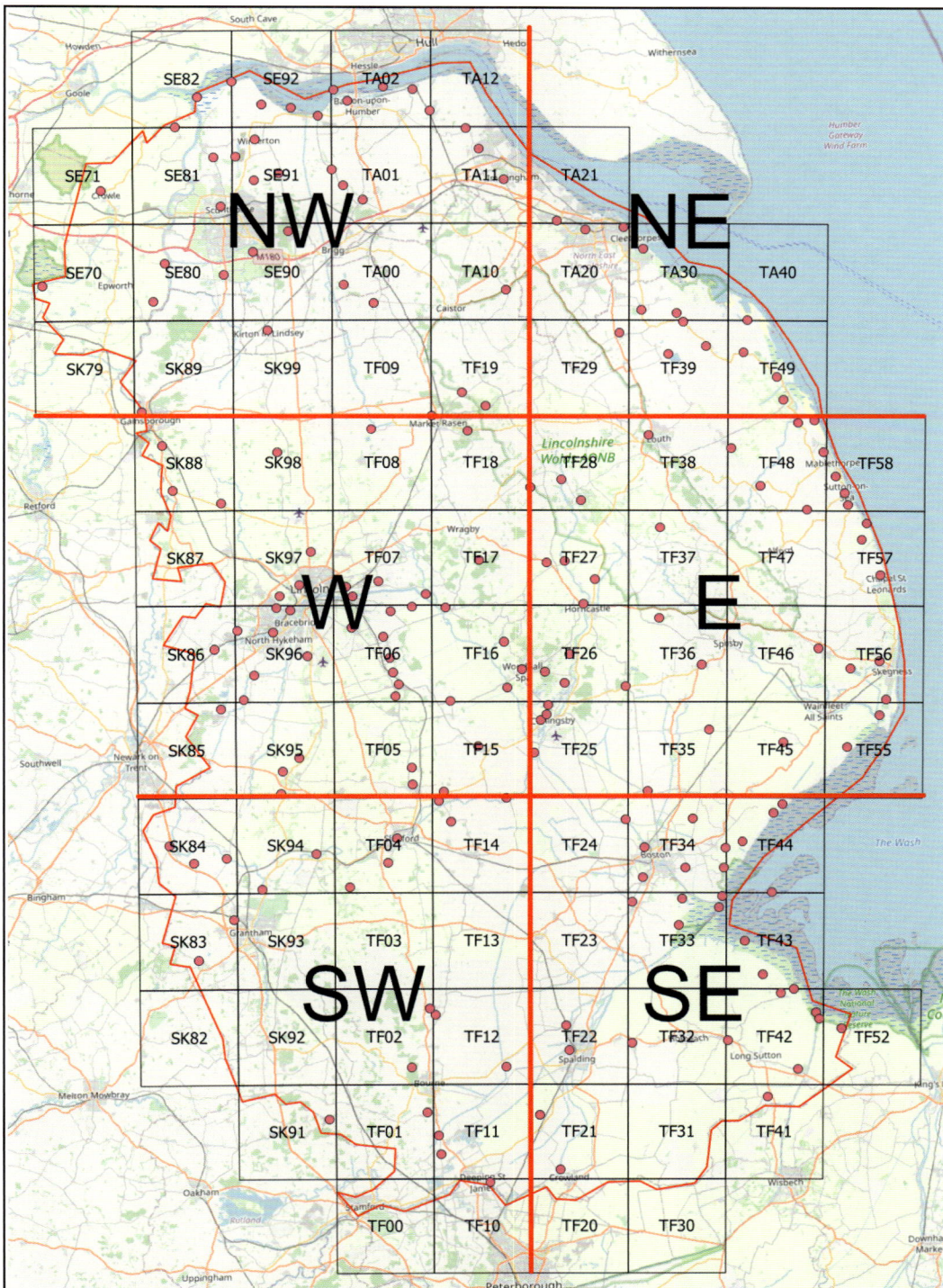

North-West

Rank	Site Name	GR	Records	Species	Rare and Scarce
6	Barton Pits	TA0023923819	45,290	253	25
10	Alkborough	SE8649723068	28,194	229	19
17	Barton upon Humber	TA0157122713	1,321	208	9
20	Messingham	SE8911604721	16,890	204	14
31	Wroot	SE7081403643	5,451	186	1
32	Goxhill	TA0985321686	7,111	182	11
34	East Halton	TA1352419892	3,791	177	2
37	Killingholme	TA1483917742	5,075	173	11
52	Scunthorpe	SE8883711782	2,861	154	7
57	Worlaby (Brigg)	TA0112513944	3,778	148	2
64	Bagmoor	SE9035916975	1,716	144	3
67	Barrow Haven	TA0524524151	2,952	139	2
76	South Ferriby	SE9861721169	2,536	134	6
80	Reads Island	SE9591122004	4,468	132	9
87	Kirton in Lindsey	SK9350698941	4,353	129	2
89	Whitton	SE9001924714	1,294	128	1
91	Winteringham	SE9295722367	1,031	126	3
96	West Butterwick	SE8318705937	493	124	1
97	Bonby	SE9999815578	1,223	123	1
101	Laughton Forest	SE8200201999	2,097	121	2
104	Normanby	SE8813216910	3,147	121	
106	Fockerby	SE8426020044	583	120	1
109	New Holland	TA0814323917	688	119	1
115	Crowle	SE7672813461	695	116	1
122	Broughton	SE9562209210	454	112	1
129	North Kelsey	TA0422001735	792	110	
133	Winterton	SE9232818760	443	107	
136	Appleby	SE9468715168	348	106	1
137	Cadney	TA0118503706	721	105	
139	Holme	SE9206607114	4,372	104	
157	Swallow	TA1754403150	1,964	97	1
170	Risby Warren	SE9222714474	606	92	
173	Elsham	TA0310412429	706	91	

This region is extremely diverse ranging from the highest point in the county at Normanby le Wold which at 168m above sea level has views across the Central Vale of the Ancholme Valley, the North Limestone Cliffs and Trent Valley to the flatlands of the Isle of Axholme. It is the largest of the six regions but quite possibly has poor general birding coverage, as only 6% of Lincolnshire Bird Club's 350 members live in the region. The whole of the Humber bank from Alkborough in the west to Immingham in the east has excellent birding sites.

The best site in the region are the flooded clay pits immediately behind the Humber bank that surround Barton upon Humber which have increasingly come under the management of LWT over the last 60 years. With the sixth largest bird list in the county the Far Ings NR is the flagship reserve of LWT in the north of the county and has excellent facilities which include a visitor centre, toilets, hides and well marked trails. The reserve has two car parks, by the Visitor Centre at TA018233 and further to the west near the best hide to see Bittern in the county at TA011229. Not every pit is in the ownership of LWT and visitors should respect signs regarding access. Just to the east of the Humber Bridge within Barton itself is the Water's Edge Country Park with similar habitats and parking at TA029231.

Barton Pits

North Killingholme

The up-and-coming site in the north-west is Alkborough Flats, a flood relief scheme at the confluence of the Trent and the Humber which was formed in 2006 and is managed by The Environment Agency. An extensive wetland of some 450ha, with lagoons, grassland, marsh and reedbeds it has the 10th largest bird list in the county and one of the fastest growing. There are two excellent hides, a marked trail and car parks at SE878220 in which disabled drivers have priority and another at the top of the hillside at SE886223.

The area around Scunthorpe, known as the "Coversands" was once an almost desert like landscape, but very little of the sandy open habitat survives and none of the interesting remnants like Twigmoor Woods, Manton Warren and Laughton Forest have easy access. The easiest to access habitats in this area are the extensive former mineral workings to the east of Messingham, now the Messingham Pits Nature Reserve of LWT. These form a mosaic of freshwater, woodland and grassland habitats. The reserve has good trails but is generally under-developed though there are hides. The car park is accessed from SE908032.

Messingham Pits

North-East

Rank	Site Name	GR	Records	Species	Rare and Scarce
2	Donna Nook	TA4197100035	83,049	277	50
7	Tetney	TA3121001082	46,621	247	25
11	Covenham Reservoir	TF3398996473	96,253	223	19
12	Humberston	TA3137407376	33,499	223	4
13	Cleethorpes	TA2941709593	14,316	217	6
21	Marshchapel	TF3544099888	3,748	204	6
26	Grimsby	TA2556509389	9,828	192	7
28	Grainthorpe	TF3774897271	6,818	191	8
35	North Cotes	TA3475900737	828	176	22
60	Immingham	TA1733914554	4,839	147	1
86	North Somercotes	TF4154096682	820	130	5
98	North Thoresby	TF2899298684	1,207	122	1
184	Great Coates	TA2268710260	790	89	5

The North-East is effectively the outer estuary of the River Humber to its mouth and the land that lies to the south of the coast rising from the Marsh into the Wolds. It is the most industrially developed part of the county and houses the large urban conurbation of Grimsby and Cleethorpes. Having said that it has a superb coastline and an attractive agricultural hinterland. Donna Nook is the best site in this region though the entire shoreline has lots of birding potential. It has to be said, and the locals mention it regularly, that the comfort of birders is not well catered for in the North-East. Hides are like hens' teeth and visitors must be prepared for long walks and no public conveniences; there are no visitor centres and very few marked trails. There is no information to be had on the ground apart from talking to other birders so visitors need all their communication devices primed and be prepared to get stuck in and find their own birds. The contrast with Spurn, which can usually easily be seen just across the estuary, could not be starker.

Donna Nook is a National Nature Reserve mainly in North Somercotes stretching 8km along the coast from Pye's Hall in the north to Howden's Pullover in the south. The easiest access is from a public car park at Stonebridge (TF420998). To the north lies a recently enclosed "managed retreat" area of former farmland. The sea wall has been breached at Pye's Hall to allow the tide in to flood the land to create new mudflats and saltmarsh. The area immediately to the south houses the control centre for the Ministry of Defence bombing range that lies along much of the foreshore. There are public footpaths along the foot of the dunes and visitors must respect the warning signs and stay out of certain areas when red flags are flying, usually Monday to Thursday and Friday morning. Certain areas are private especially around Ponderosa and the Coastguard Cottages and signs should be respected. The closure of the bombing range has allowed a very successful Grey Seal breeding colony to develop.

Donna Nook, Pye's Hall

Donna Nook to Pye's Hall

Donna Nook

The Grey seal herd which is several thousand strong is present all year round but pupping at the end of Oct to the end of Dec takes place at the foot of the dunes. It is a wildlife spectacle to compete with anything in Life on Earth and has become a popular tourist spectacle which can make access awkward due to traffic during the seal breeding season. Best to arrive early.

Donna Nook Grey Seal Pup

Quad 3 is located to the south of the bombing range control station by the southern most range tower. It is a productive area of dunes and marsh. It is accessible by foot and easiest to reach from the Howden's Pullover car park to the south at TF445958.

Moving north along the coast Tetney is another excellent and massive site, managed by RSPB but not developed for visitors. It can be accessed from the north from roadside parking at TF344024 and walking along the north bank of Louth Canal until the seawall is reached. Alternative access from the south is by walking north from Horse Shoe Point at TF381018. In between these two points is the historic (for birders) North Cotes point where several species were added to the British list in the days of "shoot 'em up" birding. The area to the south of Horse Shoe Point is Grainthorpe Marsh, a fantastically attractive area of mature saltmarsh. The whole foreshore is dangerous on a rising tide and there is a real risk of being cut off and drowned if basic commonsense and checking tide times is not employed.

Just to the north of Tetney lies Humberston, with the pools at The Fitties providing all kinds of interest. Access is from North Sea Lane, Humberston and a drive through caravan sites and an interesting 1930s chalet park to a public car park next to the Humber Yacht Club at TA337050. It is possible to walk 8km north from here along the shore to Grimsby Docks or 20km south to Saltfleet. The whole coast should soon be accessible along the new national footpath which is currently being developed.

Cleethorpes foreshore is superb for waders on migration in winter where parking is possible anywhere along the promenade. Pyewipe Marsh at Grimsby, set in the midst of an industrial landscape is another great location for waders and wildfowl and can be accessed from roadside parking at TA261105. Just inland the remains of Freshney Bog along the River Freshney, now known as Town's Holt, provides an interesting wetland contrast to the industrial coast. Access from Great Coates bridge with parking by the church at TA241090.

Covenham Reservoir lies a few miles inland from the coast and can be easily accessed from the coast road A1031 or the A16. Parking is in the north-west corner next to the sailing club at TF340962. It is a concrete embanked reservoir built in the early 1970s. The embankment is 20m higher than the surrounding Marsh and provides fantastic views to the coast and Wolds as well as over 200 acres of water that attracts a massive range of species. The perimeter walk is nearly 4km. Dense clouds of flies can be a hazard on warm sunny days throughout the year.

West

Rank	Site Name	GR	Records	Species	Rare and Scarce
23	Whisby	SK9018567417	47,531	203	7
25	Boultham Mere	SK9456471029	60,229	193	4
33	North Hykeham	SK9387867198	9,396	180	6
40	Toft next Newton	TF0393288507	13,614	171	2
42	Bardney	TF1138369778	6,558	170	7
43	Lincoln	SK9659172149	18,764	167	4
46	Bassingham	SK9085560230	2,948	163	1
47	Nocton	TF0571964563	8,467	160	1
49	Marton	SK8379182145	7,166	158	2
54	Woodhall Spa	TF1920663344	5,864	153	1
63	Fiskerton	TF0457572551	6,121	145	3
65	Anwick	TF1119850631	31,183	140	
72	Fillingham	SK9441286094	5,376	136	1
75	Aubourn	SK9195062821	2,955	135	
78	Sleaford	TF0643645760	5,739	133	1
84	Metheringham	TF0665661846	4,112	131	1
85	Norton Disney	SK8847059306	2,558	131	
94	Ruskington	TF0796951347	5,453	125	
95	Waddington	SK9731564808	2,397	125	
99	Lea	SK8277886812	426	122	1
100	Blankney	TF0626960591	2,475	122	
105	Branston	TF0183467673	16,769	120	1
112	Potterhanworth	TF0508366741	1,155	118	
114	Kirkstead Bridge	TF1765461436	1,970	116	1
116	Dorrington	TF0790253137	4,744	116	
121	Fulbeck	SK9458250442	1,697	112	1
124	Benniworth	TF1999282374	3,136	112	
140	Chambers Farm Wood	TF1477274699	3,028	103	2
141	Hartsholme	SK9418969731	5,279	102	1
142	Gainsborough	SK8076590494	516	102	1
143	Martin	TF1186260106	1,230	102	
145	Willingham Woods	TF1369188363	990	101	2
148	Welbourn	SK9645254163	3,704	100	1
149	Riseholme	SK9770375566	949	100	1
151	Branston Island	TF0940071179	570	99	1
153	Tealby	TF1550391034	2,112	99	
156	Leadenham	SK9474352793	954	98	
159	Sturton by Stow	SK8868680820	1,013	96	
160	Greetwell (Lincoln)	TF0124371922	777	96	
161	Billinghay	TF1468855320	776	96	
162	Market Rasen	TF1000090002	582	96	
164	Washingborough	TF0197970914	778	95	2
172	Stixwould	TF1733866158	319	91	1
178	Boultham	SK9565069493	2,959	90	
181	Branston Booths	TF0584069373	879	90	

The West is dominated by the City of Lincoln and the region as a whole contributes 30% of the membership of LBC. Lincoln provides surprisingly good birding with a range of good sites around the town. Whisby Nature Park is the principal site. It consists of a series of former sand and gravel workings of a kind that abound to the west and south of Lincoln. To the north east lies Market Rasen surrounded by extensive afforested former heathland centred on Willingham Forest. In between lies the upper valley of the Ancholme which is home to Toft Newton Reservoir, an attractive large waterbody in the midst of open countryside. Lincoln lies on the River Witham and the Fens extend right up into the city. Downstream there are attractive areas including Fiskerton, Branston and Nocton Fens. East of these Fens lies the unique Lincolnshire Limewoods of which Chambers Farm Wood is a prime example.

Boultham Mere

Whisby Pits

Whisby Nature Park is easy to find as it lies astride the Lincoln western bypass the A46. The main car park is off Moor Lane at SK910661. This well managed reserve has all the facilities one might wish for and it is popular with local people. Ample parking, well marked trails, a Visitor Centre, hides, cafe, toilets and shop. It is run by LWT in conjunction with Lincoln City Council. Immediately adjacent to the east is Apex Pit, the largest area of open water around the city which is always worth checking out. Access is from the Recreation Ground car park on the A1434 at SK936664. Just to the north closer to the city centre lies Hartsholme Park, an attractive area of woodland and lakes which adjoins Swanholme Lakes. Park in the main car park off Skellingthorpe Road at SK948696 and walk through both sites. Boultham Mere, marked on the OS map as "Ballast Holes" and known locally as the "Ballys" is a great local nature reserve with a small dedicated band of observers. It is awkward to access along Coulson Road. Park in Morrison's Car Park on Tritton Road and walk west. The hide is on the far western side.

Toft Newton reservoir is a complete contrast, in the middle of nowhere close to the village of Toft next Newton. It has no facilities other than roadside parking at TF039874. Follow the track and walk around the concrete perimeter. The Reservoir is owned by Anglian Water who let it to a company that runs a trout fishery so there is sometimes disturbance from boats. Permission to enter should be asked of the site manager, if possible.

Willingham Forest is a massive area owned by the Forestry Commission which has parking, a cafe and toilets. The car park lies to the east of Market Rasen off the A631 at TF132885. There is an extensive network of woodland trails to explore. Close by is the attractive LWT Linwood Warren Nature Reserve which is accessed from roadside parking on Legsby Road at TF133876. It has marked trails but no other facilities.

Fiskerton Fen LWT reserve lies next to the River Witham and has trails connected to the footpath/cycle trail on the north side of the Witham that connects Lincoln with Bardney. Good views over Branston Fen and Branston Island can be had from this footpath. The car park is at TF084719, there is a hide but no other facilities. A short roadside walk to the east leads to Short Ferry, an interesting area of wet grassland and wetland which has a hide overlooking it. Nocton Fen is worth driving round a circuit of narrow lanes and stopping to scan for wildfowl flocks feeding on the fields and hunting raptors. Follow the lane off the Branston Causeway from the junction at TF106690 through to Wasps Nest.

Chambers Farm Wood is reached via Bardney with parking at TF148738. It has well marked trails and is excellent for flowers and butterflies as well as birds.

East

Rank	Site Name	GR	Records	Species	Rare and Scarce
1	Gibraltar Point	TF5555758664	50,6914	325	78
3	Saltfleetby-Theddlethorpe	TF4881589556	37,243	274	53
8	Huttoft	TF5412778670	14,369	239	20
9	Saltfleet	TF4492194145	10,358	236	16
14	Kirkby on Bain	TF2345761971	26,863	213	16
15	Anderby	TF5368176974	6,649	211	16
22	Chapel St Leonards	TF5559273228	2,966	203	14
27	Skegness	TF5555464320	1,186	191	12
29	Mablethorpe	TF4967686121	2,459	191	4
30	Seacroft	TF5624660345	1,623	187	4
36	Wainfleet	TF5230055297	7,396	174	3
39	Manby Wetlands	TF4037186527	10,077	171	3
45	Middlemarsh Farm	TF5259563534	17,653	164	6
48	Theddlethorpe St Helen	TF4712689281	1,759	159	1
53	Friskney	TF4579055752	3,686	153	4
56	Tattershall	TF2107958059	21,338	150	2
58	Tattershall Thorpe	TF2180359647	3,301	148	
66	Louth	TF3202987880	4,451	139	3
70	Sandilands	TF5218880678	647	137	
71	Kirkby Moor	TF2146063133	2,184	136	3
73	Trusthorpe	TF5099483613	981	136	2
74	Hagnaby	TF4807780126	4,987	136	1
82	Sutton on Sea	TF5186281904	568	132	1
93	West Ashby	TF2658572749	1,774	126	
111	Donington on Bain	TF2317083179	1,999	118	
113	Stenigot	TF2511981037	2,447	117	
126	Baumber	TF2164774447	1,406	111	
128	Ruckland	TF3314478215	2,969	110	
135	Saltfleetby St Clement	TF4564091718	672	106	
144	Revesby	TF2975861614	1,229	102	
147	Withern	TF4336082609	2,285	101	
150	Burgh le Marsh	TF4931065613	584	99	1
154	East Keal	TF3741263876	1,742	99	
163	Hemingby	TF2346474649	464	96	
165	Frithville	TF3194750643	386	95	
167	Roughton	TF2398464901	813	94	
171	Horncastle	TF2541770170	717	91	1
175	Coningsby	TF2168758653	526	91	
180	Snipe Dales	TF3312368664	1,390	90	
182	Midville	TF3823057098	794	90	
183	Chapel Hill	TF2044154660	283	90	

The East comprises the North Sea coast part of The Wash, the Marsh, Wolds and Fen, stretching inland to Woodhall Spa, which lies a fraction to the west of the dividing line between east and west. It holds Gibraltar Point, the jewel in the crown of Lincolnshire birding which lies at an important migration crossroads at the northern mouth of The Wash. There are a range of interesting seawatching and birding spots all along the coast north from Gibraltar Point to Saltfleet. The newly formed Coastal Grazing Marshes Country Park has the potential to become a superb birding area and holds within it a number of good sites. Further north Manby Wetlands is a great example of how farmers have created valuable wetland sites under DEFRA support. The Wolds are neglected by birders but LWT reserves like Red Hill near Louth and Snipe Dales near Horncastle can provide great birding. Further inland the area around Kirkby on Bain has been a focus for LWT's living landscapes policy. Former gravel pits, rare surviving heather moorland and conifer plantation link together to provide great birding.

Chapel Point Observatory

Gibraltar Point is without doubt the number one birding site in the county and one of the largest at 430ha. It is our only bird observatory and the birds have been intensively studied under the auspices of LWT since 1949. At 325 species it has recorded more birds and more rares than anywhere else in the county. The Bird Observatory recording involves regular ringing, visible migration (vismig) counts, seawatching and full daily records of all birds recorded by the Observatory staff which usually consists of at least two full timers and voluntary/part-time assistants. It also has a Visitor Centre with cafe and toilets, a good selection of hides and marked trails. It lies 5km south of Skegness railway station and is most easily accessed by car from the Beach car park at TF558589 or the Visitor Centre car park at TF556580. Anyone travelling by foot should consider walking along the footpath along the edge of the saltmarsh and dunes accessed from Seacroft Esplanade at TF566609; best wear wellies or good boots. The reserve has improved enormously for birders over the last 30 years under the current management but sadly suffered a major setback in 2010 when a surge tide flooded the Field Station taking out accommodation facilities that have not been replaced in the newly built visitor centre. Despite all the birding done here the density of the dune vegetation means that much probably gets missed and more regular birders are always welcome.

Gibraltar Point

At the north end of this part of the coast lies Saltfleetby-Theddlethorpe Dunes NNR which stretches nearly 8km from Paradise Pool, Saltfleet down to Mablethorpe North End, a narrow band of dunes, marsh, saltmarsh and foreshore. It boasts the third biggest list in the county and is massively under-watched. The main access points are all famous in the birding world and from north to south with parking locations in brackets: Paradise Pools (TF456933), Sea View Farm (TF464924), Rimac (TF467917), Churchill Lane (TF477901), Brickyard Lane (TF483891) and Crook Bank (TF488882). There are no facilities and no hides but there are trails. If you see a fellow birder, better stop and exchange information because you are unlikely to see another one.

Running south from Mablethorpe there are a series of sites that have good birding and some have free parking and can be used for seawatching including Mablethorpe and Sutton on Sea which both have shelters, Huttoft Bank where you can park and view the sea from your car, Anderby Creek and Wolla Bank from which you can only watch off the beach/dunes, and the new North Sea Observatory at Chapel Point. It boasts a cafe and an excellent viewing veranda to the south side which is commandeered by seawatchers from late Aug to Nov.

The Marsh that lies between the coast and the wolds is primarily agricultural and although there are some good farmland schemes that have created new wetland and wet grassland habitats like Manby Wetlands and Middlemarsh Farm, access can be tricky and it is best to consult locals before a visit.

Saltfleet and Paradise Pool

Kirkby on Bain LWT

The Wolds is under-watched but has some real surprises like Snipe Dales NR and Country Park. It lies between Horncastle and Spilsby and has parking, toilets and trails accessed from TF330682. It is a great place for a family walk with lots of hills, hollows, woodland and glades and rough grass valley sides with scrub. For a more open woodland experience with fantastic views across the county and raptor watching potential, Red Hill Nature Reserve owned by LWT cannot be beaten. The floristically awesome Coronation Meadows have been created from arable land and in the immediate area there are a range of bridleways and country lanes that reveal the traditional wolds bird community at its best. Park at TF264808 and explore on foot.

Towards the centre of the county lies Kirkby on Bain which is arguably the best inland birding area in the county. The area between it and Woodhall Spa was originally an open heathland "waste". It was afforested with conifer plantation in the 1920s, covered in concrete for runways during World War Two and has extensive sand and gravel workings. At the heart of this lies LWT reserve Kirkby Moor which provides a glimpse of the original landscape with intact heathland accessed from Moor Lane with roadside parking at TF224629. Immediately to the west lies Ostler's Plantation, a Forestry Commission area of mature conifer forest and a mosaic of clear fell and planting with parking at TF216628. Kirkby on Bain Gravel Pits is another LWT reserve adjacent to one of the last landfill sites in the county. The reserve has a hide next to the car park at TF236609 and there is roadside parking at TF237612 opposite the entrance to the landfill site. Completing the jigsaw to the west lies a new LWT reserve at Woodhall Spa Airfield. Parking at TF206607. This is a members only site with a locked pedestrian gate opened by keycode which can be obtained from 01507 526667 during office hours. It has extensive wetland, heathland, two hides and viewing points.

Tattershall Thorpe Pits

South-West

Rank	Site Name	GR	Records	Species	Rare and Scarce
18	Deeping St James	TF1581609662	16,210	208	6
19	Baston	TF1052814613	15,379	207	6
38	Marston	SK8895343658	12,198	173	8
50	Willow Tree Fen	TF1753621889	5,668	155	4
77	Dunsby	TF1027427294	3,077	134	2
90	Thurlby (Bourne)	TF0939917058	2,470	128	
102	Langtoft	TF1077712583	2,706	121	1
107	Dunston	TF0608063068	3,214	119	1
108	Grantham	SK8968937263	991	119	1
110	Belton (Grantham)	SK9259540411	3,112	119	
117	Bourne	TF0784221775	2,085	115	1
123	Haverholme	TF1063849605	5,715	112	
125	Denton	SK8599933001	1,225	112	
134	Holywell	SK9935116342	1,749	106	2
152	Ancaster	SK9817144127	417	99	1
155	Long Bennington	SK8315445024	3,100	98	
158	Foston	SK8565843165	2,329	97	
168	Rippingale	TF0967828004	750	94	
174	Silk Willoughby	TF0543243156	555	91	
177	Ewerby	TF1187547452	408	90	2
179	Culverthorpe	TF0156940625	1,431	90	

The region is dominated by two natural areas: the South Kesteven Uplands and the Peat Fen. The main town is Grantham which has perhaps the best traditional sewage farm in the county at Marston to the north-west and a beautiful parkland with ancient oaks to the north-east at Belton House. To the east of Bourne lies the Fens with some interesting fenland remnants at LWT Baston Fen and an exciting, visionary rewilding project at nearby LWT Willow Tree Fen. In the deep south close to the Cambridgeshire border is Deeping St James whose wetlands at the LWT Deeping Lakes reserve are about as far south as you can go in Lincolnshire. It has the best list in the region.

Willow Tree Fen

Marston Sewage Treatment Works has no real facilities and no marked trails but is a top 20 site in Lincolnshire and well worth a look. It lies just to the north of the A1, There is a small car park off Barkston Road, Marston at SK892427. Walk along the path that runs along the south side of the site and look over the pools to the north. It is still a working sewage farm and is not heavily fenced but please don't be tempted to trespass. The track goes right through to the main East Coast railway line. It can be wet so wellies are a good idea. Belton Park is about five miles to the south-east. Park at SK943389 and walk up the hill and through the woods to left and right on the ridge. There is a large network of footpaths through the park and a good roam will make a fine day out.

LWT Baston Fen is a rare relict piece of fenland in South Lincolnshire and is a mix of wet grassland, reedbeds and small lagoons. There are no facilities other than a small car park and a trail. Park at TF145176. LWT Willow Tree Fen is much more extensive and ten years ago was arable fields. LWT acquired the site because of its proximity to Baston Fen and it lies between the River Glen and the Counter Drain. An excellent location for flooding, which is just as well as the vision was to restore it to fenland with wet grassland, reedbeds and lagoons. The project succeeded in spades when it attracted Lincolnshire's first breeding Cranes in 400 years. Access was closed due to the Coronavirus lockdown of Mar-Jun 2020 and the Cranes took advantage of the peace and quiet. There is a small car park at TF181213. The access arrangements and trails at this site are currently under review at the time of writing because of the hope that Cranes will nest again in future years. There is a hide and more may follow but best to check with the LWT website if you are planning a visit and are not familiar with the site.

The Deeps at Deeping St James are LWT's leading reserve in South Lincolnshire. It has a good list of species and an interesting mix of fenland habitats with a marked trail and hides for viewing the lakes and lagoons. The island in the middle of the lake has a great mixed colony of breeding herons and cormorants and is a regular overwinter roost for Long-eared Owls. Park at TF184079. Deeping High Bank which runs along the River Welland along the Crowland road to the east is worth checking out too.

South-East

Rank	Site Name	GR	Records	Species	Rare and Scarce
4	Frampton Marsh	TF3551739411	11,3753	262	26
5	Freiston Shore	TF3975642593	69,942	260	22
16	Witham Mouth	TF3959939638	12,340	211	7
24	Boston	TF3161844746	2,244	200	7
41	Gedney Drove End	TF4566629576	2,832	171	
44	Wrangle	TF4569749303	10,995	165	2
51	Nene Mouth	TF4925927615	1,893	155	4
55	Holbeach Marsh	TF4190834986	3,734	150	9
59	Crowland	TF2300110995	2,972	147	3
61	The Wash	TF4469140078	1,522	146	
62	Spalding	TF2393623629	8,119	145	6
68	Wingland Marsh	TF4959026923	434	137	1
69	Terrington	TF5182025924	3,518	137	
79	Pinchbeck	TF2367026159	5,096	133	
81	Butterwick	TF3991044701	6,754	132	1
83	Gedney	TF4020324584	1,317	132	
88	Kirton Marsh	TF3519036652	4,724	129	1
92	Moulton	TF3039324388	2,359	126	
103	Wyberton	TF3149641581	749	121	1
118	Antons Gowt	TF2976547723	856	115	
119	Dawsmere	TF4379631513	3,248	113	1
120	South Kyme	TF1756749947	814	113	
127	Hobhole	TF3657847778	539	111	
130	Kirton (Village)	TF3039038991	649	110	
131	Gedney Marsh	TF4700229995	3,855	109	
132	Leverton	TF4479748397	5,063	108	2
138	Welland Mouth	TF3925638482	247	104	3
146	Sutton Bridge	TF4743221603	268	101	1
166	Fishtoft	TF3584442614	170	94	1
169	Deeping St Nicholas	TF2090916774	336	93	
176	Tydd St Mary	TF4429818704	420	91	

This region is dominated by intensively agricultural drained fenland and The Wash, the premier site for wildfowl and waders in the UK. Birding in this region is dominated by twin RSPB reserves Frampton Marsh and Freiston Shore which lie on either side of the river Witham south of Boston. The Wash has numerous access points from which a hardy team of WeBS volunteers count the estuary's birds every month. These are not described here.

Frampton Marsh is the RSPB's flagship nature reserve in Lincolnshire. It is almost certainly the most popular site in the county with more birding visitors, especially from out of county, than anywhere else. The highly professional management team ensures a great birding spectacle is provided at all times of the year. It has "only" the fourth largest bird list of sites in the county but given that the reserve was bare arable fields less than 20 years ago, the transformation to a diverse range of freshwater habitats and the birds they attract is astonishing. Lying adjacent to The Wash is a major part of the attraction. The reserve team are proud of their ability to attract waders and more often than not they win the annual competition across RSPB reserves to record the most waders species in a year. There is also a competitive edge with Gibraltar Point with the latter usually recording more overall species in a year than the new kid on the block. The trails, hides, facilities and especially the information about bird whereabouts are all second to none. The reserve's daily update on twitter showing an aerial photo of the reserve with numbered locations of birds is an innovation that other sites should follow. It is however made possible by the dedicated team of volunteers that the reserve attracts. Frampton Marsh is four miles off the A16 south of Boston and the car park and Visitor Centre are at TF345358. There is an entrance charge of £2 for non-members.

Freiston Shore is managed by the same RSPB team as nearby Frampton Marsh. It is much less popular but it has great birding and the fifth largest list of species in the county. The site is only slightly older than Frampton Marsh. It was the first example in the county of "managed realignment". An area of rich arable land "reclaimed" from the sea in the late 1940s had a new seawall built further back inland and the old seawall was breached in 2000 to allow the sea to take back 65ha. This created a new saltmarsh and an area of brackish lagoons with shingle covered islands. A further area to the south west has been converted to wet grazing land. The reserve is accessed from a car park at TF397424.

Cut-End or Witham Mouth lies on the east side of the point where the River Witham joins The Wash. It is an excellent spot for viewing The Wash and can be great for seawatching especially at high tide when birds come right in front of the hide. The hide is run by LBC on behalf of the Environment Agency. It is a solid brick built structure with the hide on the second storey accessed by steps. It is relatively small and gets cramped with five people in it. The door is locked to reduce vandalism and a key can be obtained from the club by LBC members. Access is from a car park at the end of Cut End Lane at TF380391. Even if you don't have a key the outside of the hide can provide protection from the elements depending on wind direction and it is still a great location. The hide is a 2km walk along the river bank which on a cold windy winters day is an invigorating experience to say the least. Wrap up warm.

East Hide Frampton Marsh RSPB

The Lincolnshire Bird Club

After discussions on the desirability and possible viability of a separate county Bird Club in early 1979 in Sep that year a steering committee was formed. The Lincolnshire Bird Club held its first AGM in Mar 1981 and adopted a constitution which aimed to: "encourage and further an interest in the birdlife of the historic County of Lincolnshire; to participate in organised fieldwork activities; to collect and publish information on bird movements, behaviour, distribution and populations; to encourage conservation of the wildlife of the county and to provide sound information on which conservation policies can be based".

For almost 40 years from 1979 to 2017 the LBC had just over 300 members, despite Lincolnshire being a huge county. Since 2018 this figure has increased to over 350.

One of the very first acts of LBC was to commit itself to an ambitious project to map the breeding, passage and wintering birds of Lincolnshire in an *LBC Atlas* project based on one km squares, of which Lincolnshire comprises nearly 7,000. It was a massive undertaking, organised by Keith Atkin employing a team of hundreds of volunteers. The fieldwork was largely conducted from 1980-89 but for some scarcer species continued until 1995. During this time Keith was also hard at work researching and writing *The Birds of Lincolnshire & South Humberside* with Steve Lorand which was published in 1989. Various difficulties with completing the *LBC Atlas* project and writing it up meant that by 2007 the *LBC Atlas* editorial team came to accept it was unlikely the published *LBC Atlas* would ever see the light of day.

During this period in 2000 Anne Goodall and Keith Atkin, under the aegis of LBC published *The Status of Birds in Lincolnshire 1991-95 (SoBiL)*. This was followed up by Anne Goodall and Janet Eastmead in 2008 with *The Status of Birds in Lincolnshire 1996-2000*. Both books updated the Lincolnshire list and looked at trends in species numbers based on data collected by LBC and more widely, especially by BTO with which LBC has always had a close relationship.

Another early long term survey set up by LBC was a Winter Garden Bird Survey (GBFS) which ran from the early 1980s to 2014 when it merged with it's younger sibling, the BTO Garden BirdWatch set up in 1995. The data was used in the *SoBiL* publications and it is hoped the results will be published by LBC in a book currently in preparation.

The main focus of LBC has always been the publication of an annual Lincolnshire Bird Report (LBR). From 1979 to 1996 the LBR was produced annually. From 1997 to 2007 three reports covering years 1997-99, 2000-02 and 2003-07 were produced. In 2008, Andrew Chick became Chair, determined to resurrect the annual LBR and he has done an excellent job. From 2008-19 a high quality annual report has been produced each year.

Writing an annual bird report is a big commitment for a small team and there is only limited capacity for projects unless additional persons step up to assist. The annual report is usually printed within 15 months of the year end. LBC is fortunate to have a wide team of people who, though not keen to serve on a committee are happy to help in myriad ways. In addition the experience of past officers can always be gratefully called upon.

In addition to BTO, LBC has important relationships with the Rare Breeding Birds Panel (RBBP) to whom an annual report of Lincolnshire rare breeding birds, compiled from records received from members, is submitted each year. Also *British Birds* (BB) magazine to whom annual reports of scarce migrants are submitted and of course BBRC with whom the County Recorder liaises over the adjudication of rare bird sightings. Each of these three publish annual reports in BB which would be difficult without the assistance of bird clubs all over the UK. To keep in touch with these fellow organisations, our County Recorder is a member of the Association of County Recorders in England (ACRE). We also share data and ideas with GLNP and work with Lincolnshire Wildlife Trust (LWT) and RSPB on Lincolnshire based conservation projects and issues.

For the first 25 years or so of its history LBC communicated with members through posted newsletters. From 2003 a website was set up which included a discussion forum and photo library. Both proved popular for some years and from 2015 onwards were complemented by social media platforms like Twitter and WhatsApp both of which have been taken up enthusiastically by the club and it's members.

The LBC committee retained the wish to see the *LBC Atlas* maps produced from the original *LBC Atlas* survey, published in some form and from late 2017 having been given access to the remaining data, Colin Casey used his IT skills to resurrect an *LBC Atlas* from the records available. It was not the original editorial team's concept that was published but through persistence and determination, in 2020 the *Lincolnshire Bird Atlas 1980-99 An Historical Perspective* was produced. So many people had contributed to it over 40 years that the team which produced it felt it could only go out under the name of Lincolnshire Bird Club.

Having laid past ghosts to rest by publishing the *LBC Atlas*, the ground was cleared to write and publish a full update of Lorand and Atkin (1989).

LBC has always been run by a small central group of committee members, the principal officers of which are the Chair and County Bird Recorder. The office of President is an Honorary title bestowed by the Club when occasion arises on distinguished ornithologists living in the county. The holders of these offices since the club's founding are listed below.

Presidents		Chairpersons		County recorders	
1981 - 2009	E Simms	1979 - 1980	C Whittles M.B.O.U.	1969 - 1980	K Atkin
2010 - 2018	vacant	1981 - 1987	R Sheppard	1980 - 1987	G P Catley
2019 - 2020	Ian Newton FRS	1988 - 1991	M Davies	1987 - 1989	A G Ball
		1992 - 2007	A Goodall	1990 - 1993	G P Catley
		2008 - 2014	A P Chick	1993 - 1995	H Bunn
		2014 - 2020	P Espin	1996 - 2007 (North)	H Bunn
				1996 - 2007 (South)	S Keightley
				2008 - 2009	S Keightley
				2010 - 2019 (North)	J R Clarkson
				2010 - 2015 (South)	J Badley
				2016 - 2019 (South)	P A Hyde
				2019 - 2020	P A Hyde

Committee members

In this alphabetic list of committee members extracted from published annual Lincolnshire Bird Reports from 1979 onwards, the following conventions are followed. Persons serving as County Recorder, Chair or President are not included though they may have served in several different capacities. Any office held or sub-committee membership is placed in brackets after the name abbreviated as follows: S Secretary, T Treasurer, M Membership Secretary, Sa Sales Officer, G Gibraltar Point Representative, L LNU Representative, LWT LWT Representative, RC Records Committee, members serving at the the time of writing are marked *:

KN Barker, G Beasley (T), M Boddy, Steve Botham (RC), P Boyer, D Bradbeer, W Brooking, RJF Carr (S), CR Casey (M), P Davey, K Durose (RC), J Eastmead (S), P French (RC), W Gillatt (RC), RN Goodall (S), Matthew Harrison (RC*), Michael Harrison (M*), R Harvey (RC*), K Heath (S), R Heath, G Hopwood (LWT*), J Ingoldby (M), S Keightley (RC*), C Jennings, S Jennings, R Lambert, N Lound (RC*), I Macalpine-Leny (LNU), A Malkinson (T), K Marshall (T), J Mighell (S), J Molloy, GPF Montgomery, D Nicholson (RC), J Nickson, I Nixon, JDW Owen (T), P Porter, EJ Redshaw, S Routledge, IG Shepherd, J Siddle (RC*), A Sims (RC*), N Smith, W Sterling (Sa*), P Todd (S), N Tribe (LNU*), B Ward (RC), RK Watson (Sa), J Watt (T*), PN Watts, K Wilson (G*), KW Winfield, A Wragg (T), J Wright (S*).

Systematic List

The sequence and nomenclature adopted follows that of the British Ornithologists' Union who maintain the British List, which is the official list of wild bird species recorded in Great Britain (England, Scotland and Wales and associated waters). The List is managed by the BOU's Records Committee (BOURC).

In Dec 2017 the BOU decided that from the 9th Edition, the British List would follow the IOC World Bird List. The LBC committee agreed to do the same and the LBC and county list is based on what was the latest IOC taxonomic order at the time of writing (IOC 10.2).

Due to the potential for repetition and error, observer names which are already included in annual Lincolnshire Bird Reports are not generally included in the species accounts. An exception is made for new county firsts from 1989 onwards which are listed in a table in the section called Birds new to Lincolnshire since Lorand and Atkin (1989) with full details of date found, site and finder(s). In the case of rare and scarce species, only those that have officially been accepted by the BBRC and LBRC are shown. For completeness, in the few cases where a species that had been accepted as a first for the county but when later reviewed was found to no longer be acceptable, the record is listed along with details.

Over the years some significant changes have occurred to the county's avifauna. For example, species that were at one time a great rarity such as Little Egret, have become reasonably common and a formerly common bird, Turtle Dove, has become difficult to find. The list of species considered by the BBRC has changed many times with species being added and removed and even added again almost annually as the occurrence and abundance of rare species fluctuates.

The area that the Lincolnshire Bird Club covers is the 6,959 square kilometres of the ancient county of Lincolnshire including the areas at the coast that are in view at low tide.

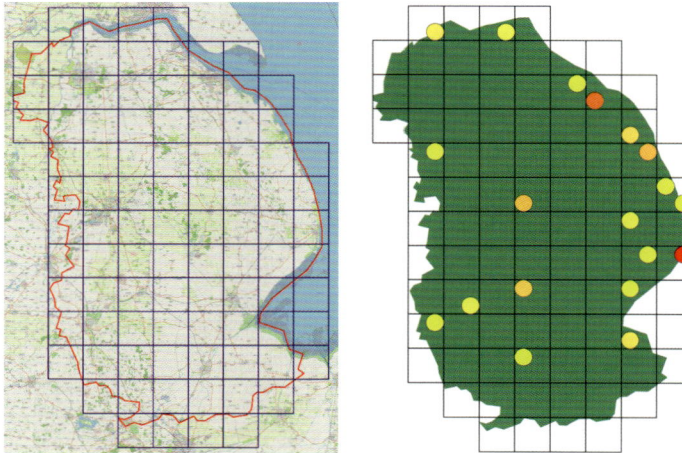

The left map shows the area covered along with the 10km grid squares that make up the LBC recording area.

Throughout the systematic list for most rare and scarce species there is a thumbnail map showing a dot at the position for each site at which the species has been recorded. The colours of the dot go from yellow for single sightings through to bright red where most sightings have occurred.

Due to the thumbnail map covering the area to low water mark some of the dots in this book may appear further inland than they actually are.

As per example, the map above to the right shows all Bee-eater records for the county with most sightings along the coast and the highest being the red dot at Gibraltar Point. The other red dot to the north represents one flock of 10 birds seen at Tetney Marsh. Each dot covers a 5km area.

As far as possible, the records for those rare and scarce species that have a thumbnail map have been verified and validated by Phil Hyde using information from BBRC and LBRC. This huge undertaking to make sure LBC records are as correct and complete as possible has taken over two years. Many hours of data analysis, cross-checking and a huge amount of correspondence by the county recorder has been required and we thank all those involved, particularly the finders of birds for being so helpful looking back in note-books from many years ago, to check that the data held is accurate.

These data have been used to create the tables of sightings accompanying the text accounts for rare and scarce species. In the list of sightings you will see that the count column has green in some cells the table and white in others. Green means that this was the first sighting and a white cell refers to the same bird being seen again on a different date or at a different site.

We have provided an image of each species taken in Lincolnshire. If we could not obtain an image of the species taken in Lincolnshire we have used a drawing, an image of a museum specimen or a painting. We must thank all the photographers who have supplied these images. Larger versions labelled with species, site, year and photographer are shown in the gallery section of this book. Without the photographs this book would have still been useful but rather bland.

River Ancholme at Bonby Carrs

At the top of each species is a green strip containing the current common name, the scientific name and then to the right icons and dots

BBRC — Refers to the BBRC

LBRC — Refers to the LBRC

RBBP — Refers to the RBBP

⬤ Refers to the BoCC Red status species

⬤ Refers to the BoCC Amber status species

⬤ Refers to the BoCC Green status species

BBRC refers to the British Birds Rarities Committee which adjudicates on the acceptance or otherwise of the rarest birds seen in the UK. The icon indicates that at the current time all records of this species must be validated by BBRC. Some species have moved in and out of BBRC consideration at various times and where this has occurred reference is made to the relevant dates in the accompanying text.

LBRC refers to the Lincolnshire Birds Records Committee which adjudicates on the acceptance or otherwise of scarcer species for which the committee feels evidence is required to accept a sighting of that species in Lincolnshire. The icon indicates that the species is currently considered by LBRC. The criteria under which these records are considered can be found on the LBC website.

RBBP refers to the Rare Breeding Birds Panel, a national body which collects information on the breeding presence of the indicated species across the UK. LBC supports the work of RBBP and contributes an annual report on Lincolnshire's rare breeding birds to it. RBBP incorporates that information into an annual report on the status of species it considers which is published in *British Birds*. This icon has only been used for species which have bred in Lincolnshire or might conceivably breed in Lincolnshire. It is not intended as a definitive indicator of bird species considered by RBBP.

BoCC refers to Birds of Conservation Concern 4: the Red List for British Birds, the latest report of this national body led by the RSPB which regularly considers the threats to Britain's birds and assigns them to red, amber or green status depending on the degree of threat; red being the highest, amber intermediate and green no current threat. Full details can be found at Eaton *et al* 2015. *British Birds* 108, 708-746.

Each species has a short "strapline" which summarises its status in Lincolnshire. These have been adapted and updated from those used in the Lincolnshire Bird Report 2018. For non breeding species the range is briefly described. Status is described on the following scale:

Vagrant	Less than one record per year
Rare	Averaging around one record per year
Scarce	Averaging one to nine records/breeding pairs per year
Fairly scarce	Averaging 10 to 99 birds/breeding pairs per year
Fairly common	Averaging 100 to 999 birds/breeding pairs per year
Common	Averaging 1,000 to 9,999 birds/breeding pairs per year
Very common	Averaging 10,000 to 99,999 birds/breeding pairs per year
Abundant	100,000 plus pairs for breeding birds and individuals for migrants and winterer

Each species has a text account describing the status of that species in Lincolnshire. For the commoner breeding, passage and wintering birds the account focuses on the last 20 years. *The Lincolnshire Bird Atlas: An Historical Perspective 1980-99* (LBC 2020), presented a wealth of information on the status of all species in Lincolnshire up to 1999 using the considerable data amassed by LBC and the local expertise of Keith Atkin and Anne Goodall. It has been used as a baseline throughout the species accounts and is referred to as the "*LBC Atlas*". Where references are made to *BTO Atlas's* that is made clear by citing the relevant years.

Where species breed relatively commonly, frequent reference is made to population estimates in 2016. These are based on the Avian Population Estimates Panel 4 (APEP4) paper by Woodward *et al* (2020) and are further explained on the BTO website. The methodology referred to on page 12 of this book has been used to derive a Lincolnshire population estimate from the APEP4 British population estimate. Any estimates quoted are for Lincolnshire only and are provided as a guide. They have not been subject to independent validification or peer review.

For species that breed regularly in Lincolnshire, BTO BBS data is referred to extensively. BTO publishes annual indexes for population changes at the UK, England and East Midlands level; the latest being Harris *et al* (2020). The Breeding Bird Survey 2019 BTO Research Report 726. Sarah Harris kindly provides Lincolnshire specific BBS data and index data for species that appear in sufficient surveys to make this statistically valid. Where possible, Lincolnshire specific data is referred to but where sample sizes are too small, data from the next level up from East Midlands to England to UK is used as appropriate. This data has also been used on the same basis to create the BBS graphs showing "smoothed" index changes of the populations of particular species. Hopefully these illustrate the direction of travel of the populations of our commoner breeding species.

For our rarer breeding birds reference is made to five-year averages of breeding pairs in Lincolnshire submitted by LBC to RBBP for the latest period available, generally 2017 or 2018.

The BTO provides a massive online resource for researchers into bird populations and extensive use has been made of BTO BirdFacts, Robinson (2005), BirdTrends Massimino *et al* (2019) and the latest annual report of the Wetlands Bird Survey Frost *et al* (2020)

The other important resources used have been the annual reports of BBRC published in British Birds up to 2019 and their online resources available to our County Recorder Phil Hyde and the annual Lincolnshire Bird Report published up to 2018 by LBC. The authors also have access to unpublished LBC records for 2019 and 2020. This book is intended to be definitive up to Dec 31st 2019 and also contains important records up to the end of 2020.

Any readers seeking to delve into past Lincolnshire Bird Reports can check the LBC website for details of the reports that are still available to purchase.

Supporting the text are a range of different features which may include a thumbnail map for rarity occurrences, a list of sightings for rarer species which have generally had 50 records or less, a breeding strip showing the breeding status of birds that breed or have bred in Lincolnshire over periods from 1968 up to the current period as explained in full on page 10 of this book, graphs depicting annual changes in the BBS index as explained above and, in some cases, other graphs of interest and, finally one or more images of the species which is either a photograph taken in Lincolnshire, a drawing based on a specimen from Lincolnshire or, for some species, a drawing based on a photograph taken elsewhere.

BOURC record categories

All of the species recorded in Lincolnshire are assigned a category by the British Ornithologists' Union Records Committee (BOURC). Each species on the British List is placed in one or more categories denoting its status on the List. The following text is a synopsis from the BOU website (https://www.bou.org.uk/british-list/species-categories/) which LBC gratefully acknowledges. Further details of the species recorded in Britain and their official category can be found on this website.

Categorisation was revised in 1997 to assist species protection under national wildlife legislation, especially of naturalised species. Category C was expanded to distinguish the different histories of introduction and naturalisation; Category D was reduced in scope, and a Category E was introduced to enable local and national recorders to monitor escaped species. Species on the Lincolnshire List are categorised according to the following definitions and to the best of current knowledge of their circumstances when recorded.

Note: The British and Lincolnshire Lists comprise only those species in Categories A, B and C. Category C6 recognises that some previously established naturalized introductions to Britain have declined (and others may do so in the future) to a level that is no longer self-sustaining, and which will ultimately lead to extinction. Further releases of such non-native species are prohibited under Section 14 of the Wildlife and Countryside Act 1981.

Category A		Species recorded in an apparently natural state at least once since Jan 1st 1950.
Category B		Species recorded in an apparently natural state at least once between Jan 1st 1800 and Dec 31st 1949 but have not been recorded subsequently.
Category C		Species that, although introduced, now derive from the resulting self-sustaining populations.
	C1	Naturalized introduced species - species that have occurred only as a result of introduction, e.g. Egyptian Goose *Alopochen aegyptiacus*
	C2	Naturalized established species – species with established populations resulting from introduction by Man, but which also occur in an apparently natural state, e.g. Greylag Goose *Anser anser*
	C3	Naturalized re-established species – species with populations successfully re-established by Man in areas of former occurrence, e.g., Red Kite *Milvus milvus*
	C4	Naturalized feral species – domesticated species with populations established in the wild, e.g. Rock Pigeon (Dove)/Feral Pigeon *Columba livia*.
	C5	Vagrant naturalized species – species from established naturalized populations abroad, e.g. possibly some Ruddy Shelducks *Tadorna ferruginea* occurring in Britain. There are currently no species in category C5
	C6	Former naturalized species - species formerly placed in C1 whose naturalized populations are either no longer self-sustaining or are considered extinct, e.g. Lady Amherst's Pheasant *Chrysolophus amherstiae*
Category D		Species that would otherwise appear in Category A except that there is reasonable doubt that they have ever occurred in a natural state. Species placed in Category D only form no part of the British List and are not included in the species totals.
Category E		Species recorded as introductions, human-assisted transportees or escapees from captivity, and whose breeding populations (if any) are thought not to be self-sustaining. Species in Category E that have bred in the wild in Britain are designated as E*. Category E species form no part of the British List (unless already included within Categories A, B or C)
Category F		Records of bird species recorded before 1800. Category F is currently under construction

Bee-eater, a typical Cat A Species

Mandarin, a typical Cat CE Species

Ross's Goose, a typical Cat DE Species

Black Swan, a typical Cat E Species

Black Grouse *Lyrurus tetrix*

Former breeder not recorded since 1930s.

Was formerly a resident on heaths and commons of the north-west, on the unenclosed moors between Lincoln and Nottinghamshire and in the Woodhall Spa area. Cordeaux (1872) thought birds had been introduced to the Frodingham district, where Scunthorpe steel works now stands, earlier in the 19th century. Lorand and Atkin (1989) state that the population was reinforced with introductions prior to the 1870s. They died out on Whisby Moor in 1842 and on Stapleford Moor in 1883. Birds lingered on in the Woodhall Spa district until about the 1870s and on Brumby Common and Twigmoor Warren until 1900.

By the 20th century birds were confined to Scotton Common. There were still several pairs present in 1920 and 18 birds were flushed together during the 1921-22 winter. Chicks were seen in 1922 and the last confirmed breeding was of two nests found there by keepers in the early 1930s. The final sighting was of a lone greyhen reported on Jun 10th 1935.

The nearest other population of Black Grouse in modern times, in the Peak District, died out in 2000 (*BTO Atlas 2007-11*). Despite a re-introduction scheme financed by Severn-Trent Water Authority which began in 2003 in the Upper Derwent Valley and released over 200 birds during a five-year period, none were left by 2011. If such a scheme were ever to be attempted or be successful in Lincolnshire the fragmentation and quality of remaining suitable habitat would have to be addressed.

1968-1972	1980-1989	2008-2011	2016-2019	Incidence	Change

Red-legged Partridge *Alectoris rufa*

A common species of arable farmland, with a resident population augmented each year by releases for shooting.

Introduced from nearby France to Suffolk in the 1770s; no record exists of how or when the "Frenchman" became widespread in Lincolnshire, but the spread seems to have occurred within 100 years. The *LBC Atlas* estimated a population of 4,000 to 8,000 pairs in the late 80s. Like many farmland species they are undergoing a decline due to agricultural intensification. Lincolnshire BBS data shows that this species declined by 54% during the period 1994-2018 and incidence data from BBS suggests the population in Lincolnshire is thinning out though it still remains widespread. It is likely that the population is now of the order of 3,000 to 4,000 pairs supplemented annually by many thousands of birds released for shooting.

1968-1972	1980-1989	2008-2011	2016-2019	Incidence	Change
				48%	-27%

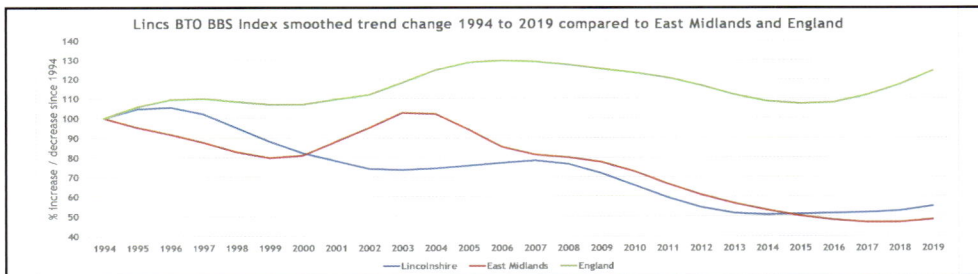

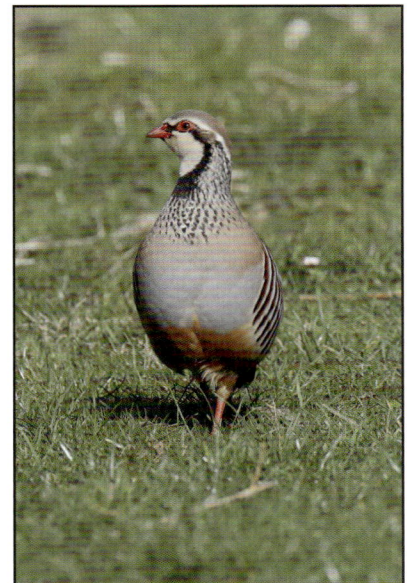

Lincs BTO BBS Index smoothed trend change 1994 to 2019 compared to East Midlands and England

Grey Partridge *Perdix perdix*

Common but declining resident especially in the south of the county.

Sadly, the English Grey Partridge population has declined by 64% since 1994 and it is now a red data species. One of our signature farmland birds and, as a primarily agricultural county, Lincolnshire remains a stronghold for this species and the BBS evidence shows that it has not declined as fast here as elsewhere. Certainly, the East Midland trend has shown a stable, even increasing, population for the last 10 years or so. BBS data for 2016 indicates Lincolnshire holds around 12% of the British population. The *LBC Atlas* estimated the population between 3,700 and 7,000 pairs and the Lincolnshire share of the APEP4 estimate suggests a current population of around 4,000 pairs. Some of the shooting fraternity in Lincolnshire understand the threat to this popular countryside bird and operate a voluntary shooting ban to help maintain it, but the recent decline over the last few years is worrying.

1968-1972	1980-1989	2008-2011	2016-2019	Incidence	Change
				23%	7%

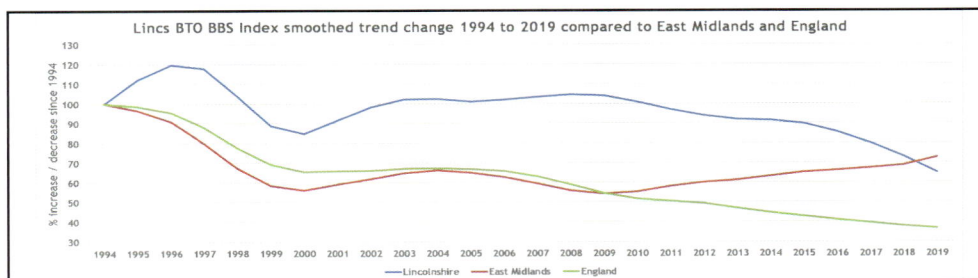

Lincs BTO BBS Index smoothed trend change 1994 to 2019 compared to East Midlands and England

Quail *Coturnix coturnix*

Scarce summer visitor Apr-Sep, rare in winter. Probably much under-recorded.

This species is a scarce and fairly local summer visitor that occurs in Lincolnshire annually. The unique breeding strategy of Quail in which the young of the year are able to migrate north and themselves breed in the same year probably accounts for "Quail years". During 2013-17 an average of eight singing males per year were reported.

Good data is available for Lincolnshire from 1996 onwards and since that year "Quail years" and the number of singing birds reported have occurred as follows: 2011 (65), 2005 (64) and 1997 (39). The worst years have been 2002 (0), 2001 (4) and 2018 (5). The average per year over the 23 years from 1996 to 2018 was 19 so the recent period looks poor and we are well overdue another "Quail year".

1968-1972	1980-1989	2008-2011	2016-2019	Incidence	Change
				0.3%	-85%

Pheasant *Phasianus colchicus*

Very common feral resident; huge numbers released for shooting annually.

While its partridge cousins are declining, Pheasant continues to become commoner. Presumably increases based on the massive numbers now being released annually. The *LBC Atlas* estimated the population in the late 1980s to be around 50,000 pairs and BBS data shows a significant increase in the breeding population of 39% during the period 1994-2018. Annual UK releases are said to be 47 million.

If Lincolnshire's share of released birds is 3% (the area of Britain which Lincolnshire covers) that is about 1.5 million, of which, it is estimated, only around 25% will be shot, leaving plenty of birds to fuel an increase and feed a multitude of foxes, corvids and other predators, like the Lincolnshire Poacher.

1968-1972	1980-1989	2008-2011	2016-2019	Incidence	Change
				91%	1%

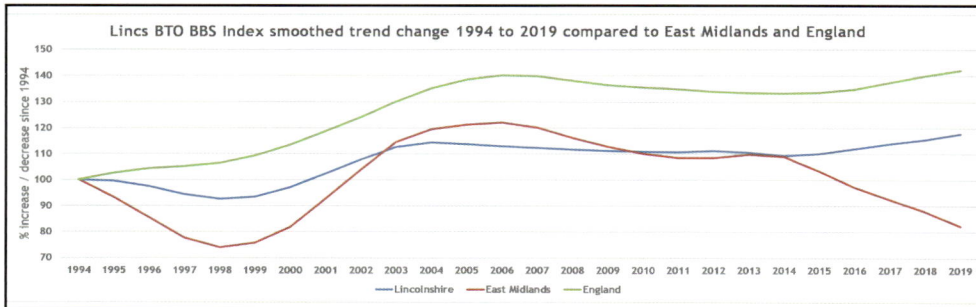

Lincs BTO BBS Index smoothed trend change 1994 to 2019 compared to East Midlands and England

Brent Goose *Branta bernicla*

Three distinct races occur. Nominate Dark-bellied Brent is a very common coastal winter visitor Sep-May, mainly to The Wash and outer Humber. Scarce but regular in summer, especially on The Wash. Occasional inland. Pale-bellied Brent *B. b. hrota*, is a scarce coastal winter visitor, in variable numbers. Black Brant *B. b. nigricans* is a former vagrant, first seen in 1982 with just three records to 1991. Now very scarce but annual in small numbers, with Dark-bellied flocks.

The Wash is of international importance for Dark-bellied Brent and peak counts between 2000 and 2019 have varied between 10,112 (2018/19) and 24,490 (2005/06). A noticeable crash occurred between 2007-08 (21,101) and 2008-09 (13,993) and every internationally important site had peak counts lower than their respective five-year averages. On the Wash, for the three winters 2016-19 peak counts have only been around 10,500. This large cyclical variation has been attributed to failed breeding seasons with increased predator pressure and poor weather on the breeding grounds.

Most Pale-bellied Brent winter in Ireland with fewer in north-east and north-west England. Few visit Lincolnshire, mostly on the Humber marshes and less so on the Wash. Annual totals are usually less than 50 but with occasional weather-dependent influxes.

The first record of Black Brant in 1982 was followed by others in 1987, 1991 and 1996 but since then there has been an unbroken run of up to three birds annually during 1999-2019 on the Wash and Humber. Movement between sites on the Wash makes interpretation of numbers difficult on occasion but there are probably two-five birds annually. Most often these are single birds but two turn up together regularly and there were two adults at Gibraltar Point (Feb-Mar 2012) accompanied by what was thought to be a hybrid, now a recurrent problem for goose counters.

Left: Brent Goose (Black Brant), Middle: Brent Goose (Dark-bellied), Right: Brent Goose (Pale-bellied)

Red-breasted Goose *Branta ruficollis*

BBRC

Vagrant. Arctic Siberia.

As with some other species of wildfowl Red-breasted Goose is a popular bird with aviculturists, and separating wild birds from those which have escaped captivity is problematic. The first accepted county record was in Oct 1978 at Covenham Reservoir. There was a second in 1984 and then several records from the Wash in 1985 probably involving the same bird first seen at Gibraltar Point in Jan 1985 and, subsequently, three other sites along the west coast of the Wash. To complicate matters two 2CY birds were seen in late Feb 1985.

The BBRC report for 1985 noted that the increase in records continues, in line with the increasing number of records in Europe, and that six British records within the space of two months is unprecedented. Six subsequent records between 2004 and 2010 have been considered to be birds having escaped from captivity but two birds at Covenham Reservoir on Oct 12th and Saltfleet on Oct 13th 2006 were accepted as wild birds. There have now been more than 80 accepted British records notwithstanding the caveat about their origins.

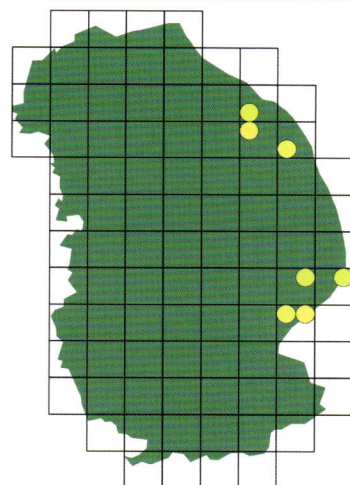

Site	First Date	Last Date	Count	Notes
Covenham Reservoir	01/10/1978		1	
North Cotes	14/11/1984	30/11/1984	1	Adult
Gibraltar Point	09/01/1985	16/02/1985	1	Adult
Leverton	13/02/1985	14/02/1985	1	Adult
Friskney	16/02/1985	17/02/1985	1	Same adult
Leverton	17/02/1985		1	Same adult
Wrangle	20/02/1985	03/03/1985	1	Same adult
Wrangle	23/02/1985		2	Second calendar year
Covenham Reservoir	12/10/2006		2	Adult
Saltfleet	13/10/2006	24/01/2007	2	Same adult

Canada Goose *Branta canadensis*

Common resident introduced from North America in the 18th and 19th centuries. The taxonomic status of Canada geese has recently changed with Canada Goose *B. canadensis* and Cackling Goose *B. hutchinsii* now being recognised as two distinct species. Only Canada Goose occurs as a feral resident in Britain.

Canada Goose is an invasive species and colonises new waterbodies rapidly. The *LBC Atlas* estimated a breeding population of around 700 pairs in the late 80s with a wintering population of some 2,500-3,000 birds. It has become more widespread since then and the breeding population has almost certainly increased though it is difficult to say by how much. Based on the latest LBRs the wintering population is around 3,000 birds. There have been no accepted records of wild Cackling Goose *B. hutchinsii*, in Lincolnshire so far. Birds treated as "escapes" are dealt with in that section.

1968-1972	1980-1989	2008-2011	2016-2019	Incidence	Change
				12%	-12%

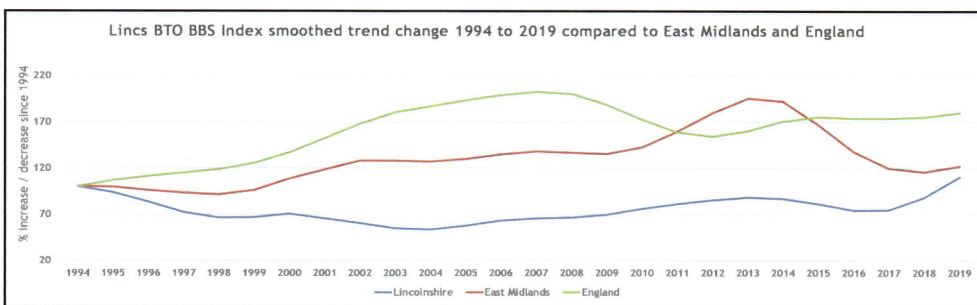

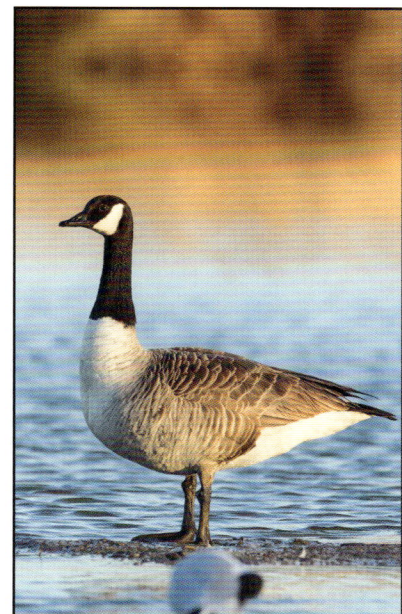

Lincs BTO BBS Index smoothed trend change 1994 to 2019 compared to East Midlands and England

Barnacle Goose *Branta leucopsis*

Fairly common localised feral population. Has bred as a feral bird since the late 1980s, with a large population established on the Humber (most nesting in Yorkshire). Wild birds are scarce but erratic winter visitors, mainly to the coast.

The true status of Barnacle Goose in Lincolnshire has become difficult to gauge with the development of a substantial feral breeding colony many hundred pairs strong based around Whitton Sands on the Yorkshire side of the Humber. These birds undoubtedly move around Lincolnshire and several pairs breed on the south bank of the Humber too. Counts of up to 2,000 birds have been made in the Alkborough Flats/Whitton area in recent winters. What are thought to be genuine birds from the Svalbard population that winters in the Netherlands and Germany still occur, escaping (increasingly rare) hard winters on the continent. These frequently accompany flocks of Russian White-fronted Geese *Anser albifrons albifrons* and tend to turn up and feed in the fields along the North Sea coast. There is ringing evidence to support this. However, the booming feral population in the Netherlands and Germany may contribute to what we perceive as "wild" influxes.

1968-1972	1980-1989	2008-2011	2016-2019	Incidence	Change
				0%	

Snow Goose *Anser caerulescens*

Possibly rare but frequent escapes from wildfowl collections cloud the true status.

The first record of a Snow Goose that may have been of wild origin was of one near Boston in 1942. Occurrences since have been almost annual from 1978 onwards. Most have involved birds at inland locations, either alone or with feral Greylag or Canada Geese when it is assumed that these birds are non-wild. Coastal records are less common, although interestingly most records from all areas have either been during the winter months or during Apr and May.

A second accepted record was of a bird seen in 2006-07 with Pink-footed Geese *A. brachyrhynchus*. It was seen at Saltfleetby-Theddlethorpe on Oct 13th and what was presumed to be the same bird was later found at Sutton Bridge/Nene Mouth Oct 26th-31st feeding with Pink-footed Geese. In 2007 one was seen near Frampton Marsh on Jan 6th and Jan 14th flying with a Pink-footed Goose flock, presumed to be the same individual.

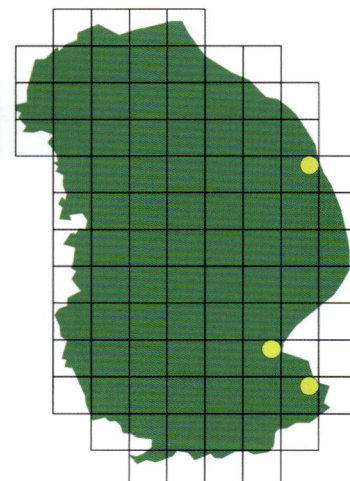

In 2011 it was estimated there were about 180 birds at large in Britain (combined numbers of wild, naturalised and other birds from WeBS and county bird reports). Wild Snow Geese are most likely to occur at traditional staging or wintering areas of Pink-footed Geese or Whooper Swans.

Site Name	First Date	Last Date	Count	Notes
Saltfleetby-Theddlethorpe	13/10/2006	14/10/2006	1	
Nene Mouth	25/10/2006	31/10/2006	1	Same
Frampton Marsh	06/01/2007		1	Same
Frampton Marsh	14/01/2007		1	Same

Greylag Goose *Anser anser*

The nominate form is now a common and widespread feral resident. Wild birds from the Scottish population are likely to be scarce but, erratic and difficult to detect winter visitors.

Over the past 40 years the breeding population of Lincolnshire has rocketed upwards from an estimated 200 pairs in the late 1980s to approaching 2,000 pairs in recent years, based on the APEP4 methodology. It has overtaken Canada Goose *Branta canadensis* and become Lincolnshire's commonest feral goose. Peak winter counts in recent winters across the county have approached 6,000 birds with Alkborough Flats in the north and Baston Fen in the south often holding large flocks of up to 1,000 birds each.

1968-1972	1980-1989	2008-2011	2016-2019	Incidence	Change
				39%	31%

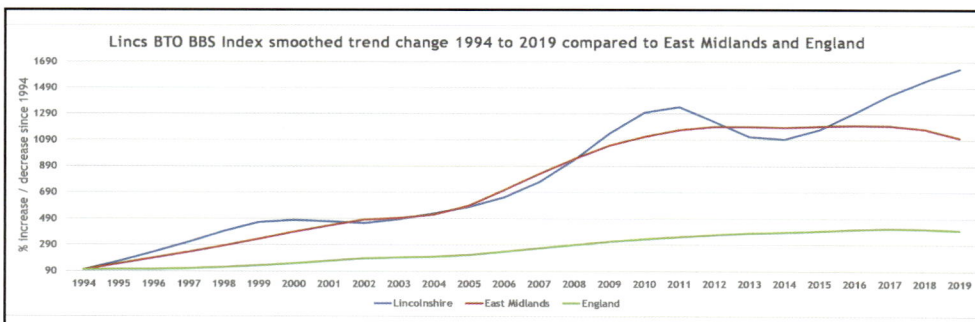

Lincs BTO BBS Index smoothed trend change 1994 to 2019 compared to East Midlands and England

Taiga Bean Goose *Anser fabalis*

Rare winter visitor, with occasional larger influxes.

Taiga Bean Geese of the nominate Scandinavian and north-west Russian subspecies are regular winterers in Britain. This species was split from Tundra Bean Goose *A. serrirostris* by the AOU (AOS from 2016) in 2007; this was adopted by the IOC and thus the BOU when they voted to change to IOC taxonomy.

Historically, and before the split, Bean Geese were not often assigned to race. After 2000 the first confirmed record of this taxon in the county was in Feb 2003 at Gibraltar Point and, with only 10 records in six of the years since, it remains a rare winter visitor. Flocks of eight were seen in 2011 and 2017.

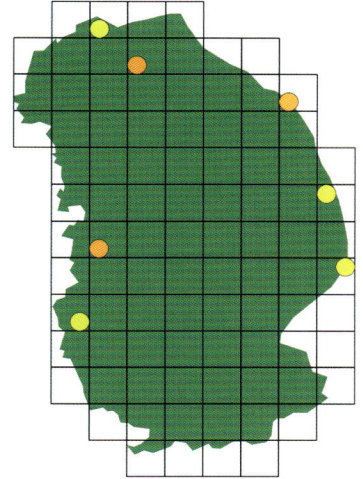

Site Name	First Date	Last Date	Count	Notes
Gibraltar Point	16/02/2003	01/03/2003	1	
Gibraltar Point	20/02/2004		1	
Donna Nook	02/01/2007		1	
Winteringham	14/01/2011	15/01/2011	1	
Donna Nook	27/01/2011	28/01/2011	3	
Worlaby (Brigg)	13/03/2011	20/03/2011	8	
Donna Nook	19/03/2011		1	
Donna Nook	11/10/2014		2	
Bassingham	21/01/2017		8	
Marston	25/01/2018	28/01/2018	1	
Anderby	26/11/2018	27/11/2018	2	

Pink-footed Goose *Anser brachyrhynchus*

A very common winter visitor during Sep-Apr, mainly to the Humber and The Wash, but there are many coastal and inland movements. A few injured birds remain in summer.

The musical notes from a skein of Pink-feet flying over in characteristic V-formation are an expected and enjoyable winter experience for most Lincolnshire folk. The county appears to be on the 'M1' between their Icelandic/Greenland breeding haunts and Norfolk wintering grounds, though many also winter in the county particularly along the Humber and The Wash. They tend to be heading south-east in autumn and north-west in spring though changes in weather can spark commuting between wintering service stations from Scotland to East Anglia. Over 30,000 per year have been regularly counted in Lincolnshire in recent winters including roosts of 29,000 on the Humber in Oct. With a British wintering population of around 500,000 (Robinson 2005) many more pass overhead, some in the daytime but many overnight. With so many birds passing through, inevitably some are injured and over-summer. These can pair up and are sometimes suspected of breeding. The evidence is sketchy, and it has never been more than an irregular occurrence.

1968-1972	1980-1989	2008-2011	2016-2019	Incidence	Change

Tundra Bean Goose *Anser serrirostris*

Scarce winter visitor, mainly coastal, in variable numbers.

In 2007, the AOU split Bean Goose *A. fabalis*, into two species; Taiga Bean Goose became *A. fabalis* and Tundra Bean Goose *A. serrirostris*. The IOC followed suit and with the adoption of IOC taxonomy by the BOU in 2017 an extra species appeared on the British List. Lorand and Atkin (1989) noted long before the split that both races, as they were then, *A. f. fabalis* and *rossicus*, had been identified in the county.

The *LBC Atlas* notes that in most years of the 1980s and 1990s that there were around 10 birds at scattered localities, usually on the coast but in years of geese influxes from the near continent there have been many more, e.g. over 150 in 2011. In most years Tundra Bean Geese turn up from the second week of Oct although occasionally early birds are recorded in Sep. The pattern in the last five years to 2019 has been of small numbers, with occasional flocks of 10 or more. A flock of 12 were on Read's Island Feb 27th 2012 and in 2017 10 or more were present in a mobile flock along the Humber Bank in what was an exceptional year for the species in the county. Most are usually coastal with a few penetrating as far as The Wash, usually at Gibraltar Point. Frost *et al* (2019) estimated that around 300 Tundra Bean Geese are present in Britain over winter.

White-fronted Goose *Anser albifrons*

A. a. albifrons Scarce winter visitor Sep-Apr, occasionally fairly common. Northern Russia, Siberia.
A. a. flavirostris rare winter visitor. Greenland (LBRC).

Historically Caton Haigh noted that it was almost unknown on the north-east coast until 1917, after which it became a regular visitor. The *LBC Atlas* noted that during 1980-99, in good years, more than 500 could be found wintering with the largest flocks of up to 100 usually being found on the coast. Most of our winterers are the nominate race arriving in small but regular numbers from Oct onwards with birds arriving from the near continent during cold weather movements. In recent times flocks of 10-20 have been regularly recorded across the county and at inland locations where more wet grassland has become available. A family party of six at Frampton Marsh in 2008 was the first record for that reserve since its inception and small numbers occur at other inland sites such as Baston & Langtoft GP, Deeping Lakes NR, Messingham SQ and Tattershall Lakes. The year 2011 saw some impressive cold weather movements into the county with maxima of 47 at Alkborough Flats Jan 1st, 60 at Donna Nook and 18 at Frampton Marsh Jan 6th. The autumn-winter of 2011 also saw significant numbers with maxima of 122 at Alkborough Flats, 50 at Donna Nook, 36 at Frampton Marsh, 74 at Gibraltar Point, and 57 at Huttoft Bank.

The Greenland race, *A. a. flavirostris*, was first noted Dec 5th 1950 at Croft Marsh, Gibraltar Point but it has always been a very irregular visitor. Since the late 1990s it has retained its status as a rare visitor with records of single coastal birds in six years to 2019, the exceptions being a flock of six at Alkborough Flats and nearby Whitton Sands Jan-Feb 2012, three at Marston STW Nov-Dec 2015 and two at Theddlethorpe St. Helen Jan 2016.

Lesser White-fronted Goose *Anser erythropus*

Vagrant. Scandinavia.

One record of an adult male bird on Jan 1st 1943 at Nene Mouth on The Wash which flew in and joined a pinioned pair of the same species in a local wildfowl collection. It stayed just one day.

These birds are highly migratory Arctic-nesting geese breeding from Scandinavia eastward to eastern Siberia. A massive population decline across its entire range occurred during the 19th and 20th centuries, with an abrupt decline since c.1950 in Finland and other parts of Scandinavia, the most likely source of British wintering birds. There was also concurrent evidence of winter decreases in south-east Europe (Marchant and Musgrove 2011). This accords with the long-term decline in British records since 1950 with only four since 2000. The classic 'carrier' species for most of the British records is Eurasian White-fronted Geese *A. anser albifrons* and this was the case for the most recent record (Essex, 2017) although they have also been found with flocks of Taiga Bean Geese *A. fabalis*.

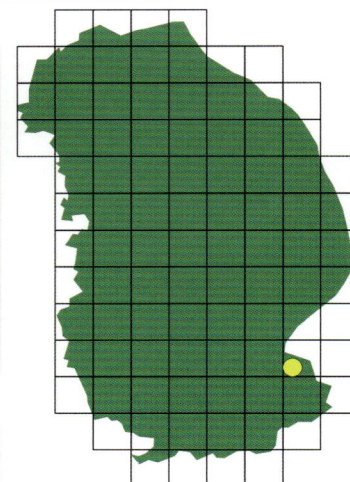

Site Name	First Date	Last Date	Count	Notes
Nene Mouth	01/01/1943		1	Adult

Mute Swan *Cygnus olor*

Fairly common resident and partial migrant.

Mute Swan is very widespread as a breeder in Lincolnshire occurring at low density on ponds, lakes, gravel pits, rivers and many of our larger drains and canals. The *LBC Atlas* estimated a breeding population of some 300 pairs and the current population is likely to be higher, based on BBS trends. Remarkably the winter population estimated at 1,000 in the late 80s is pretty much mirrored by the size of the wintering population in recent years. The largest wintering flocks tend to be in the Fens from Crowland in the south up to Branston Booths near Lincoln.

1968-1972	1980-1989	2008-2011	2016-2019	Incidence	Change
				19%	5%

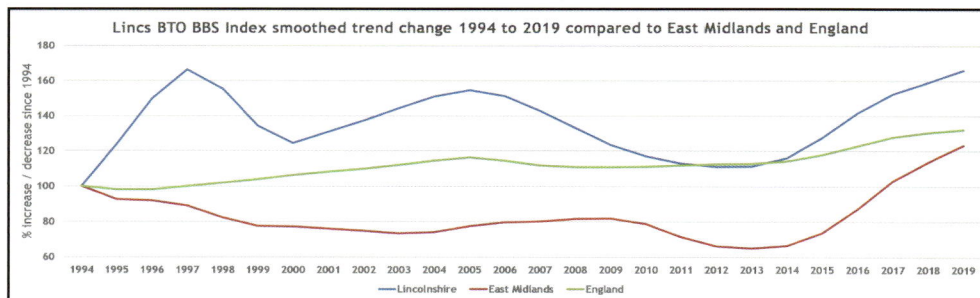

Lincs BTO BBS Index smoothed trend change 1994 to 2019 compared to East Midlands and England

Bewick's Swan *Cygnus columbianus*

Scarce passage migrant and very scarce winter visitor, mainly Oct-Mar, exceptional in summer. Amber List.

This species and Whistling Swan (see below) are races of Tundra Swan *Cygnus columbianus*, with Bewick's Swan *C. c. bewickii* the familiar British representative.

The 8th census of Bewick's Swan in Britain in Jan 2015 yielded a total of 4,371 birds, which showed a decline of 38% compared to 2010. Over the longer term, the 10-year trend 2007-08 to 2017-18 from BTO WeBS data showed a 71% decrease. The *LBC Atlas* notes that migrant flocks of 100 or more were seen in Lincolnshire in the 1980s and 1990s but in the last 10 years totals across the county reported in LBRs have rarely risen above 50. Small migrant flocks may be encountered almost anywhere in the Fens with some areas like Bardney-Branston Booths-Nocton Fen regularly holding a small wintering population consorting with the much more numerous Whooper Swan *Cygnus cygnus*.

Bewick's Swan (Whistling) *Cygnus columbianus columbianus*

Vagrant. North America.

One record on Jan 22nd 1998 at Nocton Fen. Only the second British 'Whistling Swan', with previous records 1986-90 in Hampshire and (mainly) Somerset referring to a returning individual. There have been no other British records. Elsewhere (allowing for duplication) there have been three confirmed records in the Republic of Ireland and five in the Netherlands. It remains a true vagrant in Europe. This sub-species breeds on the tundra of Arctic North America, also extreme NE Siberia. They winter in western and coastal eastern USA and N Mexico.

Whooper Swan *Cygnus cygnus*

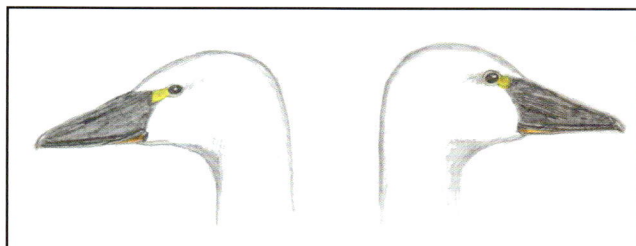

Fairly common passage migrant and winter visitor, with a notable increase since the late 1990s. Occurs mainly Oct-Apr, exceptional in summer.

WeBS reports indicate that the English wintering population of Whooper Swan grew by 425% over the 25 years to 2017-18 and 56% of that growth was in the last 10-year period. The *LBC Atlas* reports that up to the late 1990s at least, large flocks above 50 were rarer for Whooper than Bewick's Swan *C. columbianus*. A useful chart in LBR 2015 shows peak winter numbers from 1978-79 to 2014-15. Annual peak counts were below 50 up to 1994-95 and reached 100 in 1998-9, 200 in 2005-6 and 400 in 2014-15. In Jan 2019 a massive 864 were counted across all sites. It is possible some of these birds were moving between sites. The main locations currently are the coast from Donna Nook to Anderby, Nocton/Branston Fens, Frampton Marsh and Wroot. Mar sees large movements of birds relocating back towards Iceland. The largest recorded was a flock of 500 seen flying over Willow Tree Fen on Mar 13th 2016. Summering of presumably injured birds is becoming more frequent but there have been no suggestions of breeding pairs. In Jan 2014 two Whooper Swans in a small herd near Frampton Marsh carried colour rings subsequently shown to have been ringed in the same catch in Iceland Aug 2013.

Egyptian Goose *Alopochen aegyptiaca*

A scarce feral resident.

First bred in the early 1980s at Grimsthorpe Park in the south-west but numbers grew slowly. The bulk of the county population is now concentrated in the Woodhall Spa Airfield/Tattershall/Kirkby on Bain area but recent increases have seen a spread of records, including breeding, more widely across the county. It remains fairly local and the breeding population is probably in the order of 25 pairs. The largest counts tend to be in late summer, when with young, peak counts across the county can be over 60 birds. Records tend to be received from around 35 sites each year indicating birds move about quite a bit, especially in winter.

1968-1972	1980-1989	2008-2011	2016-2019	Incidence	Change
				0.6%	

Shelduck *Tadorna tadorna*

Common resident, partial migrant and winter visitor, with concentrations mainly in the Humber, on the north-east coast and in The Wash. Fairly common as a breeding species in estuaries and coastal areas, scarcer inland.

The *LBC Atlas* estimated the breeding population in the late 1980s at around 700 pairs, split between The Wash (400), Humber (200) and inland (100). The winter population was put at 10,000 to 15,000 birds. BBS suggests that the breeding population has declined but the sample is too small to draw firm conclusions. WeBS reports suggest that the 5-year winter moving average total for the Humber and The Wash from 2014-15 to 2018-19 is 4,700 and 2,250 respectively, a total of 7,000 birds suggesting a fall in the Lincolnshire wintering population. Pyewipe Marsh at Grimsby is a particularly important site. In Jul 2020, 6,650 Shelduck were counted there, over 10% of the UK population in one flock. The highest count on the Humber since 1990-91.

1968-1972	1980-1989	2008-2011	2016-2019	Incidence	Change
				11%	3%

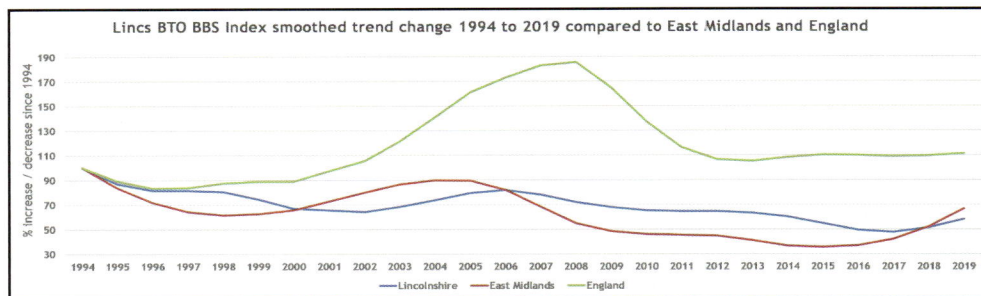

Lincs BTO BBS Index smoothed trend change 1994 to 2019 compared to East Midlands and England

Ruddy Shelduck *Tadorna ferruginea*

Very scarce. Birds are recorded in most years, mostly in late summer/early autumn. Most if not all are assumed to come from feral stock in the Netherlands.

This species has a very chequered history in Britain and it currently has Category B/D/E status; category B refers to a species recorded in an apparently natural state at least once between Jan 1st 1800 and Dec 31st 1949, but which have not been recorded subsequently (in the wild state). This is slightly odd given that there have been invasion years of this species and the first county record was of one shot at Humberstone Fitties in Sep 1892, which was an invasion year. There may have also been wild birds in 1898, 1919 and more recently in summer 1994 when an influx occurred in Britain and in north-western Europe. In 1994 there was one on the coast at Saltfleetby Jan 2nd and 4th, four Barton Pits Jul 29th and one Pyewipe Oct 1st.

All British records post-1950 have been given Category D status (there is reasonable doubt that they have ever occurred in a natural state). Small flocks are regular in the county and form an occurrence pattern unique among wildfowl of suspect (captive) origin with most arriving Jul-Aug and most likely the result of feral birds arriving from Europe, especially the Netherlands, post-breeding; flocks of up to 11 on the Humber bank have been noted in recent years.

Observers should eliminate the possibility of the similar Cape Shelduck *T. cana* when confronted with a possible Ruddy Shelduck; hybrids between these two species have also been noted in the county.

Mandarin Duck *Aix galericulata*

Scarce and increasing feral resident or visitor and local escapee.

In the 1980s the *LBC Atlas* produced one confirmed breeding record near Tallington, and it was suspected these birds might be local escapes. Since that time Mandarin has spread widely from southern England into the river valleys of the Pennines of Derbyshire and Yorkshire and the Yorkshire Moors. Breeding has become more regular in the south-west of the county at Grimsthorpe and Belton Parks and as far north-east as Kirkby on Bain. It is likely the spread will continue, and dispersing birds can appear on water bodies anywhere in the county. In 2018 they were reported from 14 sites.

1968-1972	1980-1989	2008-2011	2016-2019	Incidence	Change
				0.3%	-14%

Garganey *Spatula querquedula*

Scarce passage migrant and very scarce summer visitor. Exceptional in winter.

The *LBC Atlas* suggested that 10 pairs per annum occurred in the late 1980s, but most were never confirmed as having bred. Using perhaps a slightly tougher standard RBBP records show that an average of five pairs per year bred during the period 2013-17. An early migrant, pairs often turn up from late Mar and through into May, but usually only stay for a few days. They are generally reported from around 25 sites a year. The most consistent sites where breeding has been confirmed in recent years have been Frampton Marsh and Middlemarsh Farm, near Skegness. Records after the end of Oct are exceptional. In the 10 years to 2018 the only winter record was one seen at Alkborough Flats on Jan 14th and 27th 2018.

1968-1972	1980-1989	2008-2011	2016-2019	Incidence	Change
				0%	

Blue-winged Teal *Spatula discors*

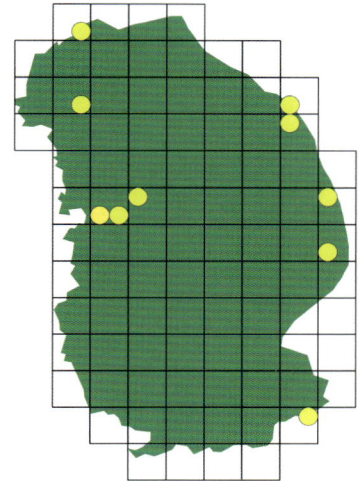 BBRC
British Birds Rarities Committee

Rare. North America.

Blue-winged Teals are fairly common in wildfowl collections and some suspicion is often attached to sightings, but true vagrancy is evidenced by ringing recoveries between Canada and Europe including three to Britain and Ireland, one of which was in Suffolk.

The first record was of an immature drake that arrived at a wildfowl collection at Sudbrook Manor, near Grantham in 1941, its identification being established for certain once it developed its full plumage. The seasonal distribution of 11 records involving 13 individuals matches that of British records as a whole, with most appearing in autumn but a smaller peak in spring.

The female at Messingham SQ in 1997 was apparently paired with a drake Shoveler *Anas clypeata*. The three together at Boultham Mere in 2013, two males and a juvenile female, was the most remarkable of all the county records to date.

Site Name	First Date	Last Date	Count	Notes
Sudbrooke	10/09/1941	23/10/1941	1	Male
Lincoln	22/04/1947		1	Male
Wisbech STW	12/09/1978	24/09/1978	1	
Wisbech STW	12/09/1978		1	
Huttoft	26/09/1982	30/09/1982	1	Immature male
Messingham	11/03/1997	13/04/1997	1	female
North Somercotes	28/09/1997		1	First winter
Boultham Mere	13/09/2013	12/11/2013	3	
Middlemarsh Farm	29/06/2014	02/07/2014	1	
Donna Nook	12/04/2015	28/04/2015	1	
Alkborough	11/08/2018		1	Second calendar year male

Shoveler *Spatula clypeata*

Fairly common winter visitor and passage migrant, and scarce breeding species.

The *LBC Atlas* suggested a breeding population of around 50 pairs and a wintering population of 150 birds in the late 1980s. RBBP records show an average of 38 pairs per year bred during the period 2013-17. Breeding reports are only received from a handful of sites each year and it is possible that the recent figures are underestimated. Frampton Marsh usually leads the way with around 25 pairs a year. The wintering population is now around 1,000 birds indicating a substantial increase over the last 30 years.

1968-1972	1980-1989	2008-2011	2016-2019	Incidence	Change
				0%	

Gadwall *Mareca strepera*

Fairly common though localised breeding species and winter visitor. Numbers have increased since the 1980s.

Gadwall only started to summer in Lincolnshire from the beginning of the 1970s, having formerly been a winter visitor from Germany. They began breeding more frequently in the 1980s and the *LBC Atlas* estimated the breeding population at 50 pairs and a wintering population of 400 birds in the late 1980s, growing to 850 in winter by 1996. It has since become much more widespread in the 10 years to 2019. BBS data for England suggests the population has tripled since 1994 and it is likely that the breeding population is now in the region of 150 to 200 pairs. LBR reports indicate the wintering population is now over 1,200 birds.

1968-1972	1980-1989	2008-2011	2016-2019	Incidence	Change
				6%	257%

Salt Marsh at Freiston Shore RSPB

Falcated Duck *Mareca falcata*

 BBRC

Vagrant. East Asia.

Field identification is often straightforward but determining their provenance as either wild or of captive origin can be extremely problematic and this species is no exception. The assessment of this species' position on the British List has lasted for more than 30 years but after an extensive review of British records of Falcated Duck by the BOURC, the species was added to Category A of the British List in Feb 2019. The index record was determined to have been a drake present at Welney, Norfolk Dec 9th-27th 1986; it moved to Pitsford Reservoir, Northants, from Feb 15th-Apr 5th 1987 and returned the following autumn and winter to Welney on Aug 20th-Oct 8th, then moving to Northamptonshire from Dec 12th. The first accepted county record thus becomes the drake present at Kirkby GP Feb 19th-21st 1995.

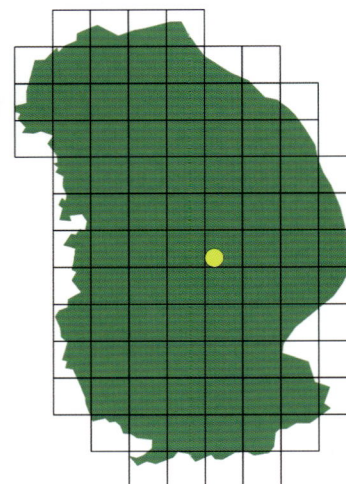

Falcated Duck breeds mainly in eastern Russia, in southern Siberia, Transbaikal, Amurland and Ussuriland. Outside Russia it also nests in Japan, parts of Mongolia and northeast China but it is nowhere common. The most recent estimate puts the global population at 89,000 birds with around 78,000 of these in China outside the breeding season. The species has also been recorded as a vagrant in the Indian Subcontinent, Central Asia, the Middle East, Europe and North America (Carboneras and Kirwan 2020).

Site Name	First Date	Last Date	Count	Notes
Kirkby on Bain	19/02/1995	21/02/1995	1	Male

Wigeon *Mareca penelope*

Common or very common winter visitor and passage migrant, especially to The Wash and the Humber. Scarce in summer.

The British wintering population of Wigeon is currently around 260,000 birds and Lincolnshire holds around 18,000 of them. Birds arrive from Iceland, Scandinavia and Russia from Oct through to Dec. The *LBC Atlas* estimated a wintering population of 10,000 in the late 1980s but suggested it could reach 25,000 in severe winters, which have been few and far between in recent years. One of the finest ornithological experiences of the Lincolnshire winter is the sight and sound of 5,000 plus Wigeon taking to the air at Frampton Marsh.

Most leave by Apr but with so many birds around it is not surprising that a few summer each year. In 2019 up to 28 birds were present across 10 sites in Jun. Pairs have been suspected of breeding in five of the last 20 years to 2019, but no confirmed breeding has been proved.

1968-1972	1980-1989	2008-2011	2016-2019	Incidence	Change
				0.3%	

American Wigeon *Mareca americana*

LBRC

Rare. North America.

First recorded in Feb 1974 there have now been 20 records in total involving 17 birds, 16 of them males, one a female. Most have turned up during the first and last quarters of the year. Like other rare wildfowl they appear to easily survive after a transatlantic crossing and associate with Eurasian Wigeon *M. penelope*. Hybrids with that species are reported, one example of which was a male resembling an American Wigeon at Covenham Reservoir in Oct-Nov 2016; the exact parentage was not established though.

There had been at least 341 British records from 1950 up to the end of 2001 when it ceased to be considered by BBRC.

Site Name	First Date	Last Date	Count	Notes
Covenham Reservoir	12/02/1974	24/02/1974	1	Male
Kirton in Lindsey	14/09/1990		1	Same male
Messingham	14/09/1990	01/12/1990	1	Male
Whisby	21/10/1991	01/11/1991	1	Male
Toft next Newton	14/03/1996		1	Male
Reads Island	05/10/1997		1	Male
Marston	06/04/2000	07/04/2000	1	
Marston	12/01/2002	27/01/2002	1	Adult male
Covenham Reservoir	30/09/2003	11/10/2003	1	
Covenham Reservoir	12/10/2003		1	Adult male
Covenham Reservoir	03/01/2004		1	Same adult male
Kirkby on Bain	02/10/2004		1	Adult male
Freiston Shore	27/02/2007		1	Adult male
Bardney	05/03/2007		1	Female
Carlton & Manby Washlands	01/04/2009		1	Adult male
Freiston Shore	13/07/2011	18/07/2011	1	Adult male
Fiskerton	09/02/2013	03/04/2013	1	
Kirkby on Bain	14/12/2018		1	Adult male
Kirkby on Bain	07/01/2019	27/01/2019	1	Same adult male

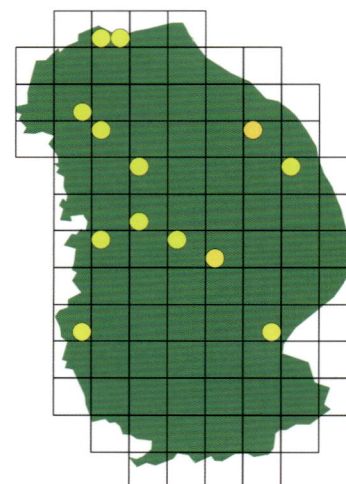

Mallard *Anas platyrhynchos*

Common resident, passage migrant and winter visitor.

Our commonest resident duck. In the late 1980s the *LBC Atlas* put the breeding population at 7,500 pairs and the wintering population at 20,000 birds. Recent evidence suggests shooting interests release tens of thousands of Mallard a year in the county. Where they come from, how many are shot and their impact on the native population is not known. BBS data shows that during the period 1994-2018 the breeding population has fallen by 10%. The estimated population for Lincolnshire was 6,000 pairs in 2016.

The current position of the wintering population is more concerning. Peak winter counts across the county in recent years reported in LBR suggest a winter population of some 4,000 birds, a fall of 80% since the 1980s, compared with a fall of 33% for England reported by WeBS Online for the period 1992-93 to 2018-19. The reasons for this fall are not clear but one possibility is that birds from Russia and Scandinavia are wintering further east as the climate changes.

1968-1972	1980-1989	2008-2011	2016-2019	Incidence	Change
				69%	0%

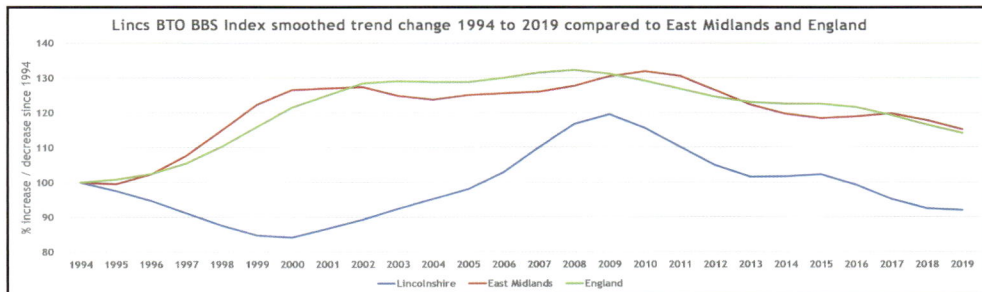

Lincs BTO BBS Index smoothed trend change 1994 to 2019 compared to East Midlands and England

Pintail *Anas acuta*

Fairly common passage migrant and winter visitor. Very scarce in summer.

The wintering Pintail population is currently in the order of 500 birds focused primarily in The Wash at Frampton Marsh and Freiston Shore and on the Humber at Alkborough Flats. Small flocks occur regularly inland but rarely get into double figures. Birds increasingly summer and in 2018 three such pairs were noted at three separate sites, but there was no other evidence of breeding. Hollings and the RBBP (2010) confirmed breeding took place in 2007 at an undisclosed site in the county where seven ducklings were seen. The only instance of confirmed breeding in the last 29 years to 2019.

1968-1972	1980-1989	2008-2011	2016-2019	Incidence	Change
				0%	

Teal *Anas crecca*

Common migrant and winter visitor. Scarce in summer, and very scarce and localised as a breeding species.

The *LBC Atlas* put the wintering population of Teal at between 4,000 and 7,800 birds and the breeding population at around 25 pairs in the late 1980s. Few reports of confirmed breeding are received, and the true breeding status of Teal in Lincolnshire is difficult to ascertain.

The wintering population is much better counted. The rolling five-year mean population over the period 2014-15 to 2018-19 reported on WeBS Online is around 8,600, split between The Wash (3,600), the Humber (2,900) and inland sites (2,100). The origin of wintering Teal in the county have been revealed by ringing studies with recoveries of Lincolnshire-ringed birds in Belarus (three), Belgium (one), Bulgaria (one), Estonia (two), Greece (one), Hungary (one), Italy (one), Morocco (one), Norway (five), Poland (four), Spain (one) and Turkey (one). Foreign-ringed birds have been recovered (mostly shot) in the county from Iceland in the north-west to far eastern Russia.

1968-1972	1980-1989	2008-2011	2016-2019	Incidence	Change
				0.2%	

Green-winged Teal *Anas carolinensis*

Very scarce. North America. First record in 1984 and a regular visitor in last 10 years.

Including the first record in 1984 there have now been 48 records of Green-winged Teal, but it is difficult to know how many individuals have occurred, since a number of records refer to birds occurring elsewhere a short time before, or at the same site in previous years. All of them have been drakes with no conclusive records of females. There have been 18 records at Alkborough Flats 2006-19 in every year apart from 2013, with two there in 2008. Another eight records have come from sites along the Humber Bank. How many individuals this involves is purely speculative given that the BTO ringing data shows that the longevity record for Eurasian Teal *A. crecca* is more than 18 years.

Away from the Humber Bank, there have been seven records at Gibraltar Point and several on inland freshwater bodies. Most birds are found in autumn and over the winter, but there is also a marked spike of arrivals during spring passage.

The earliest in autumn was at Alkborough Flats Oct 18th 2011 and the latest in spring one at Deeping Lakes NR on Jun 3rd 2018.

Red-crested Pochard *Netta rufina*

Scarce winter visitor and very scarce breeding species.

Originally a vagrant in Lincolnshire from southern and eastern Europe, a pair bred at Tetney in 1937. Lorand and Atkin (1989) noted that the species had spread north-west as a breeder as far as Denmark and The Netherlands by the 1940s. The precise origin of the Lincolnshire population is unknown, but a breeding population is now established. It is possible that both wild and escaped birds are involved. Records increased through the 1980s and a pair bred in the Baston/Langtoft/ Deeping area in 1996, which has since become the main focus of the Lincolnshire breeding population and especially the wintering population. In recent years this has been around 100 birds based on reports in LBR. Reports indicate eight pairs bred in 2018 and six in 2019.

1968-1972	1980-1989	2008-2011	2016-2019	Incidence	Change
				0%	

Pochard *Aythya ferina*

Fairly common but declining winter visitor, and a scarce breeding species found mainly in gravel and clay pits.

Pochard has shown fluctuations in both breeding and wintering populations over the last 30 years. The *LBC Atlas* put the wintering population at 2,000 birds in the late 1980s increasing to 2,500 by 1997. Over the same period the breeding population doubled from 25 to 50 pairs. More recently the wintering population has fallen to around 500 birds and RBBP records show an average of 22 pairs per year bred during the period 2013-17, a halving over 20 years. WeBS Online data shows that the English Pochard wintering population peaked in 1995-96 and has since declined by two thirds up to 2018-19. The decrease is attributed to Russian birds wintering further east due to climate change.

1968-1972	1980-1989	2008-2011	2016-2019	Incidence	Change
				6%	-47%

Ferruginous Duck *Aythya nyroca*

BBRC

Rare. Eastern Europe.

The first county record is of at least one bird (unsexed) at Newsham Lake on the Brocklesby Estate Feb 3rd-24th 1929. It remains a rare bird in the county with 17 more records since then involving 16 individuals. Of those birds which were sexed only the bird in 1959 was considered a female. Eight have occurred during both Jan-Mar and Sep-Dec, and there are two summer records in Jul. All have been on inland freshwater bodies with the exception of one at Goxhill Marsh in 1960. Six long-stayers were present 19-31 days, the remainder nine days or less with six being one-day birds.

There have been 685 British records 1950-2018 and Ferruginous Duck was removed from the list of species considered by BBRC after 2005, remaining so up to 2016. However, due to its growing scarcity during that period, it was re-instated from 2017 onwards. It should be noted that there are difficulties in assessing some records with complications including escapes from captivity, birds from the German reintroduction scheme launched in 2011, and associated hybrids with other *Aythya* species.

Site Name	First Date	Last Date	Count	Notes
Brocklesby	03/02/1929	24/02/1929	1	
Holywell	25/12/1957	27/12/1957	1	Male
Baston	21/11/1959		1	Female
Goxhill	04/09/1960	26/09/1960	1	Adult
Wisbech STW	22/07/1972	10/08/1972	1	Male
Boston	16/03/1974	24/03/1974	1	Male
Covenham Reservoir	04/11/1974	05/11/1974	1	Male
Chapel St Leonards	11/07/1976		1	Male
Chapel St Leonards	24/01/1980	30/01/1980	1	Male
Killingholme	05/03/1981	06/03/1981	1	Male
Kirkby on Bain	18/10/1998	05/11/1998	1	Adult male
Kirkby on Bain	10/12/1998		1	Same
Barton Pits	16/02/2001	18/03/2001	1	
Huttoft	25/02/2007		1	
Covenham Reservoir	11/09/2008		1	
Bagmoor	28/10/2008		1	
Whisby	05/01/2009	31/01/2009	1	
Glentworth	06/10/2009	11/10/2009	1	Eclipse male

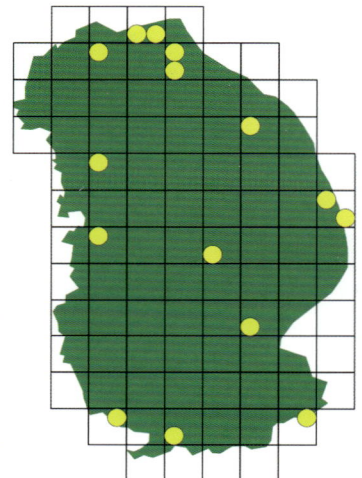

Ring-necked Duck *Aythya collaris*

Very scarce visitor. North America.

The first county record was a male at Messingham in Jun 1979, returning in Oct 1979 and also seen at nearby Barrow Haven into Nov 1979; what was likely to have been the same bird was seen in Feb and Apr 1980. There was an eight-year gap before the next record in 1988 but after that the species has turned up regularly to 2019. Most of these have been males but there have been at least four females in 2000, 2003, 2006 and 2017; the female in 2000 arrived with a 1CY male. Determining the number of birds involved is difficult but a reasonable estimate is of between 15-20 birds. The gravel pits in the Kirkby/Tattershall area and the Humber bank clay pits have been regular locations for Ring-necked Duck in recent years. Many of these birds have been long stayers but also very mobile between sites. A 1CY drake arrived at Gibraltar Point on Dec 28th 2006 staying until Apr 26th 2007. Another male, an adult, arrived Jan 15th 2012 at Kirkby GP and toured other water bodies at West Ashby and Donington on Bain before finally being at Kirkby GP on Apr 23rd 2012.

The species was removed from the BBRC list in 1994 and there have been 854 British records 1958-2018 (White and Kehoe 2020).

Tufted Duck *Aythya fuligula*

Common/fairly common breeding species on gravel pits and similar water bodies, and common winter visitor.

The *LBC Atlas* put the wintering population at around 2,500 birds and the breeding population at around 1,250 pairs in the late 1980s. The latter had increased dramatically from around 50 pairs in the 1960s. WeBS Online indicates that the Lincolnshire wintering population is still around 2,500 birds, but the breeding data provides conflicting information. BBS suggests there has been a small long-term decline from 1994 to 2019, and incidence has fallen over the last 10 years. The estimated population for Lincolnshire in 2016 was in the region of 550 pairs which looks low. The real position currently, is probably somewhere between 550 and 1,250 breeding pairs.

1968-1972	1980-1989	2008-2011	2016-2019	Incidence	Change
				12%	-17%

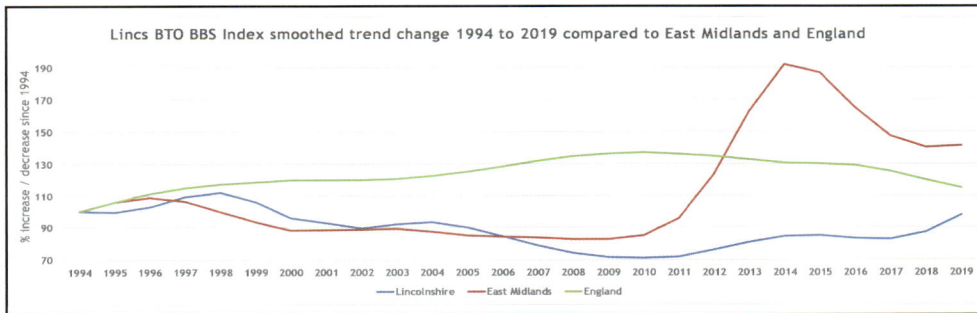

Lincs BTO BBS Index smoothed trend change 1994 to 2019 compared to East Midlands and England

Scaup *Aythya marila*

Scarce passage migrant and winter visitor to estuaries and coast; numbers have declined. Rare in summer; bred in 1944.

Cordeaux (1872) stated that Scaup was the commonest duck wintering on the Humber. Flocks of more than 1,000 were recorded there in the late 19th century. The *LBC Atlas* refers to flocks of several hundred wintering in the 1980s. In recent years WeBS Online reports show that the wintering population has averaged around 35 birds per year to 2018-19 with a mean of 13 on The Wash, 12 on the Humber and 11 at inland sites. The maximum British wintering population in recent years has been 1,954. It is the most irregular breeder imaginable; a female was found with seven ducklings on Tetney Marsh in 1944. No breeding has been suspected since, though birds do sometimes summer on the Humber Bank clay pits.

1968-1972	1980-1989	2008-2011	2016-2019	Incidence	Change

Sea Bank at Frampton Marsh RSPB

Lesser Scaup *Aythya affinis*

BBRC

Vagrant. North America.

Not recorded in Britain until 1987, it was another eight years before one was found in Lincolnshire, a 2CY male at Barton-upon-Humber, in Feb 1995. Most of the Humber duck flock were flighting onto one of the east side pits at Barton to bathe and roost in the late afternoons. On Feb 13th there were 270 Pochard *A. ferina*, and two drake Greater Scaup *A. marila*, and on Barrow Mere, 300 Pochard, and a few Goldeneye *Bucephala clangula*, Ruddy Duck *Oxyura jamaicensis* and two more Greater Scaup together with a rather odd-looking duck resembling a Tufted Duck, but fast asleep. It looked reminiscent of a Lesser Scaup *A. collaris*, and soon after it awoke its identity was confirmed. Aged as a 2CY bird due to the extensive area of brown feathering in the plumage. There have been four others since, three males and one female, the most recent being a 2CY male at Freiston shore in Mar 2013.

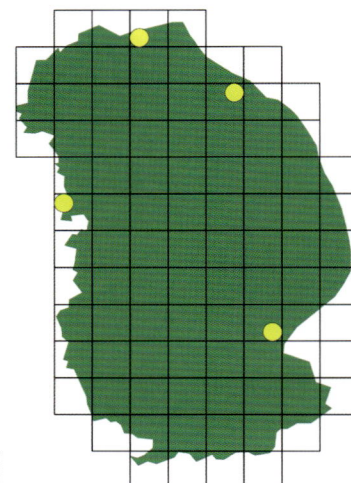

Once the unique features identifying this Nearctic duck became more widely known the number of British records gradually increased through the 1990s and 2000s with a peak of 27 in 2007 and it was recorded every year between 1992-2020. It was removed from BBRC consideration in 2015 but added back again in 2020.

Site Name	First Date	Last Date	Count	Notes
Barton upon Humber	13/02/1995	16/02/1995	1	Second calendar year male
Torksey	14/05/1998		1	
Cleethorpes	05/04/1999	16/04/1999	1	Female
Barton Pits	28/04/2004	30/04/2004	1	Adult male
Freiston Shore	19/03/2013	14/04/2013	1	Second calendar year male

King Eider *Somateria spectabilis*

BBRC

Vagrant. Scandinavia.

This high Arctic duck has been hard to find offshore in the county despite the presence of large flocks of Common Eider *S. molissima*, wintering in The Wash. The first record was of an eclipse male found off the Witham Mouth on Sep 5th 2009 where it remained until Oct 9th. It moved slightly north later in its stay and was last seen off Leverton on Oct 18th. Many local observers saw this bird from land, but several boat trips were hastily arranged to get better views.

Ageing proved difficult, but it most resembled a second calendar year bird. When first found, it had completely moulted its primaries and secondaries. A boat trip into The Wash on Oct 4th observed its first flight, so it had taken a maximum of thirty days to complete a full wing moult. This was the first county record. The second record of an adult female in Apr 2012 was also found by a boat trip. It later relocated along The Wash and was seen in Jul from Gibraltar Point.

Scotland is still the best place to see this species and BBRC statistics show that there have been 194 British records 1950-2018.

Site Name	First Date	Last Date	Count	Notes
Freiston Shore	05/09/2009	18/10/2009	1	Male
Wrangle	21/04/2012		1	Adult female
Gibraltar Point	01/07/2012	07/07/2012	1	Same, Adult female

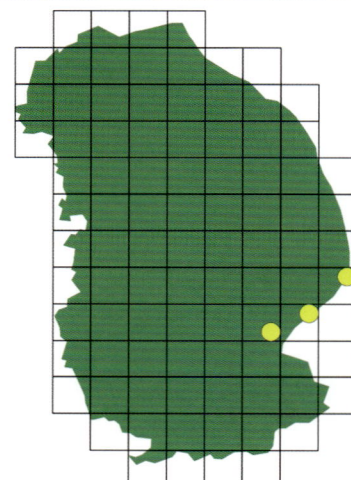

Eider *Somateria mollissima*

Present offshore throughout the year: fairly common on passage and in winter, scarce in summer. Mainly found in The Wash and rare inland. Under-recorded because most of the population stays well offshore; however, numbers have recently declined.

The UK wintering Eider population is in decline, the WeBS 25-year trend showing a 26% fall to winter 2018-19. Most birds in Lincolnshire winter on The Wash. LBR 2017 reported a record low peak count of 34 from the Lincolnshire Wash coast in Feb 2017. The recent peak for The Wash had been 3,721 in winter 2011-12 and the previous low was 226 in 2016-17. In winter 2018 there were 896. Birds can move around and are often far out to sea, utilising sandbanks at low tide. During a 10-year low tide count of Roger and Toft Sands, four km offshore from Wainfleet Flats and only accessible by boat, 350 were counted roosting on sandbanks in Jan 2020. Summer counts which have always been much lower than winter ones have also fallen. There was a low of only two in Jul 2018 but so far there has not been a month with no Eider recorded.

Velvet Scoter *Melanitta fusca*

Scarce passage migrant and winter visitor. Rare inland and in summer.

The *LBC Atlas* reported that in the mid 1990s flocks of up to 100 had been seen exceptionally, but nothing like that has been reported in recent years. Over the five years to 2018 LBR reports indicate that annual numbers ranged between 29 (2015) and 124 (2017). The largest flocks were both in Jan 2017 and peaked at 44 off Saltfleetby-Theddlethorpe and 21 off Tetney. Some of these birds stayed into Mar. Typically, no birds are seen from May until passage begins, usually in Aug, rarely in Jul. Passage picks up through autumn until Nov when wintering flocks sometimes stay through until Mar. During this five-year period there were two reports of inland birds. A single one-day bird at Covenham Reservoir in Nov 2016 and two at Toft Newton Reservoir in 2014 that arrived on Nov 16th and stayed though to Dec 13th.

Common Scoter *Melanitta nigra*

Common offshore passage migrant and winter visitor, scarce in summer but autumn movements begin in Jun-Jul. Very scarce inland.

Regular all along the coast from autumn until spring, the *LBC Atlas* reported that up to the mid-1990s flocks of several hundred and on rare occasions more than a thousand could be seen along the coast. There is now good evidence from visible migration counts of a westerly migration along the Humber, presumably to the Mersey, in the autumn and more recently in 2020 from nocturnal migration recordings, of a wide front of spring migration across both the country and the county from west to east in Mar. As large flocks can be found well out to sea it is difficult to accurately estimate the wintering population and WeBS is unable to monitor long term trends for this species from onshore counts. The volume of passage and the number of birds wintering varies significantly from year to year as reported in LBR and viewed from the shore. In the five-year period to 2018, 2014 was very poor and 2018 was good with a peak of over 2,700 in Oct and Nov. The largest flocks counted tend to be off Saltfleetby-Theddlethorpe where there were 2,000 on Nov 16th and 1,500 on Nov 18th, and there were 1,650 off Friskney in the Wash in Feb 2018.

Long-tailed Duck *Clangula hyemalis*

Scarce passage migrant and winter visitor, rare in summer and inland.

The *LBC Atlas* reported up to 50 at Witham Mouth in the 1980s and 1990s but nothing like that number has been seen in the five years to 2018. Total annual counts inferred from LBR reports range from 10 (2015) to 43 (2016). The largest flocks reported were in Nov 2016 with 17 at Witham Mouth on Nov 7th and 13 at Sutton on Sea on Nov 16th. Coastal records split between 33 from Jan-May and 83 from Sep to Dec.

Birds reported inland ranged from five in 2014, four in 2016, two in 2017 and one in 2015 and 2016. Covenham Reservoir used to be a regular inland location. This is another species whose offshore wintering pattern makes it difficult to monitor long term trends using shore-based counts.

Bufflehead *Bucephala albeola*

Vagrant. North America. Sometimes escapes from captivity.

There is just one accepted record at Covenham Reservoir on 27th April 2012, age and sex uncertain; this was a county first and at the time the 15th British record. Known escapes from captivity included a female which frequented Huttoft Pit 2002-05, and more recently another female, an adult, at Baston and Langtoft GP on Mar 30th 2019. This latter bird was ringed and had been seen elsewhere in the East Midlands.

There have been 16 accepted British records 1950-2018 including three males in the spring of 2004. All but three of the previous records have been since the well-watched drake at Colwick Country Park Mar 17th-26th 1994. Nov appears to be the best month for vagrant Buffleheads with four records. As ever, wild origin is often difficult to prove in unringed birds but the surge in records of other Nearctic duck species such as American Wigeon *Mareca americana*, Green-winged Teal *Anas carolinensis* and Ring-necked Duck *Aythya collaris* from the late 1990s provides some evidence for wild origin, although some might disagree.

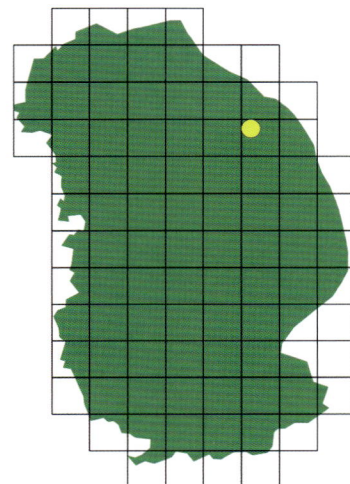

Site Name	First Date	Last Date	Count	Notes
Covenham Reservoir	27/04/2012		1	Juvenile male

Goldeneye *Bucephala clangula*

Fairly common but somewhat localised winter visitor, rare in summer.

This handsome duck is declining as a wintering bird in England with WeBS data showing the 25-year trend down 43% to 2017-18. This decline is reflected in Lincolnshire too. The five-year average of wintering birds shown by WeBS Online to 2013-14 was 812 (split between Humber 490, The Wash 91, and inland 231) while the five-year average to 2018-19 was 583 (split between Humber 291, The Wash 74, and inland 218). The key area for this species in Lincolnshire is the Humber clay pits. Some of the top counts from there reported in LBR in the last five years have come from the Goxhill area with 666 at Barrow-Goxhill in Jan 2016 and 340 there in Dec 2018. One of the most consistently counted sites for Goldeneye in the county has been Covenham Reservoir. Robinson *et al* in LBR 2014 reported numbers from 1969-70 to 2014-15 showing that after construction numbers stayed reasonably constant from 1983-84 ranging between 60-100 with occasional spikes in cold weather to around 150 in 2012-13. Since then numbers have stayed within the range reported. Birds arrive there from Oct and numbers peak in Feb-Mar, birds after Apr when the main departure takes place are rare. Coastal passage occurs through autumn and sometimes larger movements are noted. An example of this was on Nov 21st 2015, when up to 80 birds were seen moving north at locations between Gibraltar Point and Mablethorpe.

Smew *Mergellus albellus*

Very scarce winter visitor, with occasional larger influxes in severe weather.

Smew has always been very scarce in Lincolnshire with occasional influxes in hard winters. Over the 10 years to 2018 LBR reports ranged from two in 2015 to 31 in 2012, with an average of 13, although the average over the last five years has been only five. The second-best year was 2011 with 30. Both 2011 and 2012 were effectively winter influxes with ice and freezing weather. The largest single flock was of five on the unfrozen waters of the Witham south of Kirkstead Bridge on Feb 16th 2012. Reports from the coast on the sea are unusual. On Jan 11th 2013, three redheads were reported drifting south off Gibraltar Point. The *BTO Atlas 2007-11* suggests wintering in Britain may be declining while numbers in Sweden are increasing because of "short-stopping", i.e. a shorter migration resulting in a wintering distribution closer to breeding areas in Finland and Russia as winters become milder.

Goosander *Mergus merganser*

Fairly common passage migrant and winter visitor, mainly inland. Rare in summer.

In contrast to its close cousin Red-breasted Merganser *M. serrator*, Goosander is doing well in Lincolnshire. It differs in that most birds winter on a wide range of inland waters rather than the sea. It was reported from over 50 sites in 2018. WeBS data indicates the rolling mean five-year wintering population of Lincolnshire to 2018-19 was 124 (Inland 119, The Wash two, Humber three) while in the previous five-year period to 2013-14 it was 149 (Inland 143, The Wash two, Humber four). The 25-year long term trend for England is down 6% according to Frost *et al* in the 2018-19 WeBS report. LBR reports show that in the last five years the largest flocks come from the lakes around Lincoln where there were 95 in Feb 2017 and 58 in Dec 2015. Willow Tree Fen had 66 in Dec 2018. In this five-year period there were no records in Jul and one in Aug. In late Jul 2020 a non-birder tweeted a photo of two female or immature Goosanders said to be on the R. Freshney nr Grimsby. Not good enough to confirm the first breeding record for Lincolnshire but enough to heighten suspicion and to encourage searching for possible breeding birds on north Lincolnshire rivers in 2021.

Red-breasted Merganser *Mergus serrator*

Scarce passage migrant and winter visitor, mainly coastal and especially in The Wash. Very scarce inland and in summer.

Another declining wintering duck, Red-breasted Merganser was added to the list of species considered by RBBP from 2018 onwards. Though it has bred in neighbouring Yorkshire there is no record of it ever breeding in Lincolnshire. WeBS data indicates the rolling mean five-year wintering population of Lincolnshire to 2018-19 was 84 (The Wash 81, Humber three) while in the previous five-year period to 2013-14 it was 166 (The Wash 159, Humber seven). Clear evidence of a recent fall in line with the 25-year long term trend for England which is down 43% according to Frost *et al* in the 2018-19 WeBS report. Peak winter flocks on the sea in recent years have come from Gibraltar Point with 87 in Feb 2014, 31 in Nov 2015, 47 in Feb 2016, 59 in Jan 2017 and 27 in Jan 2018. Birds can occur in all months of the year but are very scarce from Jun-Aug.

Ruddy Duck *Oxyura jamaicensis*

Former scarce feral breeder and winter visitor, now rare.

This introduction from North America was admitted to the British list under Category C in 1971, the first county record having occurred at Burton Pits in 1964. A small breeding population became established in the Humber Clay pits from 1984. By 1989 there were 12 pairs across the county which had grown to 30 breeding pairs and 66 wintering birds by 1999. The UK population was thought to be 6,000 in 2000 and some had begun migrating to Spain where they allegedly hybridised with the endangered White-headed Duck. Spanish conservationists successfully argued for a European Union wide cull. By 2007 the Lincolnshire wintering population had grown to 284 but then extirpation began. By 2010 when the last breeding was recorded "in the north of the county" the population had fallen to 49. The last record was a single on Holywell Lake on Sep 14th 2014. Probably the first species to be directly extirpated from Lincolnshire by shooting since Red Kite around 1870.

1968-1972	1980-1989	2008-2011	2016-2019	Incidence	Change
				0%	-100%

Nightjar *Caprimulgus europaeus*

Scarce summer visitor and very scarce passage migrant.

Lorand and Atkin (1989) noted a recent decline and contraction in the range of Nightjar in the county with losses from the Market Rasen, Woodhall Spa and Lincoln areas. That decline has continued with breeding birds now found in only four localities one of which only holds a single pair. A full county survey in 1992 found a total of 59 singing males of which 39 were in Laughton Forest. This area held 32 males in 2005 and has continued to hold the bulk of the county breeding birds to 2019. Changes in forestry practice at Laughton with a move from clear felling to selective harvesting in recent years has reduced the area available to breeding Nightjars and there was a maximum of 20 singing males found in a full survey in 2015. Numbers in the Market Rasen area are similarly affected by forestry practices affecting habitat availability while the population at Crowle is on managed heathland and thus more stable. Breeding birds usually arrive from the first week of May, occasionally at the very end of Apr and most have departed breeding sites by early Sep. The county population is now in the region of 30-40 singing males and is thus somewhat critical. Passage migrants occur occasionally on the coast and may be found in the most unexpected places, such as gardens, during the post-fledging dispersal of juveniles in Sep.

1968-1972	1980-1989	2008-2011	2016-2019	Incidence	Change
				0%	

Alpine Swift *Tachymarptis melba*

Rare. Southern Europe. Recorded from Mar-Aug with a single record from Oct.

First recorded in the county on Apr 23rd 1964 at Sturton Park near Horncastle. There have been a further 20 confirmed records with the earliest on Mar 20th, the latest on Oct 24th. Gibraltar Point has been the best site to find one of these huge swifts with nine records. Not all of these birds have arrived in the best of health. The bird at Healing on Aug 6th 1971 was found in a weak condition, later dying, and the bird seen at Gibraltar Point on Apr 24th 1987 was later found dead at Seacroft on May 2nd. These birds may be tracked as they travel across the country and such was the case with one first seen at 07:00h Apr 27th 2003 at Gibraltar Point. It flew south with two Common Swifts *Apus apus*, and was later seen at two localities in Norfolk and then at Minsmere, Suffolk in the evening where it stayed for a few days.

The species was ex-BBRC in 2006, unsurprisingly given that in the peak years for this species more than 20 were being recorded and between 1950-2005 there were 484 UK records.

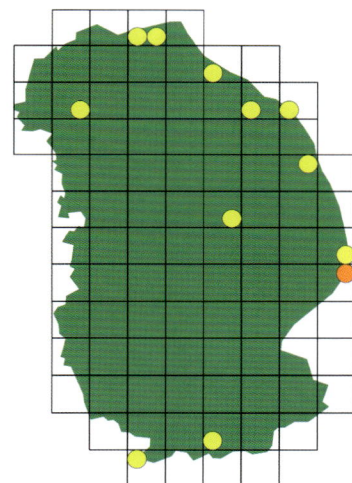

Site Name	First Date	Last Date	Count	Notes
Horncastle	23/04/1964		1	
Tetney	24/10/1969		1	
Healing	06/08/1971		1	
Gibraltar Point	11/06/1974		1	
Donna Nook	20/07/1975		1	
Messingham	16/06/1979		1	
Gibraltar Point	31/08/1985		1	
Gibraltar Point	24/04/1987		1	
Seacroft	02/05/1987		1	Same found dead
Stamford	20/03/1990		1	
Goxhill	11/06/1993		1	
Skegness	19/05/1996		1	
Gibraltar Point	27/04/2003		1	
Gibraltar Point	23/10/2006		1	
Gibraltar Point	12/04/2009		1	
Barton Pits	01/05/2009		1	
Gibraltar Point	06/07/2011		1	
Saltfleetby-Theddlethorpe	11/05/2012		1	
Gibraltar Point	03/08/2013		1	
Gibraltar Point	28/05/2017		1	
Crowland	22/03/2018		1	
Gibraltar Point	24/04/2018		1	

Swift *Apus apus*

Very common but declining summer visitor and passage migrant.

It will be obvious to many that the large screaming flocks of hundreds of Swifts of yesteryear so common in our town and villages are in decline, though they remain widely spread. The *LBC Atlas* conservatively estimated the population at 10,000 pairs in the late 1980s. The estimated British population was 59,000 pairs in 2016. The BBS index chart below reflects a steady and continuous fall of just over 70% in birds counted over the period from 1994 to 2019, though the BBS is not necessarily a good measure as Swifts tend to be away from nest sites feeding when BBS counts are made.

The coast can experience large movements of feeding birds in summer in the right weather conditions. On the morning of Jun 29th 2020, a stunning count of 45,844 birds was made at Gibraltar Point. Exactly where these birds were breeding is unknown, but the total represents nearly half of the British breeding population.

Most breeding birds depart by the end of the first week of Aug but migrants are frequent into Sep and even later; such birds always receive detailed scrutiny.

1968-1972	1980-1989	2008-2011	2016-2019	Incidence	Change
				36%	-8%

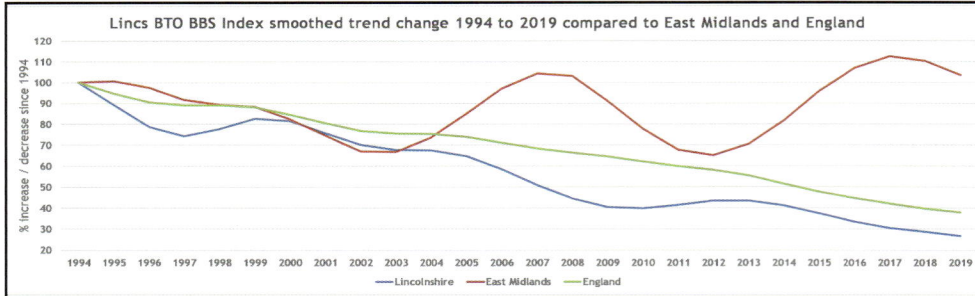

Lincs BTO BBS Index smoothed trend change 1994 to 2019 compared to East Midlands and England

Pallid Swift *Apus pallidus*

BBRC British Birds Rarities Committee

Vagrant. Southern Europe.

A real birder's bird, finding a Pallid Swift in late autumn is literally a feather in the cap for the finders. Just three have been found in the county all between Oct 23rd and Nov 11th, all one-day birds. The first was found in Oct 2004 in unusually good weather conditions cruising overhead in clear blue skies at the north end of Skegness. Viewing conditions were excellent and a full and complete description was easily made. It disappeared off to the north after a few minutes leaving the three finders in a state of some elation. The second and third records didn't arrive until 14 years later and then within days of each other. Once again, the diligence of the observers including photographs of the third bird made BBRC acceptance a formality.

First recorded in Britain in 1978, three to four Pallid Swifts arrive in Britain on average each year usually associated with warm sectors of low-pressure systems in late autumn. There have been occasional 'invasion' years as was seen in 1999 (12), 2001 (12), 2004 (16) and 2018 (15); there have been 117 records since 1978.

Site Name	First Date	Last Date	Count	Notes
Skegness	23/10/2004		1	
Gibraltar Point	07/11/2018		1	
Goxhill	11/11/2018		1	

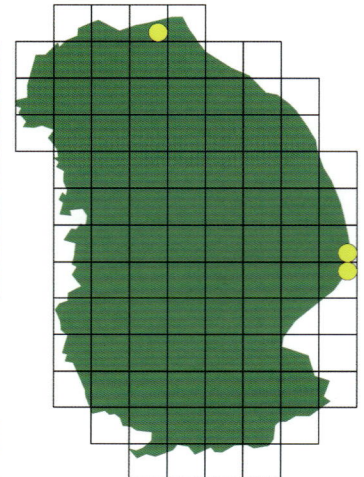

Pacific Swift *Apus pacificus*

BBRC British Birds Rarities Committee

Vagrant. East Asia.

One fortunate observer was in the right place at the right time to see and photograph the sole county record at Saltfleetby-Theddlethorpe in Jun 2013. The bird had earlier been seen moving slowly south over the Spurn peninsula before being lost to view there at 08:50h. It reached the reserve at 15:20h arriving from the south. This astounding observation proved to be the seventh British record. When the 2013 records were examined by BBRC it became clear that all four records of Pacific Swift in 2013 most likely related to just one individual roaming from Suffolk to Yorkshire, and back to Suffolk via Lincolnshire over a period of 19 days. Away from Britain, all European reports of Pacific Swift fall between May 10th and Aug 19th.

In Sweden, the repeated occurrence of an individual at the same inland site on May 15th 2013 and on May 10th-11th and May 30th 2014, gives support to the theory that vagrant Pacific Swifts may well return to Europe in subsequent years. It would be nice if the next county record stayed around for a bit longer.

Site Name	First Date	Last Date	Count	Notes
Saltfleetby-Theddlethorpe	12/06/2013		1	

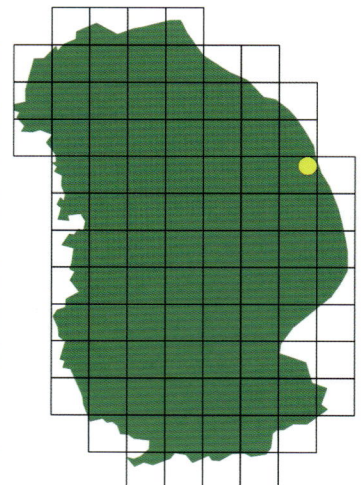

Little Swift *Apus affinis*

Vagrant. North Africa.

Two county records which occurred on successive days in Jun in 1998 and 2002. The first turned up early afternoon, with a group of around 500 Common Swift *A. apus*, feeding low over reed beds as approaching heavy rain clouds threatened. It was present until early evening only. The second at Gibraltar Point was found early morning feeding over the Freshwater Mere and present for minutes only, in the company of a few Common Swifts. It was apparent after further searching that there was a southerly movement of Common Swifts underway and that the Little Swift had moved south with them.

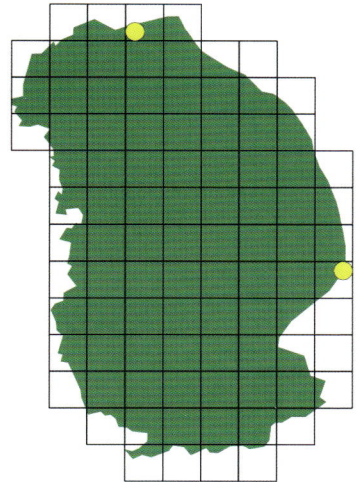

There have been 26 British records 1950-2018. The nearest birds breed in southern Spain and spring overshoots between Apr and Jun account for 19 of these. The most recent record though turned up in Nov 2018 being found by an observer who had been watching Pallid Swift *A. pallidus*, at the same site on that date. All Swifts in Nov should be scrutinised closely.

Site Name	First Date	Last Date	Count	Notes
Barton upon Humber	26/06/1998		1	
Gibraltar Point	25/06/2002		1	

Great Bustard *Otis tarda*

Vagrant. Former resident, not bred since early 19th century

The Lincolnshire Wolds were very different when Great Bustards roamed the county in the 18th century. Widespread rolling vistas with much less cultivation, few trees and no hedges. Think of the plains of Extramadura. But the enclosures from the 1760s onwards, agricultural intensification and increased shooting pressure brought about their demise by 1810. Since then there have been seven records in the county all of them historic. The two most recent records were single females in Dec 1902 and, predictably, both were shot. Other immigrants at the time, also shot, turned up in Wales and two in Ireland. These were thought to be genuine vagrants despite the earlier release of some in 1900 in an early attempt to reintroduce the species. Most Spanish birds are known to remain in Spain all year round so the source of vagrants to Britain may be further to the east from Germany and into Russia.

A brief word should be said about the Great Bustard Group, a charity set up in 1998 to reintroduce the species once again into Britain. Initially birds were obtained from Russia where eggs were collected from nests likely to fail because of agricultural operations. This practice ceased when genetic research by the University of Chester identified Spanish birds as being closest to the old British Great Bustard population and the first Great Bustards reared from Spanish eggs were released on Salisbury Plain from 2013 onwards. Wing-tagged birds do wander and have been seen as far afield as Staffordshire. In 2018 RBBP reports show that eight females nested. The population was estimated at 70 birds with an even sex ratio though only 15 females were of breeding age. There were no imports of birds or eggs in 2018.

How long the population takes to reach Category C status remains to be seen.

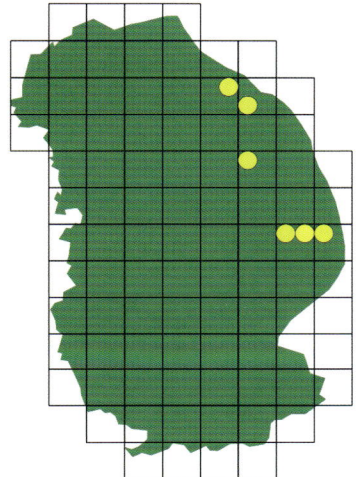

Site Name	First Date	Last Date	Count	Notes
Louth	01/01/1818		1	
Lincolnshire	01/01/1837		1	
Candlesby	01/01/1860		2	Male and female
Halton Holegate	10/04/1866		1	
Addlethorpe	01/01/1890		1	Male
Grimsby	15/12/1902		1	Female
Tetney	29/12/1902		1	Female

Macqueen's Bustard *Chlamydotis macqueenii*

Vagrant. Central Asia.

In 2004 following the recommendations of the BOURC taxonomic sub-committee Macqueen's Bustard *C. macqueenii* was elevated to full specific status. All five British records of Houbara Bustard *C. undulata* were accepted as *C. macqueenii* and so Houbara Bustard was removed from the British List. The Lincolnshire List has thus 'lost' Houbara Bustard due to this taxonomic update and 'gained' Macqueen's Bustard; the latter record on Oct 7th 1847 was the first British record.

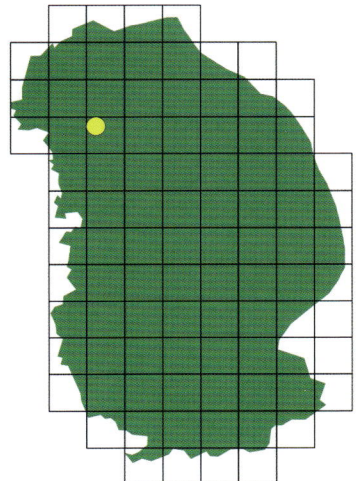

The Central Asian populations are strongly migratory, leaving their breeding grounds in Aug-Oct on trans-Himalayan migration and are thus prone to vagrancy. They have occurred in many European countries west to France and Britain as well as to the Middle East (Lebanon). However, the species has suffered severe declines from over-hunting, with an estimated 4,360 birds killed by local and visiting hunters in winter 1982-83, and 4,955 in 1984-85 in Pakistan, for example. In addition to hunting, other threats include intensive agricultural practices, human disturbance and habitat degradation through livestock overgrazing.

A repeat record would seem to be unlikely.

Site Name	First Date	Last Date	Count	Notes
Kirton in Lindsey	07/10/1847		1	Shot

Little Bustard *Tetrax tetrax*

BBRC British Birds Rarities Committee

Vagrant. Southern Europe.

The first record for the county was of one shot in a turnip field on Welbourn Heath, south of Lincoln, on Jan 30th 1854. A further five individuals have occurred, four of which were also shot: two in the 19th century and three in the 20th, with none since 1955. The female bird shot in 1933 was examined by H. F. Witherby and assigned to the eastern race *T. t. orientalis*; modern taxonomy has since evidenced this species as being monotypic.

The lack of recent records presumably reflects the declining population trend in continental Europe and matches the frequency of occurrence in Britain as a whole: some 183 records are known up to 1949 but only 29 since then up to 2019, and only five in the most recent 20 year period.

The last two records in 2015 and 2019 were both in Yorkshire so there may be some reward in searching the rolling fields of the Wolds in autumn especially.

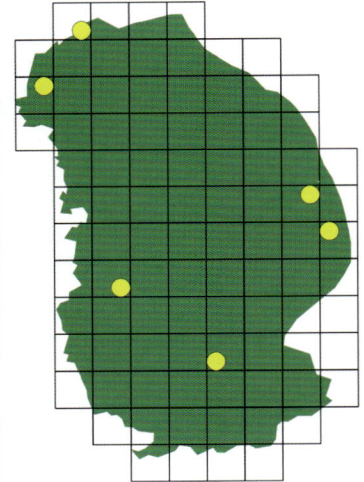

Site Name	First Date	Last Date	Count	Notes
Welbourn	30/01/1854		1	
Bilsby	01/01/1856		1	Date Jan 1856
Belton & Westgate (Axholme)	01/01/1890		1	Female date Jan 1890
Walcot	22/01/1913		1	Female
Addlethorpe	22/11/1933		1	
Gosberton	30/12/1955		1	

Great Spotted Cuckoo *Clamator glandarius*

BBRC British Birds Rarities Committee

Vagrant. Southern Europe.

Just two county records: an immature bird in May 1971 at Anderby Creek and an adult in Jul 1974 at Donna Nook. Both were one day birds, in contrast to some recent British records. The three which occurred in 2019 stayed for periods of four, 15 and 29 days respectively for example.

With 48 accepted British records 1950-2018 this species averages less than one record per year. Great Spotted Cuckoo is a summer visitor to the Mediterranean Basin, they breed regularly in Portugal, Spain and western Italy, irregularly in the Balkans and Turkey east to Asia Minor. They arrive early, sometimes even in January, departing south to its wintering grounds in sub-Saharan Africa in mid-summer. Most records in Britain occur in early spring, with Mar and Apr the peak months for arrival, and most of these are sub-adults overshooting north. Juveniles from Europe wander widely in late summer and autumn.

There is a geographical bias to records towards the south and south-west in spring (Feb-May) and towards the east coast in autumn (Jul-Oct). Cornwall and Norfolk are the best counties in which to see them, with 10 records each.

Site Name	First Date	Last Date	Count	Notes
Anderby	09/05/1971		1	Immature
Donna Nook	01/07/1974		1	Adult

Yellow-billed Cuckoo *Coccyzus americanus*

BBRC British Birds Rarities Committee

Vagrant. North America.

It's a slightly staggering statistic that this east coast county has had as many records - two - of this American vagrant as it has had of Great Spotted Cuckoo *Clamator glandarius*. The first record was of one found dead (and unringed) at Welton le Marsh in Oct 1978; interestingly another was trapped and ringed at Spurn Point three days before. Perhaps a bigger surprise was the discovery of a live individual at Rauceby Warren Oct 18th 1987 which stayed just two days. There was also another British record at this time on Lundy Island, Devon, two days after this one.

Between 1950-2018 there were 51 records and it remains a rare bird in Britain with less than one record per year. The month of Oct accounts for 70% of British records and a significant percentage of this species arriving in Britain are either dying or found dead.

Those which survive may remain for 10-12 days and are frequently observed feeding on either caterpillars or (in the south-west) stick insects.

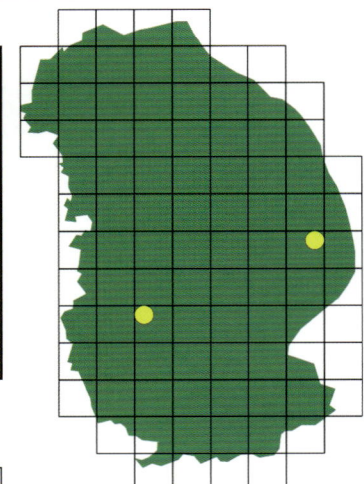

Site Name	First Date	Last Date	Count	Notes
Welton le Marsh	30/10/1978		1	Found dead skin Lincoln museum
North Rauceby	18/10/1987	19/10/1987	1	

Cuckoo *Cuculus canorus*

Fairly common but declining summer visitor.

This popular harbinger of spring is missing for an increasing number of folk in Lincolnshire. The *LBC Atlas* put the population in the late 1980s at 1,000 to 1,500 pairs. The BBS index chart shows that it declined by over 70% in the county between 1994 and 2019. The estimated Lincolnshire population in 2016 was 200 pairs, but there may still be a few more than that. They still seem to do well in coastal dunes scrub where there are plenty of caterpillars and less application of pesticides. Recent advances in satellite tracking have highlighted that the British Cuckoo population winters in the Congo basin and returns each spring via Spain. In autumn Cuckoos from northern parts of Britain follow a different migration route via Italy while the southern population is more likely to go via Spain. The population in the north of Britain seems to be doing much better, suggesting that something is going wrong on the autumn Spanish migration route and causing Lincolnshire Cuckoos to decline. Continuing research will hopefully shed more light on the precise reasons for the decline. Hopefully it may yet be reversed.

1968-1972	1980-1989	2008-2011	2016-2019	Incidence	Change
				12%	-39%

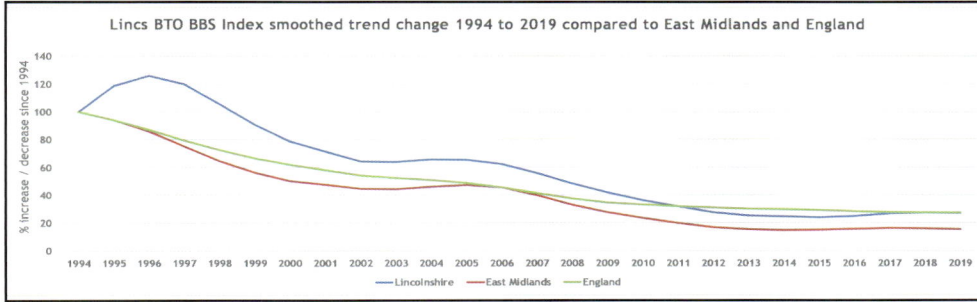

Lincs BTO BBS Index smoothed trend change 1994 to 2019 compared to East Midlands and England

Early morning at Kirkby Moor LWT

Pallas's Sandgrouse *Syrrhaptes paradoxus*

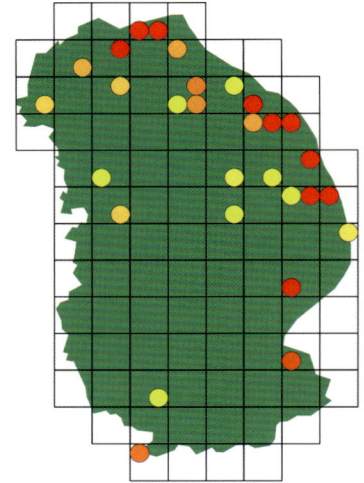

Vagrant. Central Asia.

Arguably one of the most sought-after rarities for every British birder, but by and large these are birds from a bygone era. This is more than evident from BBRC statistics – 6,848 records prior to 1949, only seven since 1950. In the species' heyday, Lincolnshire had its share of records of Pallas's Sandgrouse during the irruptions in the late 19th century. All of the county records were during the five years 1863, 1888, 1889, 1890 and 1899. The irruption of 1863 and the first record of a flock of at least 40 at Saltfleetby-Theddlethorpe is an astonishing thought. Three other double-figure flocks were recorded elsewhere in that year with the last record on Aug 1st. The next irruption 25 years later in 1888 was more widespread and began on May 18th with birds still present into early Nov. Records came from 23 locations with the highest count being flocks of 40 at Mablethorpe Aug 30th and Grainthorpe on Nov 8th. After a gap of almost two months more birds arrived on the coast with 100 at North Somercotes which moved around the sand dunes between there and Saltfleetby. Another flock of 30 birds was found in the northern Wolds and the last birds were five at Brigg May 4th. The last two occurrences in 1890 and 1899 were lesser affairs although flocks of 30 on Jan 1st 1890 and Feb 26th 1899 were still very significant records of course.

These immense numbers are unthinkable today. There were also breeding attempts in Britain and clutches of eggs and young were found in 1888 and 1889. It seems unlikely we shall see such spectacular movements again. In more modern times three were seen in Britain in 1969: Foula (Shetland) May 26th-31st, and two in Northumberland, a male shot at Seahouses Sep 5th and another at Elwick Sep 6th. In Finland that year there were four to six birds in May and singles in May and Jun in The Netherlands. The last British records were two on the Isle of May on May 11th 1975 and one on Shetland May 19th-Jun 4th 1990.

Site Name	First Date	Last Date	Count	Notes
Saltfleetby-Theddlethorpe	01/04/1863	01/07/1863	40	
Alford	01/05/1863		1	
Great Coates	01/05/1863		1	
Grimsby	15/05/1863		1	
Old Leake	25/05/1863		20	
Huttoft	01/08/1863		20	
Irby upon Humber	18/05/1888		10	
Swallow	22/05/1888		4	
Epworth	23/05/1888		5	
Tetney	25/05/1888		2	
Cawkwell	28/05/1888		2	
Stamford	31/05/1888	01/06/1888	12	
Tetney	01/06/1888	03/10/1888	15	
Bourne	02/06/1888		1	
Fulstow	02/06/1888		8	
High Toynton	02/06/1888		1	
Lincoln	02/06/1888		5	
Asserby	04/06/1888	09/06/1888	20	
Cabourne	04/06/1888		1	
Horkstow	14/06/1888		21	
Scunthorpe	15/06/1888	09/07/1888	7	
Barton on Humber	18/06/1888	30/06/1888	20	
Authorpe	08/07/1888		1	
Holbeach Marsh	20/07/1888	28/07/1888	2	
Cuxwold	15/08/1888		5	
Mablethorpe	30/08/1888		40	
North Cotes	13/09/1888	10/10/1888	20	
North Somercotes	10/10/1888		2	
Goxhill	23/10/1888		20	
Grainthorpe	08/11/1888	11/11/1888	40	
North Somercotes	01/01/1889	01/03/1889	100	
Ingoldmells	26/01/1889		3	
Lincolnshire	26/01/1889		30	
Brigg	04/05/1889	27/07/1889	5	
Barton Pits	01/01/1890		30	
Killingholme	24/05/1890		8	
Lincolnshire	26/02/1899	19/05/1899	30	
Holbeach Marsh	16/03/1899		13	

Saltfleetby-Theddlethorpe Dunes NNR

Rock Dove (Feral) *Columba livia 'feral'*

Common and widespread resident.

Popularly known as the Feral Pigeon and a familiar site across our cities, towns and farmsteads. Old abandoned buildings of all kinds are favoured nesting and roosting locations. The *LBC Atlas* estimated the population at 5,000 pairs in the late 1980s and the BBS index chart shows a roller coaster ride since then with an increase, then a decrease. Is it just coincidence that the decrease accelerated in 2005 just as Peregrine *Falco peregrinus*, colonised and started to breed and spread in Lincolnshire? Or has the drive to develop old buildings eliminated breeding sites? The estimated population for Lincolnshire in 2016 was 3,700 pairs.

1968-1972	1980-1989	2008-2011	2016-2019	Incidence	Change
				25%	-7%

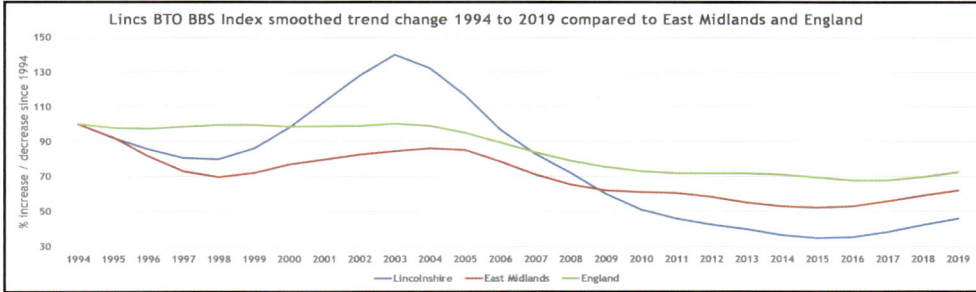

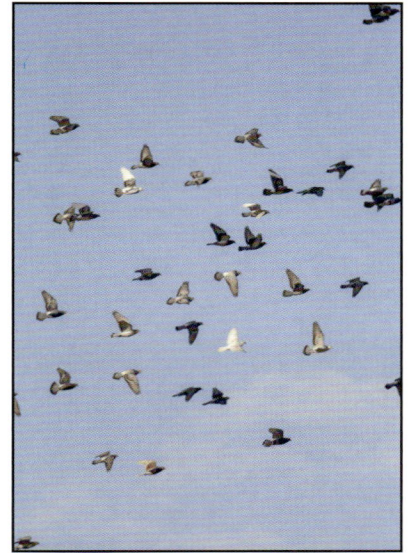

Lincs BTO BBS Index smoothed trend change 1994 to 2019 compared to East Midlands and England

Stock Dove *Columba oenas*

Common and increasing resident and winter visitor.

The *LBC Atlas* put the population at 10,000 pairs in the late 1980s. The Lincolnshire BBS index showed a steady decline from 1994 to around 2009 and then an increase since then. The incidence data suggests it has become more widespread and the estimated population was around 11,000 pairs in 2016. As a bird of arable farmland, one might expect it to do well in Lincolnshire and it does. Characteristically heard in the spring making its excitable call or seen as a pair flying away, it flocks up in winter and can be subject to movements in some years. In Nov 2018 there was an exceptional count of 364 birds moving south at Gibraltar Point, the highest since a flock of 550 in Nov 2009 at Donna Nook.

1968-1972	1980-1989	2008-2011	2016-2019	Incidence	Change
				47%	20%

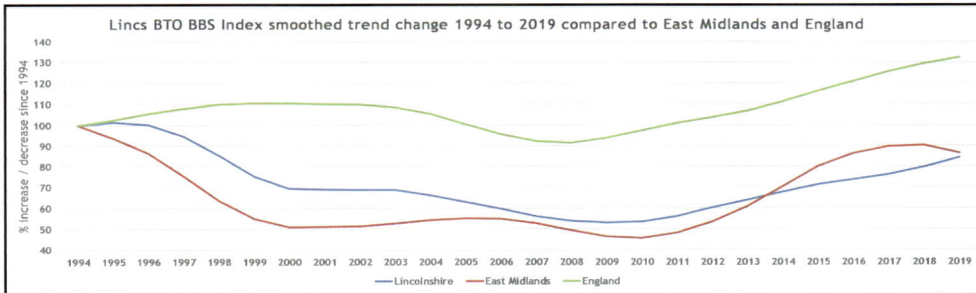

Lincs BTO BBS Index smoothed trend change 1994 to 2019 compared to East Midlands and England

Woodpigeon *Columba palumbus*

Abundant resident, passage migrant and winter visitor.

One of the top three commonest birds in Lincolnshire, and almost certainly the most widespread. Found everywhere, all year round, but perhaps less numerous in the Fens in winter. The *LBC Atlas* estimated the breeding population at around 78,000 pairs in the late 1980s. BBS data shows a significant change of 41% increase in the breeding population during the period 1994-2019. There is still controversy about the precise status of movements. Are they entirely within the UK or do continental birds get here? In an analysis of a four-year period from 2012-15 in LBR 2015 movements south at Gibraltar Point occurred between mid-Oct and mid-Nov with a maximum of 6,160 south on Nov 8th 2015. The numbers moving vary considerably from year to year.

1968-1972	1980-1989	2008-2011	2016-2019	Incidence	Change
				100%	1%

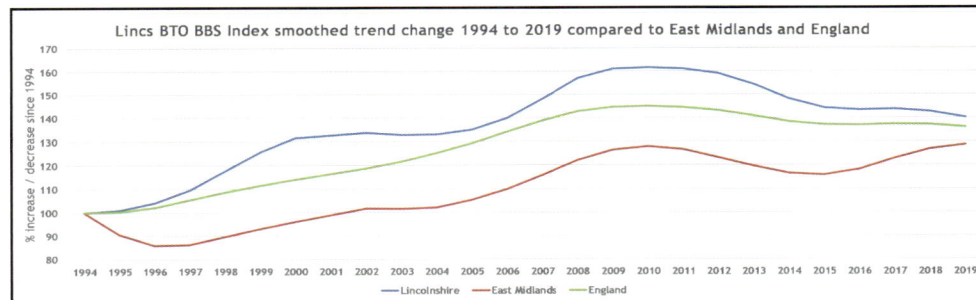

Lincs BTO BBS Index smoothed trend change 1994 to 2019 compared to East Midlands and England

Turtle Dove *Streptopelia turtur*

Common in the late 1980s now scarce and declining summer visitor and passage migrant.

This species was already declining when the *LBC Atlas* estimated the Lincolnshire population at between 3,700 and 7,300 pairs in the late 1980s. The superb *LBC Atlas* cover illustration of a pair of Turtle Doves by Neil Smith, superimposed on a map of their distribution at that time, is a telling indictment of species destruction by agricultural intensification. The BBS index chart shows the catastrophic change that has taken place from 1994-2019 earning Turtle Dove the title of Britain's fastest declining bird. Recognising this decline the species was considered by the RBBP from 2018 onwards in which year in Lincolnshire there were a maximum of 37 breeding pairs reported. The decline is popularly attributed to the shooting of migrant birds in southern Europe in both spring and autumn. However re-wilding at the Knepp Estate in Sussex where the breeding population has surged, suggests increased agricultural intensification, eliminating the "weed" seeds on which Turtle Doves feed their young, is the more likely culprit. The RSPB is working with farmers in East Anglia to introduce seed mixes which may stem, and hopefully start to reverse, the decline. Urgent action is definitely required if the Lincolnshire population is to be saved.

1968-1972	1980-1989	2008-2011	2016-2019	Incidence	Change
				0.9%	-91%

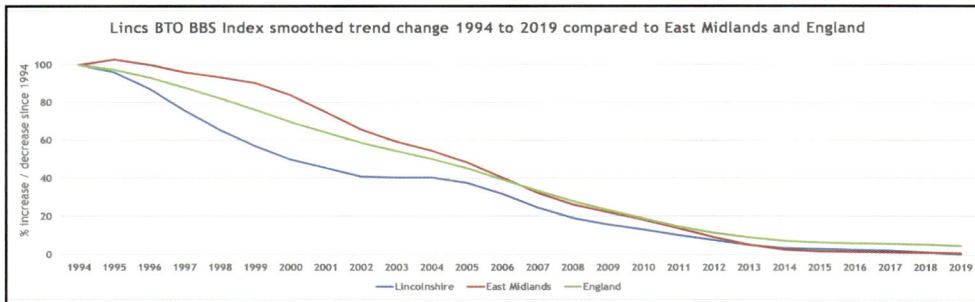

Lincs BTO BBS Index smoothed trend change 1994 to 2019 compared to East Midlands and England

Collared Dove *Streptopelia decaocto*

Very common resident.

This species famously spread across Europe like wildfire from Turkey in the first part of the 20th century, colonising Lincolnshire from 1957 onwards. The LBC Atlas put the population at around 30,000 pairs in the late 1980s and 33,000 by 1997. The BBS index chart below shows a roller-coaster ride from 1994, peaking in Lincolnshire and the East Midlands around 2008-09, then declining, in our county, back to 1994 levels. The estimated Lincolnshire population is currently 20,000 which may be a more accurate reflection of the current situation than the suggested LBC Atlas figure of 30,000. It remains a very common and widespread resident despite some suggestion of local declines. There is no connection at all between the success of Collared Dove and the decline of its migratory cousin the Turtle Dove *S. turtur*. The question remains of where was the first record for Britain? See the discussion in the 'Escapes' section.

1968-1972	1980-1989	2008-2011	2016-2019	Incidence	Change
				50%	-9%

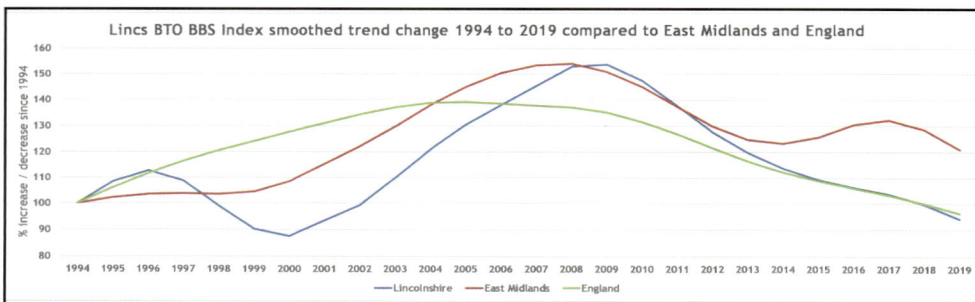

Lincs BTO BBS Index smoothed trend change 1994 to 2019 compared to East Midlands and England

Water Rail *Rallus aquaticus*

Scarce resident fairly common passage migrant and winter visitor.

Much under-recorded, Water Rails tend to be confined as breeding birds in Lincolnshire to reed swamp vegetation like *Phragmites* reedbeds. That makes them difficult to see and most observers record them from their distinctive calls. The relative paucity of reed beds in the county limits their distribution. RBBP records show an average of 42 pairs per year bred during the period 2013-17. It ceased to be considered by RBBP as a rare breeding species from 2018 onwards. By that time a better understanding suggested the British population was above the 2,000 pairs threshold that merited detailed consideration. The *LBC Atlas* suggested that in the late 1980s the population was around 10 to 20 breeding pairs. At that time a key site for the species in the county was the Humber Bank Clay Pits. They remain important but have recently been eclipsed by the Alkborough Flats Environment Agency managed retreat project at the confluence of the Trent and Humber which has substantially increased the amount of reed bed in Lincolnshire. In 2020 up to 80 territories were recorded here.

Winter influxes are not unusual and at such times Water Rails can be found in all kinds of aquatic habitats across the county.

1968-1972	1980-1989	2008-2011	2016-2019	Incidence	Change
				0%	

Corncrake *Crex crex*

Very scarce passage migrant. Former breeder no confirmed breeding since 1920.

Cordeaux (1872) indicated that Corncrake numbers varied greatly from year to year depending on wetness and the productivity of hay meadows. In dry years there were sometimes none. Breeding ceased by 1920 as hay making became mechanised. No territories holding males recorded from 1991-2007 but since then eight males have been recorded in seven of the last 11 years to 2018. These may arise from the nearby reintroduced Nene Washes population. All other records relate to spring and autumn migrants.

1968-1972	1980-1989	2008-2011	2016-2019	Incidence	Change
				0%	

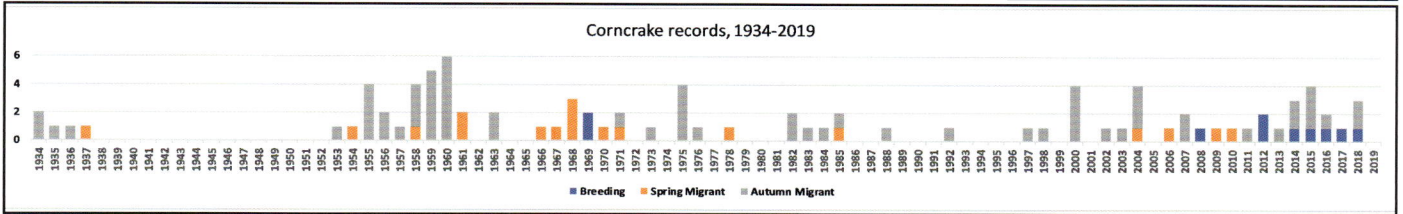

Corncrake records, 1934-2019

■ Breeding ■ Spring Migrant ■ Autumn Migrant

Sora Rail *Porzana carolina*

Vagrant. North America.

A report of a Spotted Crake *P. Porzana*, by a visiting couple on Mar 5th 2006 prompted a closer look from a member of staff at Gibraltar Point who, disbelievingly, discovered a possible Sora Rail. The bird promptly disappeared along the ditch where it was first sighted, but happily was re-found by mid-afternoon and its identity confirmed as the county's first and (not surprisingly) only record of a Sora Rail. It remained on site until Mar 18th delighting the many visitors. The bird was trapped on its penultimate day and confirmed as a 2CY female.

The occurrence of such a bird away from south-western Britain is of course not without precedent and this follows one found at Attenborough NR, Nottinghamshire in Dec 2004. With fewer than 20 British records including just eight since 2000 this remains a great rarity.

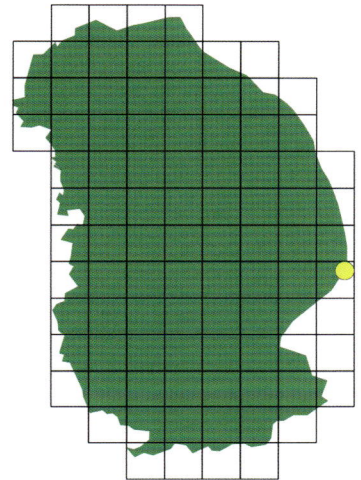

Site Name	First Date	Last Date	Count	Notes
Gibraltar Point	05/03/2006	19/03/2006	1	Trapped

Spotted Crake *Porzana porzana*

Rare migrant, though historically a breeding species up to the latter half of the 19th Century.

Probably widespread in the Fens but increasingly scarce into the late 19th century. Wet areas around Ashby near Scunthorpe, Tetney and North Cotes were said to be regular breeding locations. Blathwayt (1915) considered pairs could have been breeding as late as 1915. Until 1900 it was a regular passage migrant chiefly in Sep-Oct but became much scarcer with habitat loss and shooting, with plenty seen at local taxidermists. Between 1930-55 it was recorded only four times. Lorand and Atkin (1989) noted that there were 32 records 1963-88. There was then a run of blank years from 1989-94 followed by 21 records 1995-2019. At just under one record per year it remains a rare migrant in the modern era. Finally, calling males held territory at Baston Fen NR in 2006 and again in Apr-May 2012, only the second and third probable breeding records in the last 30 years.

1968-1972	1980-1989	2008-2011	2016-2019	Incidence	Change
				0%	

Covenham Reservoir

Moorhen *Gallinula chloropus*

Very common resident and partial migrant.

Moorhen has always been a very common resident and the *LBC Atlas* estimated a population of 15,000 to 20,000 pairs. They are associated with the smallest waterbodies which helps make them very widespread. The BBS index indicated the population was fairly stable across England until around 2008-09. At that point something happened, and the population may have fallen by as much as half in Lincolnshire in the last 10 years. More so than the rest of the East Midlands or England. The population estimate for Lincolnshire in 2016 was 7,000 pairs. The reasons for this fall do not appear to be well understood. The winter population also appears to be slightly down. WeBS data indicates that a mean of only 534 Moorhen per annum were counted on Lincolnshire WeBS sites over the winters of 2014-15 to 2018-19 compared to 667 between 2003-04 and 2007-08. Of course, the population is very widely spread and does not congregate in flocks the way many species of waterbirds do, making counting them more difficult.

1968-1972	1980-1989	2008-2011	2016-2019	Incidence	Change
				32%	-24%

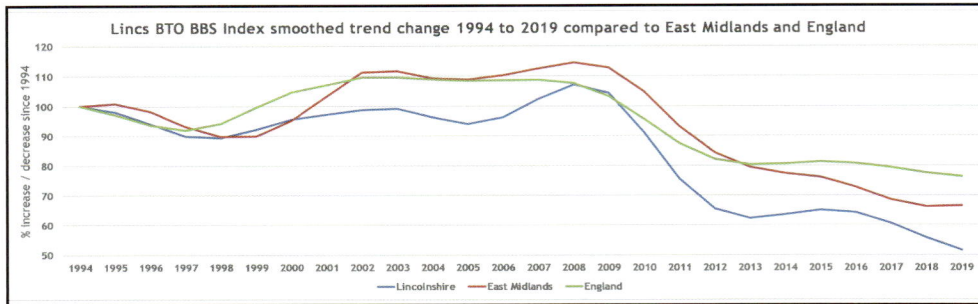

Lincs BTO BBS Index smoothed trend change 1994 to 2019 compared to East Midlands and England

Coot *Fulica atra*

Common resident, passage migrant and winter visitor.

The *LBC Atlas* estimated a population of 1,750 breeding pairs in the late 1980s and a wintering population of 7,000 birds. The BBS index suggests the breeding population declined a little from 1994 improved again and then declined more from 2010-19 ending up 40% down since 1994. The estimated population in 2016 was 400 pairs which is likely to be on the low side. The five-year mean wintering population to 2018-19 from WeBS Online is 2,600.

Perhaps not all sites are counted but this suggests a fall in the wintering population too. Another resident waterbird with a declining population. Perhaps our waterways are not as clean as we would hope?

1968-1972	1980-1989	2008-2011	2016-2019	Incidence	Change
				14%	-9%

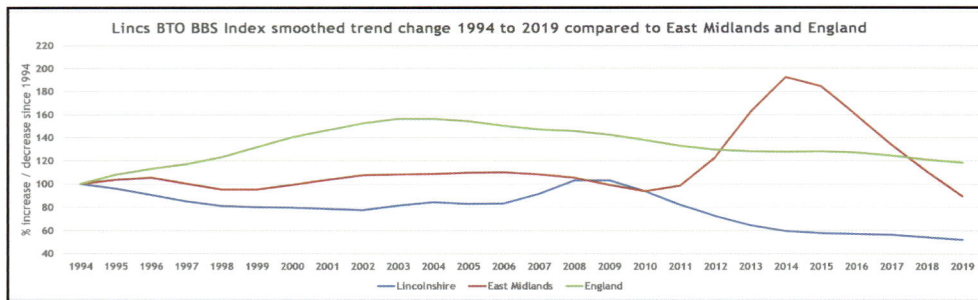

Lincs BTO BBS Index smoothed trend change 1994 to 2019 compared to East Midlands and England

Western Swamphen *Porphyrio porphyrio*

Vagrant. Southern Europe.

A Western Swamphen was found on a reed-fringed pool at Minsmere, Suffolk on Jul 31st 2016 where it remained until Aug 5th. This was the first occasion on which this species has been recorded in Britain and incredibly what was thought to be the same bird was then found in Lincolnshire at Alkborough Flats on Aug 31st-Nov 23rd 2016 and again briefly on Jan 4th 2017.

The species had been declining in the Mediterranean and a reintroduction scheme began in Spain in the 1980s and as numbers increased the species spread into France where it first bred in 1996. Despite a setback after severe weather in Feb 2012 numbers recovered and in 2016 there were a number of sightings in central and northern France.

One found at Deux-Sèvres on Jul 19th 2016 was only c.300km due south of Minsmere; could this have been the same bird?

Site Name	First Date	Last Date	Count	Notes
Alkborough	30/08/2016	23/11/2016	1	
Alkborough	04/01/2017		1	Same

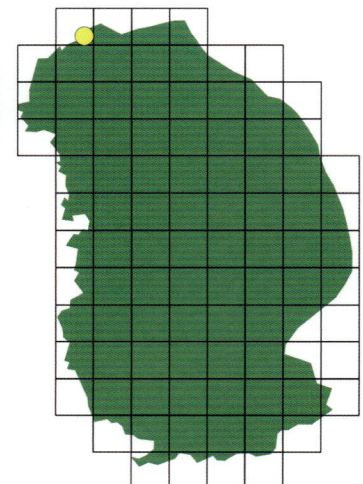

Little Crake *Porzana parva*

 BBRC

Vagrant. Eastern Europe.

There are no modern confirmed records of this species in the county and just two historical records. Cordeaux recorded the first in 1869 at Great Coates as he watched the bird down to a few feet and the second in 1910 was, unfortunately, killed by a dog and preserved in Lincoln Museum. Small crakes thought to be this species were seen at Tetney, Oct 1888, 1907 and 1930; at Gibraltar Point Aug 1953 and at Wrangle Aug 1978. None of these was able to be specifically identified however and are not included in the county statistics.

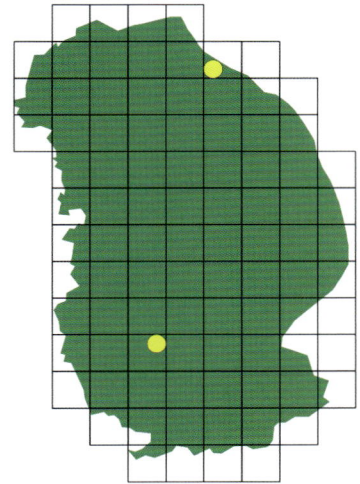

The secretive nature of this species is likely to mean they are easily missed. The BBRC *Report on rare birds in Great Britain in 2018* published a Cambridgeshire record of a female which had been sound recorded only, and notes that the last time one was actually seen in Britain was in 2015.

Records have tended to be of adults in spring and juveniles in autumn and the species is only detected in Britain about every two years. However, with an increasing number of wetland restoration projects in Britain and on the near continent there may be hope for a record in the near future.

Site Name	First Date	Last Date	Count	Notes
Great Coates	09/10/1869		1	
Spanby	01/01/1910		1	Female

Crane *Grus grus*

RBBP

Scarce but increasing migrant, recorded most months. Former breeding species and a common migrant to 16th/17th centuries. Bred in 2020 for first time in 400 years.

Lorand and Atkin (1989) give a good account of the history of Crane in Lincolnshire. They have bred in adjacent South Yorkshire, Cambridgeshire and Norfolk in recent years but up to 2019 had only appeared as a scarce migrant in Lincolnshire. This changed in 2020 when, out of the blue, a pair bred at the LWT reserve at Willow Tree Fen and in Aug one chick successfully fledged, the first to do so for more than 400 years.

Cranes are more familiar as scarce but regular migrants in the county and have turned up in most months of the year. Most occur during Mar-May with far fewer in the autumn and the majority are seen flying over coastal migration hotspots. In recent springs the first birds have generally been noted in Mar, an exception being 2012 when eight flew over Bourne on Feb 17th. Most times one to five birds have been seen but occasionally larger flocks with seven (May 2004, Apr 2013), eight twice (Dec 2006, Oct 2013), nine (Apr 2017) and 11 (Apr 2018). The last birds of the year are seen in Oct-Nov in smaller numbers and occasionally singletons have overwintered.

1968-1972	1980-1989	2008-2011	2016-2019	Incidence	Change

Little Grebe *Tachybaptus ruficollis*

Fairly common resident, partial migrant and winter visitor, the latter especially noticeable in coastal regions.

Little Grebe can be a secretive species, usually given away by its whinnying trill of a call. Not formerly as widespread as one might imagine given the availability of waterbodies in Lincolnshire. The *LBC Atlas* put the breeding population at 200 to 400 pairs in the late 1980s. It has clearly spread since then as the last *BTO Atlas 2007-11* showed a net gain of birds breeding in 23 10km squares. This spread has been mirrored across much of England.

The estimated population for 2016 was 400 pairs. The *LBC Atlas* put the wintering population at 200 birds and WeBS Online shows the five-year mean population to winter 2018-19 at around 300 birds, split broadly 50/50 between the estuary fringes and inland. The largest counts of the year tend to be in Aug-Sep with Frampton Marsh leading the LBR tables in the last few years with a peak of 65 in Aug 2018.

1968-1972	1980-1989	2008-2011	2016-2019	Incidence	Change
				10%	209%

Red-necked Grebe *Podiceps grisegena*

Scarce and more recently, very scarce passage migrant and winter visitor, Sep-Apr; rare in summer.

Found along the coast and inland, numbers vary from year to year. Lorand and Atkin (1989) mention significant cold-weather influxes which occurred in Jan-Feb 1865 and in Jan-Feb 1937. More recently there was an influx of over 100 birds in mid-Feb 1979. In more modern times LBR reports indicate that the average in the last five years to 2018 was eight (range five to 12) and in the five years to 2013, 12 (range six to 18).

The best year over the 10-year period with 18 was 2012, a relatively hard winter. Records were split about 60/40 between coast and inland and across the whole 10-year period there were 18 long-stayers (birds that stayed on an inland water for three days or more), and three years with no long-stayers. Of the 18, 12 were at Covenham Reservoir.

There were no summer records in the period. Red-necked Grebe is too scarce for WeBS to produce an index and WeBS data indicates the rolling mean five-year UK wintering population to 2018-19 was 19 while in the previous five-year period to 2013-14 it was 24.

Great Crested Grebe *Podiceps cristatus*

Fairly common resident, partial migrant and winter visitor. Suggestion of a recent decline from submitted records.

Plume hunters virtually eliminated this attractive grebe as a breeding bird in Lincolnshire by the end of the 19th century but protection (brought about by the formation of what is now the RSPB) allowed a slow renaissance of the population, charted by Lorand and Atkin (1989). The population continued to expand in Lincolnshire and by the late 1980s the *LBC Atlas* estimated that 350 pairs were breeding with 300 wintering. The last *BTO Atlas 2007-11* showed a fairly stable position in Lincolnshire, mirrored across much of England. The estimated Lincolnshire population for 2016 is 200 pairs. Anecdotal evidence from LBR records suggests that the recent population decline may well be correct. The wintering population appears to be generally doing well, and WeBS Online shows the five-year mean population to winter 2018-19 at around 400 birds, split broadly 35/65% between the coast and inland. Covenham Reservoir has tended to be the leading wintering site for this species in Lincolnshire over the last 30 years and the fluctuations in the peak counts there from 1970-2014 recorded by Robinson *et al* in LBR 2014, make for interesting reading.

1968-1972	1980-1989	2008-2011	2016-2019	Incidence	Change
				4.4%	84%

Slavonian Grebe *Podiceps auritus*

Very scarce/ scarce passage migrant and winter visitor, mainly Aug-May.

The *LBC Atlas* reported that in the 1980s and 1990s there were usually less than 10 birds per year and that exceptionally, in hard winters, influxes of up 25 had occurred. In the five years to 2018 LBR reports averaged 20 per year with a range of eight (2017) to 29 (2016). There was no specific influx as such but more reports came from coastal areas. Of the 97 birds reported, 81 were on or off the coast and 16 inland of which 10 were long stayers at a wide variety of locations throughout the county. There were no summer records in the period. Two particularly popular birds were at Cleethorpes Country Park from Jan 29th to Feb 9th 2016. WeBS data indicates the rolling mean five-year UK wintering population to 2018-19 was 846, that for Lincolnshire was three.

Black-necked Grebe *Podiceps nigricollis*

Rare breeder, scarce passage migrant and winter visitor.

Bred in the Fens and was said to be particularly common around Spalding but gone by the early 19th century with the drainage of the region. For many years it was a vagrant but from 1954 it became a rare winter visitor and migrant. The *LBC Atlas* records show that pairs started to linger in the 1980s, but no breeding was confirmed until 1998 when three pairs produced seven young at two gravel pit sites. Nesting birds were targeted by egg collectors and since then details of the three sites where it bred up to 2018 have been kept confidential. From 1998 one to three pairs bred each year. Then in 2009 five pairs bred. RBBP records show an average of five pairs per year bred during the period 2013-17. The best year was 2014 when seven pairs raised two young. Sadly in 2018 there was just one pair. In 2019 and 2020 young were successfully raised at Frampton Marsh under the protection of the RSPB.

Migrant and wintering birds appear in small numbers from late Aug mainly at inland sites such as Covenham Reservoir and Toft Newton Reservoir, and again from late Feb-early Mar. Spring flocks of up to seven birds have been recorded at these two sites, but only one to two in most winters. Similarly, coastal sites record a few at both seasons but they are much scarcer on the sea than their congeners, Slavonian Grebe *P. auritus*.

1968-1972	1980-1989	2008-2011	2016-2019	Incidence	Change
				0%	

Reads Island, Humber Estuary

Stone-curlew *Burhinus oedicnemus*

Rare summer visitor. Former breeder. No confirmed breeding since 1904.

Formerly bred on the Wolds and the heaths of the north-west and south-west of the county but became progressively scarcer from the 1870s. Last bred at Manton Warren near Scunthorpe in 1904. A brief appearance by one there in Apr 2005 set the heart racing. In 1989 a lone female, ringed as a chick in the Brecks in 1986, attempted, without success, to incubate two eggs at Fulbeck Airfield. She reappeared in spring 1990 but did not lay.

Lorand and Atkin (1989) noted 20 records between 1950-67, mainly coastal migrants. From 1968-88 there was less than one bird per year with a blank run from 1979-87. There were just five records in the 1990s but then three in 2000 with singles in 10 of the years afterwards up to 2019.

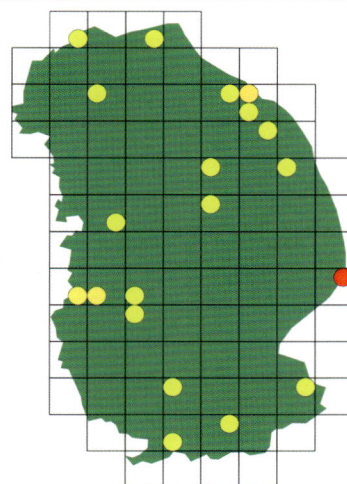

1968-1972	1980-1989	2008-2011	2016-2019	Incidence	Change
■	■		■		

Site Name	First Date	Last Date	Count	Notes
North Cotes	02/03/1935		1	
Gibraltar Point	07/04/1950		1	
Gibraltar Point	08/04/1950		1	
Gibraltar Point	11/05/1950		1	
Gibraltar Point	15/05/1950		1	
Gibraltar Point	22/09/1950		2	
Goxhill	25/09/1953		1	
Gibraltar Point	24/09/1957		1	
Lincoln	02/04/1958		1	
Cleethorpes	11/05/1958		1	
Gibraltar Point	08/06/1958		1	
Gedney Drove End	16/08/1958		1	
Beckingham	10/09/1958		3	
Humberston	15/06/1960		4	
Biscathorpe	27/07/1964		1	
Gibraltar Point	03/12/1967		1	
Sotby	04/05/1968		1	
Cranwell	24/05/1968		1	
Gibraltar Point	25/11/1978		1	
Gibraltar Point	16/07/1988		1	
Fulbeck	09/05/1989	13/07/1989	1	Female ringed Brecks 1986
Fulbeck	10/04/1990	05/06/1990	1	Same female
Gibraltar Point	15/05/1993	16/05/1993	1	
Gibraltar Point	31/05/1993		1	
Gibraltar Point	11/05/1994		2	
Gibraltar Point	31/05/1994		1	
South Rauceby	15/04/1996		1	
Gibraltar Point	27/04/2000		1	
Langtoft	01/06/2000		1	
Cowbit	02/09/2000		1	
Marshchapel	28/05/2003		1	
Gibraltar Point	06/05/2006		1	
Greetwell (Scunthorpe)	13/04/2007	14/04/2007	1	
Saltfleetby St Peter	24/04/2008		1	
Alkborough	18/05/2009		1	
Dunsby	06/09/2010		1	
Gibraltar Point	06/05/2011		1	
Gibraltar Point	21/05/2012		1	
Gibraltar Point	13/07/2014		1	Juvenile
Holbeach St Marks	16/05/2019		1	

Heathland, Laughton Forest

Oystercatcher *Haematopus ostralegus*

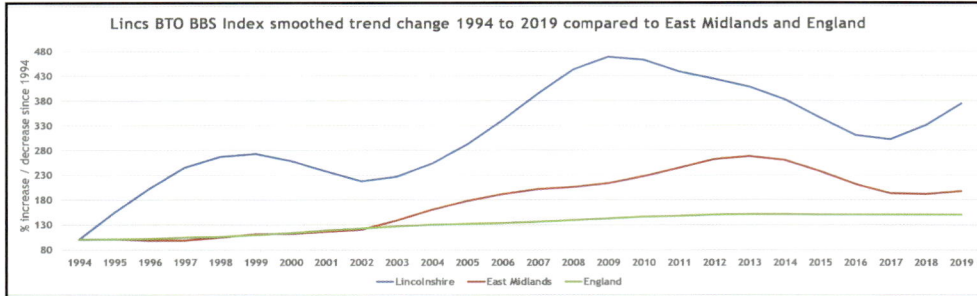

Very common coastal passage migrant / winter visitor and fairly common resident. Less common inland but now breeds in increasing numbers.

Formerly restricted to the coast as a breeding bird, Oystercatcher has progressively expanded its presence inland from the 1980s onwards. The *LBC Atlas* estimated the population at 300 breeding pairs concentrated primarily around The Wash and Humber with a few inland. The *BTO Atlas 2007-11* showed that 22 10km squares inland had been colonised, primarily in the north of the county. The BBS index chart shows that the population has fluctuated considerably over the last 10 years. It is difficult to get a grip on the current population without another full tetrad-based atlas survey. The estimated population in 2016 was 700 pairs which might be in the right area. The winter population was put at 20-30,000 in the *LBC Atlas* and data from LBR suggests the wintering population is now in the region of 13,000 in Lincolnshire. Birds are found mainly in the Humber Mouth from Cleethorpes to Saltfleet and The Wash. WeBS data indicates that the wintering Oystercatcher population in England is down 23% on the 25-year trend from 1992-93 to 2017-18. Migrant numbers in late summer tend to be higher than in winter, with a high count over the last five years to 2018 of 12,000 at Freiston Shore and 5,000 at Gibraltar Point in Aug 2018.

1968-1972	1980-1989	2008-2011	2016-2019	Incidence	Change
				8%	-21%

Lincs BTO BBS Index smoothed trend change 1994 to 2019 compared to East Midlands and England

Black-winged Stilt *Himantopus himantopus*

Rare, but increasing, summer visitor.

The first record was of two birds flying south along the shoreline at Gibraltar Point on Apr 26th 1965. The next was not until more than 18 years later in 1983 and it remained a rare bird until the mid-2000s when three turned up at Barton Pits in 2006. From 2010 onwards they have become almost annual with potential breeding birds present at Frampton Marsh in 2015, 2017 and 2018. A pair present at Frampton Marsh in May 2015 were seen mating and building a nest but did not settle. The same behaviour was reported in May 2017, but the birds were scared off by foxes. A lone bird stayed for three weeks in May 2017. With successful breeding in adjacent counties, superb habitat and new predator proof fencing at RSPB Frampton Marsh, the first Lincolnshire hatched chicks cannot be too long coming.

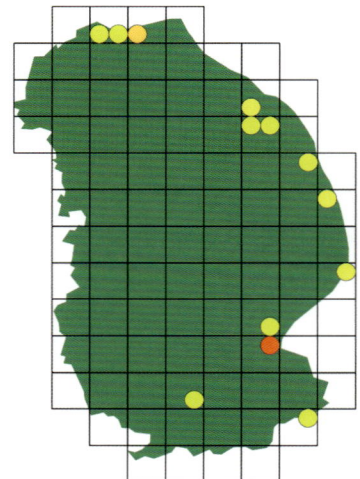

1968-1972	1980-1989	2008-2011	2016-2019	Incidence	Change
				0%	

Site Name	First Date	Last Date	Count	Notes
Gibraltar Point	26/04/1965		2	Male and female
Wisbech STW	29/04/1983		1	
Tetney	19/06/1986		1	
Barton upon Humber	06/09/1987		1	
Barton upon Humber	08/09/1987	09/09/1987	1	Same
Winteringham	10/09/1987		1	Same
South Ferriby	11/09/1987	12/09/1987	1	Same
Covenham Reservoir	06/05/1991		1	
North Cotes	21/04/1993		1	
Huttoft	04/05/1997	05/05/1997	1	Adult
Barton Pits	04/05/2006		3	First summer
Frampton Marsh	29/05/2010		2	
Willow Tree Fen	14/04/2012	18/04/2012	1	Female
Frampton Marsh	19/04/2012	18/05/2012	1	
Saltfleetby-Theddlethorpe	13/10/2012		1	
Frampton Marsh	07/05/2013		1	
Frampton Marsh	03/06/2013		2	
Willow Tree Fen	15/05/2014	16/05/2014	2	BBRC Pending, Late submission
Grainthorpe	04/05/2015	09/05/2015	1	
Frampton Marsh	27/05/2015	28/05/2015	2	
Frampton Marsh	03/05/2016	04/05/2016	1	Adult female
Anderby	05/05/2016	07/05/2016	1	Same
Frampton Marsh	11/05/2017	12/05/2017	2	
Frampton Marsh	01/05/2018	28/05/2018	1	Male
Freiston Shore	29/05/2018	31/05/2018	1	Same male
Frampton Marsh	01/06/2018		1	Same male

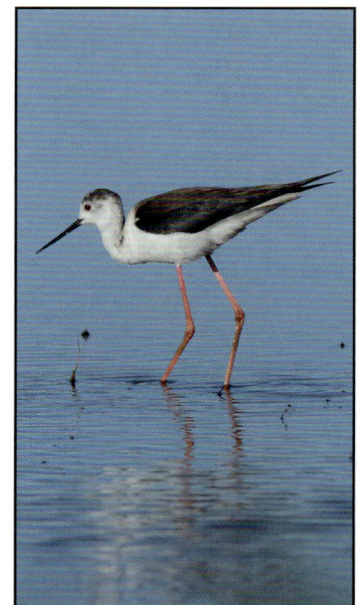

Avocet *Recurvirostra avosetta*

Fairly common coastal breeder, has colonised inland localities in the past few years. Fairly common passage migrant, scarce in winter.

Bred in the Fens around the Wash up until the early 19th century and disappeared as the Fens were drained. Only five records in the first half of the twentieth century but became more numerous in the 1980s culminating in breeding again at Welland Marsh in 1991. Scrapes created by the RSPB at Read's Island on the Humber in 1997 proved successful and held up to 250 pairs in 2010. The colony has now moved as silting up allowed foxes and badgers to access the island. RBBP records show an average of 246 pairs per year bred during the period 2013-17. Numbers can vary from year to year as birds move around looking for the best breeding conditions. An analysis in LBR 2018 shows that in the 11-year period 2008-18 birds bred at 28 different sites but only around 12 sites in each year. The wintering population is increasing with up to 100 birds in recent years.

1968-1972	1980-1989	2008-2011	2016-2019	Incidence	Change
				0.3%	

Lapwing *Vanellus vanellus*

Common but declining breeding species, and very common passage migrant and winter visitor.

The Lapwing and its demonstrative display flight is the "archetypal" wader of the Lincolnshire countryside and the symbol of LWT. Sadly, our breeding population is in long term decline. The *LBC Atlas* estimated 7,000 pairs in the late 1980s, but the population was thought to have halved between then and 1997. The Lincolnshire BBS index shows there was a rebound up to 2009 followed by a further and greater decline. The population estimate for 2016 was 2,400 pairs. One bright spot is that, over the last 15 years, DEFRA habitat improvement schemes have demonstrated across the county that grassland areas can recover their Lapwing populations if water levels are raised and sufficient wet muddy edges for young to feed are provided. The winter population in the 1980s was thought to be up to 1,300,000 birds, with widespread flocks of 1,000 to 5,000 birds. In the last five years LBR reports suggest that the top 10 sites reported held an average of 21,000 birds in Jan with considerable variation from winter to winter. In this period Alkborough Flats has usually had the highest counts with a peak of 8,560 in Dec 2018.

1968-1972	1980-1989	2008-2011	2016-2019	Incidence	Change
				27%	-41%

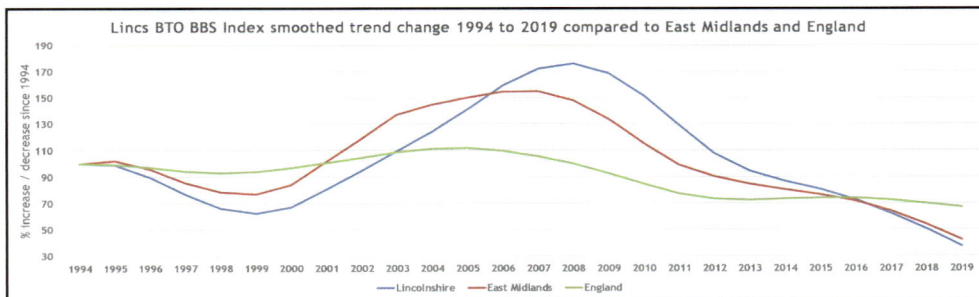

Lincs BTO BBS Index smoothed trend change 1994 to 2019 compared to East Midlands and England

Sociable Plover *Vanellus gregarius*

Vagrant. Central Asia.

A stunning adult in summer plumage graced the fields around the Kirkby on Bain GP May 30th-Jun 2nd and again Jun 12th 1993. When sitting down, it was at times difficult to pick up as its cryptic colouration made for ideal camouflage against the sandy soil of the fields it frequented. Despite being seen to fly off high north on Jun 2nd it was briefly relocated on Jun 12th. It was presumed to be the same individual which had previously been in Derbyshire, Apr 17th, and Norfolk Apr 21st-23rd.

There have been 40 British records 1950-2018 but only eight since the Kirkby bird with the last in 2008 on the Isles of Scilly. The species has suffered a catastrophic decline in numbers across its breeding range. In northern Kazakhstan a decline of 40% was reported during 1930-60, with a further 50% decline during 1960-87, together with a massive contraction in its range. The bird is considered to be globally threatened.

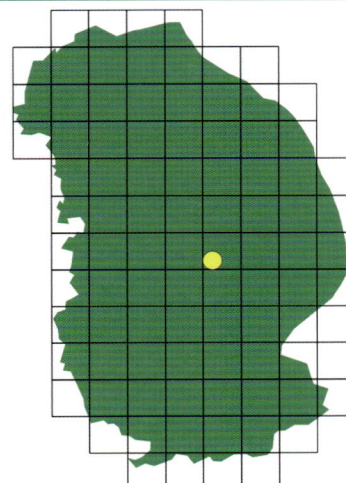

Site Name	First Date	Last Date	Count	Notes
Kirkby on Bain	30/05/1993	02/06/1993	1	
Kirkby on Bain	12/06/1993		1	Same

Golden Plover *Pluvialis apricaria*

Very common passage migrant and winter visitor, occasional in summer.

Golden Plover is a widespread wintering bird in Lincolnshire found in every 10km square during the *BTO Atlas 2007-11*. The Humber is the most important wintering site for the species in Britain. WeBS data indicates that the rolling five-year mean Lincolnshire wintering population to 2018-19 was around 47,000 of which 32,000 were around the Humber and 13,000 around The Wash and only 4,000 inland. It is possible that this total is understated because many birds feed in fields across the county and don't get counted. Observer experience suggests that this behaviour is declining as agricultural intensification proceeds and the food value of arable fields declines for birds. There is also considerable movement and flocks can be very mobile. Some individual flocks can be massive with Alkborough Flats and Read's Island holding flocks of 26,000 birds in Nov 2014 and Nov 2016. Other sites reporting flocks of more than 10,000 birds on the Humber include Cleethorpes-Grainthorpe (10,000 Nov 2014), on The Wash at Frampton Marsh (15,000 Jan 2015) and Gibraltar Point (10,000 Nov 2014). The annual WeBS report for 2018-19 indicates that the long term 25-year trend for Golden Plover wintering in England is up 13%.

Pacific Golden Plover *Pluvialis fulva*

Vagrant. Arctic Siberia and west Alaska.

Pacific and American Golden Plovers were long considered conspecific, together known as "Lesser Golden Plover", but were treated as separate species by the BOURC from 1986.

The first county record was in Jul-Aug 1986 at Tetney and North Cotes when, amazingly, it was accompanied by an American Golden Plover, *P. dominica*. The detailed description of this bird by the finders in LBR 1986 was testament to the careful observations required to separate this species from its sister taxon. At the time this was the fifth British record.

Seven further county records have followed involving eight birds with two together in Jul 2002 and the last in 2015, all of them on the Humber estuary. There have been 97 British records in all 1950-2018 and it remains much the rarer of the two "Lesser" Golden Plovers with about three per year.

Site Name	First Date	Last Date	Count	Notes
North Cotes	27/07/1986	29/07/1986	1	Adult
Tetney	03/08/1986	08/08/1986	1	Same adult
Tetney	18/08/1986	19/08/1986	1	Same adult
South Ferriby	02/07/1993		1	Adult
South Ferriby	18/07/1993	19/07/1993	1	Same adult
South Ferriby	10/07/1994	11/07/1994	1	Adult
North Cotes	17/07/2002	19/07/2002	2	Both adults
Alkborough	29/01/2015		1	Adult winter

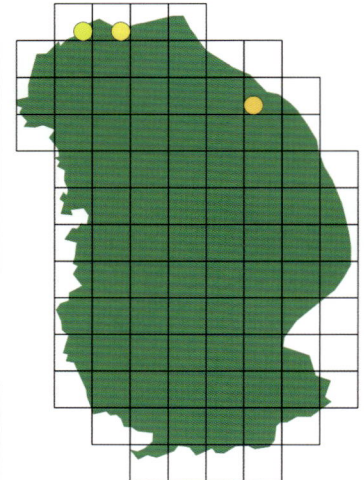

July at Frampton Marsh RSPB

American Golden Plover *Pluvialis dominica*

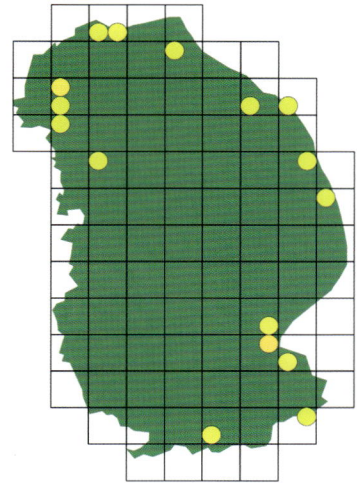

BBRC

Rare. North America. Recorded from May-Nov.

Formerly known as 'Lesser Golden Plover,' the species was split into American Golden Plover *P. dominica*, and Pacific Golden Plover *P. fulva*, by the BOURC in 1986, prior to which there had been only seven British records of Pacific. By 2005 the total had risen to 60, compared with 275 Americans.

The first county record was found at Wisbech Sewage Farm on Aug 19th 1974. There have been a further 23 confirmed records, the earliest on May 9th and latest on Nov 4th with records in all months between those extremes. Of those which were aged, 14 were adults, three were 2CY and four were juveniles. The 2CY birds were seen in May Jun and Sep while the juveniles were seen in Oct (two) and Nov (two). Most were recorded in Jul (seven) and Oct (nine).

It is noteworthy that both American and Pacific Golden Plovers do not complete their moult into winter plumage until they reach their winter quarters and retain variable amounts of black underpart feathering during their autumn migration. European Golden Plovers complete their moult by Sep-Oct. Thus both Pacific and American may retain traces of summer plumage well into the winter which can be an aid to identification.

Site Name	First Date	Last Date	Count	Notes
Wisbech STW	10/08/1974		1	
Killingholme	24/07/1982		1	Adult
Saltfleetby-Theddlethorpe	16/07/1983		1	Adult
Tetney	27/07/1986	29/07/1986	1	Adult
South Ferriby	24/07/1993		1	Adult
Holbeach Marsh	24/07/1994		1	Adult
Butterwick	08/10/1994	16/10/1994	1	Adult
Frampton Marsh	03/09/1995		1	First summer
South Ferriby	24/07/1997	13/08/1997	1	
Huttoft	09/05/1998		1	First summer
Frampton Marsh	02/10/2005		1	Adult
East Ferry	02/10/2006		1	Adult
Crowland	06/10/2006	09/10/2006	1	Adult
Crowland	28/10/2006		1	Juvenile
East Butterwick	06/09/2007	08/09/2007	1	Adult
East Butterwick	07/09/2007	08/09/2007	1	Adult
Dawsmere	14/10/2007		1	Juvenile
Fillingham	04/11/2008	05/11/2008	1	Juvenile
Freiston Shore	31/10/2010	01/11/2010	1	
Donna Nook	04/10/2013		1	Juvenile
Susworth	05/10/2013		1	Adult
East Butterwick	05/10/2018		1	Adult
Frampton Marsh	22/09/2019		1	Adult
Frampton Marsh	25/09/2019		1	Same

Grey Plover *Pluvialis squatarola*

Common passage migrant and common winter visitor. Very scarce inland.

Grey Plover tends to be confined to inter-tidal habitats in winter though it occasionally occurs inland in small numbers during passage. The annual WeBS report for 2018-19 indicates that the long term 25-year trend for Grey Plover wintering in England is down 35%. The key wintering sites in Lincolnshire are The Wash and the Humber. WeBS Online indicates the rolling five-year mean Lincolnshire wintering population to 2018-19 was around 12,400 of which 3,100 were around the Humber and 9,300 around The Wash. The Wash is the number one site in the UK for wintering Grey Plover, with the key area being Wainfleet/Gibraltar Point. The Humber is number four with the area between Cleethorpes and Grainthorpe Haven holding most birds. The largest flocks in these areas reported in LBR in the five years to 2018 have been Wainfleet with 7,345 in Mar 2014 and Marshchapel with 2,000 in Nov 2017.

High tide at Frampton Marsh RSPB

Ringed Plover *Charadrius hiaticula*

Nominate British and southern Scandinavian form common passage migrant, mainly coastal. Fairly common declining towards scarce breeder. Northern Scandinavian and Russian form *tundrae* (Tundra Ringed Plover) fairly common passage migrant.

Ringed Plover is having a bad time breeding in Lincolnshire. Its coastal breeding sites came under disturbance pressure from the 1960s onwards. In the 1970s it started to spread inland. The *LBC Atlas* estimated around 280 pairs in the late 1980s with around a third of those across 18 inland 10km squares. Seventeen out of 18 coastal squares had confirmed breeding and by the time of the *BTO Atlas 2007-11* only eight inland 10km squares were occupied, and eight out of 18 squares had confirmed breeding. There are no recent detailed estimates, but the population must have at least halved and could be below 100 pairs. Peak passage months tend to be May and Aug and the wintering population appears to have declined. The *LBC Atlas* estimated 700-800 birds wintered in the late 1980s. LBR tables indicate that numbers have been 200-300 in the five years to 2018.

1968-1972	1980-1989	2008-2011	2016-2019	Incidence	Change
				0%	-100%

Little Ringed Plover *Charadrius dubius*

Scarce summer visitor and passage migrant since 1950.

Little Ringed Plover colonised Britain in the 1930s and first bred in Lincolnshire in 1950. It is limited by available breeding habitat, most frequently flooded gravel pits and coastal lagoons. The *LBC Atlas* reported around 25-35 pairs in the early 1990s. RBBP records show that an average of 24 pairs per year bred during the period 2013-17. The data reported by RBBP from 1996 onwards shows a run of bad years in the late 1990s. Counts have ranged from four pairs (1996, 1998, 2006) to 33 pairs (2017). The recent improvement in fortunes is down to attractive breeding habitat created at Frampton Marsh which holds about half the county population. Birds arrive from Mar onwards and the last ones are usually gone in Sep.

1968-1972	1980-1989	2008-2011	2016-2019	Incidence	Change
				0%	

Fishing boat attracting gulls on the Humber

Kentish Plover *Charadrius alexandrinus*

RBBP | BBRC

Rare. Western Europe. Has bred once, in 1979.

The first record was Oct 8th 1881 with no further reports until Apr 1950 when there were three at Gibraltar Point. They have always been a rare migrant with 35 records in total. These have largely involved single birds but in 1979 there were, eventually, four when a pair bred at Seacroft, Skegness and two young were raised. No suggestion of breeding since and this must be regarded as a one-off occurrence. Though two birds at Gibraltar Point on May 17th 1980 must have raised hopes for a repeat of 1979. Since then there have been singles in 11 years, with the most recent in Aug 2011, Apr 2014 and two in May and Aug 2019.

In 2020 Kentish Plover was again added to the BBRC list needing a full description to be accepted.

1968-1972	1980-1989	2008-2011	2016-2019	Incidence	Change

Site Name	First Date	Last Date	Count	Notes
Friskney	08/10/1881		1	
Gibraltar Point	13/04/1950		3	
Gibraltar Point	14/04/1950		1	
Wisbech STW	21/05/1959		1	
Wisbech STW	26/05/1964		1	
Gibraltar Point	09/08/1967		1	
Wisbech STW	09/08/1967		1	
Saltfleetby-Theddlethorpe	15/06/1972		1	
Killingholme	29/05/1977	30/05/1977	1	Female
Gibraltar Point	21/05/1978		1	Female
Gibraltar Point	11/06/1979		1	Male
Gibraltar Point	17/06/1979		1	Same male
Gibraltar Point	27/06/1979		1	
Gibraltar Point	30/06/1979		1	Same female
Gibraltar Point	10/07/1979		4	Pair with 2 immature
Gibraltar Point	23/07/1979		1	Same Male
Gibraltar Point	17/05/1980		2	Male and female
Gibraltar Point	18/05/1980		1	Same male
Saltfleetby-Theddlethorpe	15/06/1980		1	
Saltfleetby-Theddlethorpe	29/05/1981	30/05/1981	1	Male
Saltfleetby-Theddlethorpe	05/07/1981		1	Male
Gibraltar Point	17/06/1981	20/06/1981	1	Male
Killingholme	06/06/1982		1	Male
Gibraltar Point	07/05/1983	08/05/1983	1	
Donna Nook	07/05/1984		1	
Saltfleetby-Theddlethorpe	02/06/1984		1	
Tetney	24/05/1986		1	Female
Gibraltar Point	07/05/1987		1	Female
Barton on Humber	24/12/1989	01/04/1990	1	First winter male
Gibraltar Point	25/04/1995		1	Female
Saltfleetby-Theddlethorpe	05/06/1996	06/06/1996	1	Female
Saltfleetby-Theddlethorpe	22/08/2011	24/08/2011	1	Juvenile
Gibraltar Point	19/04/2014		1	Female
Gibraltar Point	14/05/2019		1	Female immature
Gibraltar Point	05/08/2019		1	Female

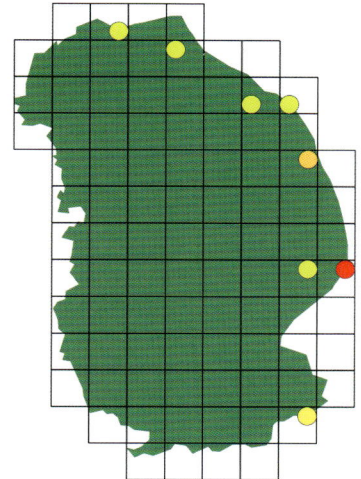

Barton Pits and Humber Bridge

Lesser Sand Plover *Charadrius mongolus* BBRC

Vagrant. Central Asia.

One record in May 2002, an adult female, this vagrant from Asia is still an extreme rarity in Britain and nearly all of them have posed an identification challenge given the variation in, and sometimes similarity to its congener, Greater Sand Plover *C. leschenaultii*. This individual was in bright summer plumage and showed well for numerous observers during its five-day stay out on the sands at Saltfleetby-Theddlethorpe. Currently IOC recognises five subspecies and this bird was of the form *C. m. atrifrons*, colloquially Tibetan Plover and a potential split from Lesser Sand Plover, from south-central Russia, Himalayas and east and south Tibet. Plumage features supporting this included a subdued head pattern with no strong blackish or dark brown component, plain orange forehead, long creamy supercilium merging into orange toned neck sides and nape collar, plain rear flanks and plain tail.

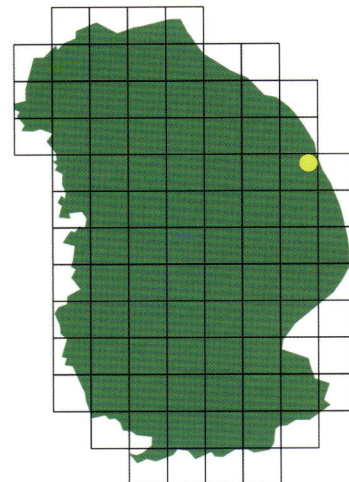

Lesser Sand Plovers are long-distance migrants and have occurred as a vagrant in several European countries, in at least four American states and two Canadian provinces. There are only six British records to date, the last being in Scotland in 2013.

Site Name	First Date	Last Date	Count	Notes
Saltfleetby-Theddlethorpe	11/05/2002	15/05/2002	1	

Greater Sand Plover *Charadrius leschenaultii* BBRC

Vagrant. Central Asia.

One record of an immature bird on Aug 7th 1981 at North Cotes Point, the fourth British record. It was considered to be the same bird which had been at Spurn for the previous nine days. It turned up in a mixed wader flock on the morning high tide but quickly flew off only to be re-found later in the day affording good views as it fed with Ringed Plovers *C. hiaticula* and Dunlin *Calidris alpina*.

These are large, long-billed and long-legged Sand Plovers and, given good views, identification should be reasonably easy. They are polytypic with three subspecies currently recognised by IOC. They breed across the arid regions of Asia from central Turkey to northwest China and Mongolia; none of the British records has been ascribed to a particular subspecies.

First recorded in Britain in 1978 when a bird was found in Pagham Harbour in Dec staying for 24 days until Jan 1st 1979, there have been 17 records in all up to 2018 with only five since the turn of the century. Of the 17 to date, 11 have been in England and six in Scotland.

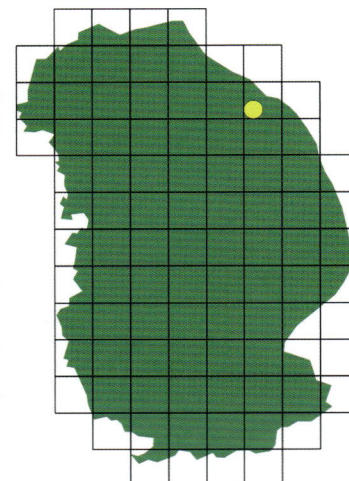

Site Name	First Date	Last Date	Count	Notes
North Cotes	07/08/1981		1	

Dotterel *Charadrius morinellus*

Rare/scarce passage migrant, mainly spring.

The *LBC Atlas* observed that in the 1980s there was an average of 20 birds per year which increased to 30 per year in the 1990s; LBR 2016 suggested that in the previous 20 years the average had been about 25. Looking at the five years to 2018 the wide range in birds counted is apparent. In 2014 there were 80 birds, the best year since a total of 109 in 1996. In 2018 there were just two. The total over the five years was 117 birds, an average of 21. Of these birds 113 were in spring between Apr-May, three were autumn juveniles and remarkably one adult bird was photographed at Frampton Marsh in Mar 2017. The rarity of this event is such that at that time there were no other winter records for Dotterel on Birdtrack from Dec-Mar for the UK. It is also noticeable that of the 117 birds, 109 of them were seen on the coast from Cleethorpes to Huttoft, Frampton Marsh had three, Gibraltar Point two and Moulton Seas End had one. A total of 115 on the coast leaving two (barely a trip) inland at Worlaby Top in May 2017. This was historically a regular stop off but has not delivered in recent years.

Whimbrel *Numenius phaeopus*

Fairly common passage migrant, mainly coastal. Rare in winter.

The *LBC Atlas* stated in the 1980s and 1990s peak migration was in May with counts rarely exceeding 100. During autumn migration in Jul-Aug peak counts of up to 250 were recorded at individual sites. WeBS does not publish a long-term trend for this species but across Britain counts are generally up. In the six years to 2019 LBR reports reflect that birds are arriving earlier in spring and records from Feb-Mar are becoming more frequent.

Peak spring migration was in Apr rather than in May in three out of six years. The spring total peak counts ranged from 138 in 2018, to 288 in 2019. Autumn migration peaked in either Jul or Aug and ranged from 180 in 2018 to 526 in 2015. The largest count in spring was reported from Frampton Marsh with 100 in Apr 2014 and in autumn, Gibraltar Point with 221 in Jul 2015.

If the ratio between spring and autumn counts is any indicator of breeding success (it may just reflect differing migration depending on weather), 2015 was a good year with a ratio of 1:2 while 2017 was a poor one with a ratio of 1:0.86.

Curlew *Numenius arquata*

Common passage migrant and winter visitor; scarce and local breeder.

The status of Curlew in Britain, whose shores hold around a quarter of the world population, is of real concern. The British population estimate of 58,500 for 2016 has fallen by nearly 70% since 1970. In Lincolnshire there is a small breeding population largely confined to the west of the county. The *LBC Atlas* confirmed them breeding in six 10km squares and estimated 30-50 pairs in the late 1980s. The *BTO Atlas 2007-11* confirmed breeding in 11 10km squares suggesting an increase in range. The current figure is likely to be 10-50 pairs.

There is a strong autumn passage particularly on The Wash in Jul-Aug. The *LBC Atlas* estimated the wintering population at 3,000-4,000 birds. During 2014-18 LBR tables indicate the average wintering population was at least 3,200 with a range from 1,900 (2017) to 3,600 (2014). These birds probably come from Scandinavia and Russia so hopefully this indicates that the eastern population is more stable than the British one.

1968-1972	1980-1989	2008-2011	2016-2019	Incidence	Change
				4%	7%

Bar-tailed Godwit *Limosa lapponica*

Very common coastal passage migrant and winter visitor. Scarce inland.

The Wash holds over 10% of the world population of this species and is Britain's number one internationally important site. The Humber comes in at number 10. Data from the WeBS survey shows rolling five-year mean counts to 2018-19 of 18,579 on The Wash and 1,331 on the Humber. For the five years to 2013-14 these figures were 16,491 and 2,097 respectively, indicating that the local population is fairly stable. Overall the long-term trend shows the UK wintering population down 17%. LBR reports peak counts over the five years to 2018 from the following sites: on The Wash 8,000 at Gibraltar Point Sep 2016 and 8,000 at Friskney Aug 2014; on the Humber 1,200 on Pyewipe Marsh, Grimsby Mar 2015. High Aug counts reflect the fact, demonstrated by ringing, that Siberian birds pass through en route to their West African wintering grounds.

Black-tailed Godwit *Limosa limosa*

Icelandic Black-tailed Godwit, *L. l. islandica* is a common passage migrant and winter visitor.

The European race *L. l. limosa* is a rare migrant and has bred, last in 1974, and around 50 pairs breed annually in Britain up to 2019 (LBRC, RBBP).

The Icelandic race *L. l. islandica*, is the commonly seen race in Britain and in Lincolnshire although there were many fewer 20 years ago. Since the 1990s the ratio of the two races in Europe has changed quite dramatically. European *L. l. limosa* used to number around 200,000 in the 1960s but has declined to around 60,000. There are no good historical counts for *islandica*, but it is estimated that there were c.50,000 in the early 2000s. However, whereas *limosa* probably outnumbered *islandica* by about 8:1, the two are now probably closer to 1:1. Most *limosa* winter in west and south-west Europe but virtually the whole of the Icelandic *islandica* population winters exclusively in Britain. The BTO WeBS data show a 228% increase in *islandica*, 1992-93- 2017-18. This has been reflected in the county and a record count of 10,505 roosted at Freiston Shore in Nov 2019. There is also a growing catalogue of colour-ringed records in the county showing the origins and movements of these long-lived shorebirds as the species is the subject of an intensive project based at the University of East Anglia.

The European race *L. l. limosa* is not regularly recorded in the county and is difficult to separate from *islandica* in the field. However, there is one recent confirmed record of *limosa* at Frampton Marsh: a bird colour-ringed bird as a chick at the Nene Washes was seen on Apr 24th 2019. Project Godwit is a partnership between RSPB and WWT aiming to supplement the Ouse Washes *limosa* population through a rear-and-release programme to help re-establish the birds at sites adjacent to the Ouse Washes.

1968-1972	1980-1989	2008-2011	2016-2019	Incidence	Change

Black-tailed Godwits on the Humber

Hudsonian Godwit *imosa haemastica*

Vagrant. North America.

The Hudsonian Godwit, known in the past as the Ring-tailed Marlin or Goose-bird, old folk names, is the least well known of the world's four godwit species. One was found at Blacktoft Sands, Yorkshire, in Sep 1981 which was a first for Britain and the Western Palaearctic. It visited Alkborough Flats for one day on Sep 15th 1981 and remains the sole county record. It returned to Blacktoft where it stayed until Oct 3rd, last seen flying off to the east. It, or another, turned up in Devon seven weeks later, from Nov 22nd 1981 to Jan 14th 1982 at least, and what was assumed to be the same bird turned up again at Blacktoft in Apr 26th to May 6th 1983.

There have only been two British further records, one in 1988 and much later in Apr-May 2015. Given the huge flocks of Icelandic Black-tailed Godwits *L. l. islandica*, regularly occurring on The Wash these days, assiduous observation of these may be the best chance of locating another in the county.

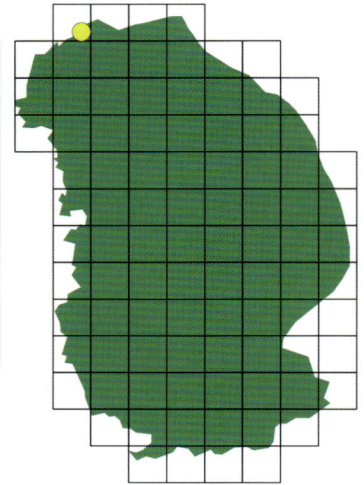

Site Name	First Date	Last Date	Count	Notes
Alkborough	15/09/1981		1	

Turnstone *Arenaria interpres*

Fairly common/common passage migrant and winter visitor. Very scarce inland.

The Wash is Britain's second most important site for Turnstone and the Humber is the 21st. WeBS data show that the rolling five-year mean counts to 2018-19 are 809 on The Wash and 272 on the Humber. For the five years to 2013-14 these figures were 787 and 347 respectively, indicating that the local population is fairly stable. Overall, the long-term trend shows that the UK wintering population is down 43%. This is thought to be a result of a northward range shift in response to climate change. LBR reports show peak site counts over the five years to 2018; on The Wash, 420 at Freiston Shore Oct 2016, and on the Humber, 285 at Goxhill Feb 2018.

Knot *Calidris canutus*

Greenland and Canadian form *islandica* abundant passage migrant and very common winter visitor. Very scarce inland.

The Wash can hold a third of the world population of this race of Knot during peak passage and is Britain's number one internationally important site. The Humber comes in at number four. WeBS data reports rolling five-year mean counts to 2018-19 at 177,869 on The Wash and 22,500 on the Humber. For the five years to 2013-14 these figures were 147,606 and 29,847 respectively, indicating that the local population is falling slightly. Notwithstanding that it is Lincolnshire's commonest wader, the long-term trend overall shows the UK wintering population down 20%. LBR reports peak counts over the five years to 2018 from the following sites: on The Wash 105,000 at Gibraltar Point Sep 2014 and on the Humber 20,000 at Cleethorpes Nov 2018. The sight of a flock of Knot wheeling at roost from Cleethorpes promenade or Mill Hill at Gibraltar Point is one of the finest experiences in British birding.

Ruff *Calidris pugnax*

Fairly common passage migrant and scarce winter visitor. Bred to 19th century.

Ruff originally bred in the Fens and Isle of Axholme but died out by 1882. In the last 29 years to 2018, four instances of probable breeding have occurred in 2010, 2016, 2017 and 2018 at two separate sites. Frampton Marsh has reported lekking behaviour in all the four years mentioned, with males and females present for over a week. While encouraging, there has been no evidence of nesting behaviour as yet. The wintering and migratory population varies from year to year and in the five years to 2018 peak monthly table counts from LBR show an average wintering population of 80 birds ranging from 18 (2014) to 155 (2015). Peak monthly spring migration in Apr-May averaged 100 birds ranging from 60 (2015) to 152 (2014). Peak monthly autumn migration from Aug-Oct averaged 220 birds ranging from 184 (2018) to 571 (2017). The peak monthly count in the five-year period at all sites was 257 at Alkborough Flats in Oct 2017.

1968-1972	1980-1989	2008-2011	2016-2019	Incidence	Change
				0%	

Broad-billed Sandpiper *Calidris falcinellus*

BBRC

Rare. Scandinavia.

The first county record was found at Wisbech Sewage Farm in May 1959 and subsequently there have been 15 more. The nominate European race *falcinellus* breeds in the boreal forest bogs of northern Scandinavia and into Arctic Russia. They migrate through the eastern Mediterranean, Black and Caspian Seas to winter in the Persian Gulf, western India and Sri Lanka, with small numbers in coastal east Africa. Departure from the Fennoscandian breeding grounds, where there are estimated to be some 25,000 breeding pairs, peaks in Jul for adults and Aug for juveniles.

Of the 11 spring records the earliest has been May 12th and the latest Jun 9th. The autumn records (five) have all been between Jul 15th and Aug 4th. More than 250 have been found in Britain to date with more than 70% of those being found in May-Jun, mainly on the east coast. Looking through flocks of Ringed Plover *Charadrius hiaticula* and Dunlins *C. alpina* in spring, as they speed northwards is probably the best way of finding this species.

Site Name	First Date	Last Date	Count	Notes
Wisbech STW	18/05/1959		1	
Saltfleetby-Theddlethorpe	26/05/1982		1	
Saltfleetby-Theddlethorpe	19/05/1984		1	
North Cotes	29/05/1984		1	
Gibraltar Point	26/07/1990	27/07/1990	1	
Reads Island	06/06/1992		1	
Donna Nook	15/07/1995	19/07/1995	1	
Gibraltar Point	04/08/1997		1	Adult
Reads Island	29/05/2000	31/05/2000	1	
Alkborough	16/05/2008	18/05/2008	1	Adult
Alkborough	16/05/2014		1	
Gibraltar Point	29/07/2014		1	Adult
Frampton Marsh	09/06/2015	12/06/2015	1	Second calendar year
Frampton Marsh	12/05/2016	14/05/2016	1	Second calendar year
Frampton Marsh	24/07/2016	28/07/2016	1	Second calendar year
Frampton Marsh	19/05/2019	20/05/2019	1	Adult

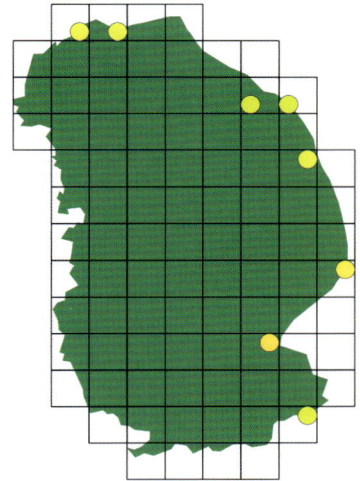

Sharp-tailed Sandpiper *Calidris acuminata*

BBRC

Vagrant. Eastern Siberia.

There have been two county records of this subtly plumaged medium-sized *Calidrid*, an adult in 1982 at Killingholme Pits on The Humber and a juvenile in 1985 on The Wash. The account of the Killingholme bird in LBR 1982 notes a behavioural characteristic of this species and that "...it bore a strong resemblance to a small crake as it moved with head and neck held forwards...and its deliberate jerky leg movements".

Confusion with Pectoral Sandpiper *C. melanotos*, occurred in the early British records and was discussed by Britton (1980) in a paper in *British Birds*.

There have been 28 British records to 2018, the last in 2012 in Scotland and it remains a much sought after shorebird. The well-watched coastal sites in the county are overdue a third record.

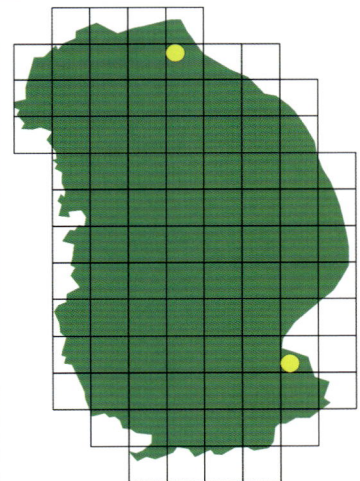

Site Name	First Date	Last Date	Count	Notes
Killingholme	18/09/1982	20/09/1982	1	Adult
Holbeach Marsh	24/08/1985		1	Juvenile

Stilt Sandpiper *Calidris himantopus*

BBRC

Vagrant. North America.

Three county records of this charismatic Arctic wader, two being in the 1960s at the shorebird magnet that was Wisbech Sewage Farm and the third in 2018 at the current shorebird magnet that is Frampton Marsh. The first individual was trapped before it had been seen in the field and as the finders noted "(it) could not readily be identified"; unsurprising given the field guide literature of the day. After release it flew to the muddy banks of the River Nene, just inside the Cambridgeshire border, and it remained in the area for nearly three weeks. The two subsequent county records were also of relatively long stayers, both remaining for around two weeks.

There have been 38 British records 1950-2018 with two records in eight years in that period and one with three (2008). It remains a rare shorebird with one record every one to two years.

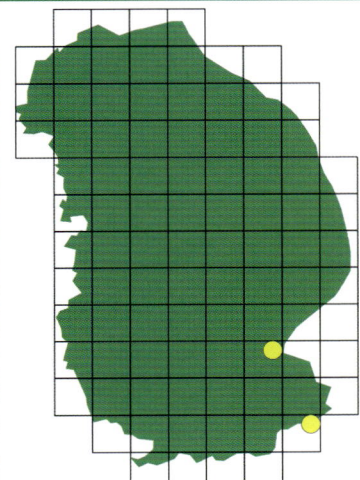

Site Name	First Date	Last Date	Count	Notes
Wisbech STW	19/07/1963	07/08/1963	1	
Wisbech STW	12/08/1965	26/08/1965	1	
Frampton Marsh	21/08/2018	04/09/2018	1	Adult

Curlew Sandpiper *Calidris ferruginea*

Passage migrant, scarce in spring and scarce/fairly common in autumn. Rare in winter.

Analysis of LBR reports for the five years to 2018 shows that spring migration is light with total peak monthly counts across the county in Apr-May ranging from five to 10. Autumn totals from Jul-Sep tend to vary much more and ranged from 27 in 2017 to 322 in 2016. Frampton Marsh holds all the record counts for this species in recent years with a peak of 258 in Aug 2016. Other sites with good counts include Gibraltar Point with 32 in Jul 2018 and 32 at Freiston Shore in Aug 2016. The *LBC Atlas* reports the exceptional passage of Aug 1969 when Wisbech Sewage Farm had 400 birds and Donna Nook 265. Birds in Nov are usually very scarce, but Frampton Marsh had 30 lingering in 2016. There were only two other reports in Nov and there were no winter records between Dec-Mar.

Temminck's Stint *Calidris temminckii*

Very scarce/scarce passage migrant.

White and Kehoe (2020) report an average of 95 Temminck's Stints per year in Britain during 2010-18 with no trend either way since 1990. In Lincolnshire the *LBC Atlas* reported an average of three to four birds per year in the 1980s and 1990s. LBR reports for the 10 years to 2018 suggest around 94 birds though totals are difficult to distinguish as Frampton Marsh gets so many. The mean of nine per year suggests that Lincolnshire is hosting more birds than previously, a result of Frampton Marsh providing such good wader habitat. Over the 10-year period county totals ranged from one in 2018 to 15 in 2015. Spring passage accounted for 78% of records and autumn 22%. The best spring was 2017 with 12 birds (10 at Frampton Marsh) and the worst 2018 with none. Autumn migration provided four blank years and a maximum of seven in 2016 of which five were at Frampton Marsh.

Sanderling *Calidris alba*

Common passage migrant and winter visitor. Scarce inland.

The Wash is Britain's number one internationally important site for Sanderling. The Humber comes in at number 17. WeBS data reports show that the rolling five-year mean counts to 2018-19 were 9,052 on The Wash and 509 on the Humber. For the five years to 2013-14 these figures were 3,916 and 632 respectively, indicating that the local population has increased, though many of these birds are passing through rather than wintering. Overall, the long-term trend shows that the UK population is up 36%. The open sandy beaches from Skegness to Mablethorpe are poorly monitored by WeBS and likely hold several hundred more wintering birds. LBR reports peak counts over the five years to 2018 from the following sites: on The Wash 10,121 at Gibraltar Point Aug 2018 and on the Humber 450 at Saltfleetby-Theddlethorpe in May 2017. There have been numerous ringing recoveries, and sightings of colour-ringed birds show that passage birds are en route from high Arctic breeding grounds to winter quarters in Mauritania and Ghana.

Dunlin *Calidris alpina*

Southern form *schinzii* a common passage migrant; north-east Greenland form *arctica* a passage migrant in unknown numbers; Scandinavian/Russian form *alpina* a very common passage migrant and winter visitor, mainly coastal. Bred to late 19th century.

One of our commonest migrant and wintering waders it hung on as breeding species at Scotton Common until the 1890s. Lorand and Atkin (1989) and the *LBC Atlas* referred to a possible breeding record from Read's Island in 1958 which on its own is no longer considered sufficient evidence to sustain a status of irregular breeder in the 50 year period to 1989. The *LBC Atlas* put the wintering population in the late 1980s at 40,000 on The Wash and 10,000 on the Humber. In the last few years the numbers reported in LBR have been slightly less than half that. The largest numbers occur on passage in Jul-Aug when small numbers can turn up at many inland sites. The biggest count in the five years to 2018 was 16,300 at Gibraltar Point in Aug 2018.

1968-1972	1980-1989	2008-2011	2016-2019	Incidence	Change

Purple Sandpiper *Calidris maritima*

Very scarce/scarce passage migrant and winter visitor, mainly coastal.

The *LBC Atlas* suggested that average numbers were around 11 birds per year in the 1980s and six to seven in the 1990s. With no natural rocky shores and very few artificial ones, Lincolnshire has little to offer for Purple Sandpipers. The 25-year long term trend for UK wintering birds is down 50% according to Frost *et al* 2020 in the Waterbirds in the UK report, 2018-19. In the five-year period to 2018 LBR reported around 38 birds or seven to eight per year, very much in line with previous Lincolnshire experiences. Annual totals ranged from three in 2015 to 12 in 2017 and 2018. Of the 38, 11 were winter birds, five in spring and 22 in autumn. The only winter long stayers, as in many previous years, were at Cleethorpes. Two reports were of inland birds, both at Covenham Reservoir.

Baird's Sandpiper *Calidris bairdii*

Vagrant. North America.

First recorded at Wisbech Sewage Farm Jul 22nd to Aug 6th 1963 there have only been four records since, most recently at Frampton Marsh in Jul 2013. Strangely for a vagrant wader, three of the five were adults and one a juvenile (1979) with one unaged (1966).

Baird's Sandpipers have been recorded on 285 occasions in Britain between 1950-2018. Overall, there are about seven British records per year although since 2000 there have been double figure counts in 2004 (11), 2005 (12) and 2016 (11).

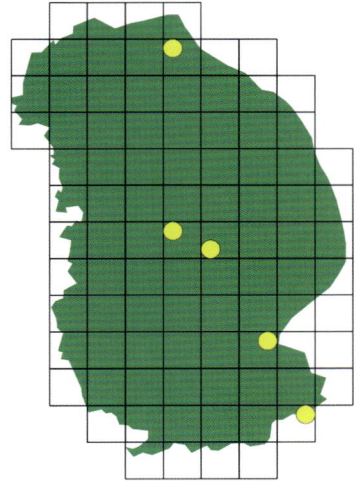

Site Name	First Date	Last Date	Count	Notes
Wisbech STW	22/07/1963	06/08/1963	1	
Bardney	01/09/1966	08/09/1966	1	
Killingholme	19/09/1979	20/09/1979	1	
Kirkby on Bain	19/09/2005	27/09/2005	1	Adult
Frampton Marsh	24/07/2013	26/07/2013	1	Adult

Little Stint *Calidris minuta*

Passage migrant, very scarce in spring and scarce/fairly common in autumn with rare/very scarce winter records.

The general feeling is that while Little Stints have increased in winter and in spring, autumn numbers are down. LBR reports show that in the five years to 2018, birds wintered in four years out of five and averaged about six birds per year with none in 2018 and 16 in 2016. Peak spring reports averaged about 10 per year with four in 2014 and 16 in 2015, while peak autumn reports averaged 38 per year ranging from 13 in 2018 to 67 in 2016.

The *LBC Atlas* reported that exceptional influxes of over 100 had occurred in the past but there has been nothing like that in recent years. The largest counts around the Wash were of 40 at Frampton Marsh in Aug 2016 and on the Humber, 15 at Alkborough Flats in Sep 2015.

The Coast of Lincolnshire

White-rumped Sandpiper *Calidris fuscicollis*

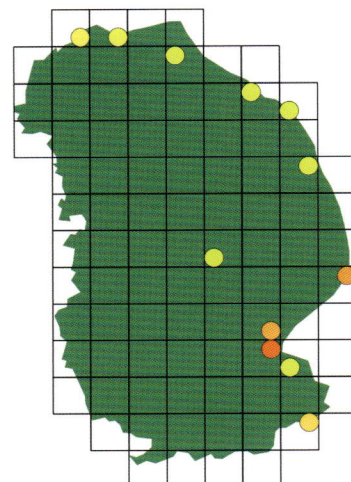

Very scarce migrant. North America.

With around 35 individuals up to 2019 this is the second commonest North American shorebird recorded in the county after Pectoral Sandpiper *C. melanotos*. Inevitably the first five records came from Wisbech Sewage Farm, the first in 1955 being trapped. The majority of those birds aged have been adults, 17 out of 20, and only three have been in spring, May 1999, 2006 and 2015. The autumn dates are widely spread between Jul 3rd and Nov 23rd. There have been some long staying individuals; the records of the wandering juvenile seen at several sites on The Wash in 2013 spanned 34 days, and 12 have stayed for at least seven days. There has been a surge in records since 2006 due, in part, to the availability of suitable habitat at places such as Alkborough Flats and Frampton Marsh from about this time. Almost half have been in The Wash between Gibraltar Point and Holbeach Marsh. The only inland record, apart from those at Wisbech Sewage Farm, one at Kirkby on Bain gravel pits on Jul 14th 2012.

Site Name	First Date	Last Date	Count	Notes
Wisbech STW	13/11/1955	17/11/1955	1	
Wisbech STW	24/10/1964	01/11/1964	1	
Wisbech STW	10/08/1970	19/08/1970	1	
Wisbech STW	07/08/1971	19/08/1971	1	
Wisbech STW	28/07/1973	10/08/1973	1	
Frampton Marsh	16/09/1973		1	
Killingholme	26/07/1976	27/07/1976	1	
Welland Mouth	25/08/1985		1	Adult
South Ferriby	05/08/1993	15/08/1993	1	Adult
Reads Island	07/08/1998		1	Adult
Holbeach Marsh	16/05/1999		1	
Saltfleetby-Theddlethorpe	14/05/2006		1	
Freiston Shore	08/10/2006	10/10/2006	1	
Gibraltar Point	03/07/2007		1	
Alkborough	23/11/2008	30/11/2008	1	
Gibraltar Point	08/08/2010		1	
Alkborough	26/09/2011		1	Adult
Frampton Marsh	07/07/2012	25/07/2012	1	Adult
Kirkby on Bain	14/07/2012		1	
Donna Nook	07/10/2013		1	
Gibraltar Point	14/10/2013		1	Juvenile
Frampton Marsh	01/11/2013		1	Juvenile
Gibraltar Point	02/11/2013		1	Same juvenile
Frampton Marsh	04/11/2013		1	Same juvenile
Alkborough	09/11/2013		1	Juvenile
Frampton Marsh	09/11/2013		1	Same juvenile
Frampton Marsh	12/11/2013	22/11/2013	1	Adult
Gibraltar Point	28/07/2014	30/07/2014	1	Adult
Frampton Marsh	02/08/2014	04/08/2014	1	Adult winter
Gibraltar Point	06/08/2014	08/08/2014	1	Adult summer
Gibraltar Point	12/08/2014		1	Same
Frampton Marsh	29/05/2015	12/06/2015	1	Adult summer
Frampton Marsh	28/07/2016	09/08/2016	1	Adult
Gibraltar Point	22/08/2017		1	Adult
Humberston	30/08/2017	12/09/2017	1	
Gibraltar Point	27/09/2018		1	Adult winter
Frampton Marsh	20/07/2019	21/07/2019	2	Adult
Frampton Marsh	22/07/2019	27/07/2019	1	Same adult
Freiston Shore	28/07/2019		1	Same adult
Freiston Shore	30/07/2019	04/08/2019	2	Same adult
Freiston Shore	05/08/2019		1	Same adult
Freiston Shore	06/08/2019	07/08/2019	2	Same adult
Freiston Shore	08/08/2019	09/08/2019	1	Same adult

Buff-breasted Sandpiper *Calidris subruficollis*

Very scarce migrant. North America. Recorded from May-Nov.

One of several species added to the Lincolnshire list by G. H. Caton Haigh, who shot the first Buff-breasted Sandpiper on the fitties at North Cotes on Sep 20th 1906. The next was not until 1975 on the Wash, but there have now been 33 records involving around 32 birds; most in the autumn but with four in the spring, all in late May-early Jun. The bulk of autumn birds arrive in Sep with extreme dates of Jul 14th to Oct 22nd.

Autumn birds can be very mobile making it difficult to assess total numbers. For example, two adults were seen at Horseshoe Point on Aug 14th 2011 followed by a juvenile at Donna Nook on Sep 13th. This bird was then presumed to have moved to Saltfleet and then Saltfleetby-Theddlethorpe on Sep 14th where a second juvenile joined it. Later, two juveniles were found at Alkborough on Oct 2nd. So, a minimum of four, perhaps six birds arrived that autumn. To put this in context, the last week of Sep 2011 was notable for the large numbers which arrived in Britain and Ireland when as many 60 as may have arrived including an incredible flock of 28 at Tacumshin (Wexford) on Sep 28th.

Buff-breasted Sandpiper ceased to be considered by BBRC in 1983. White and Kehoe (2020) in the report on scarce migrants in Britain gave the total number of British records, 1958-2018 as 1250 with 32 in 2018.

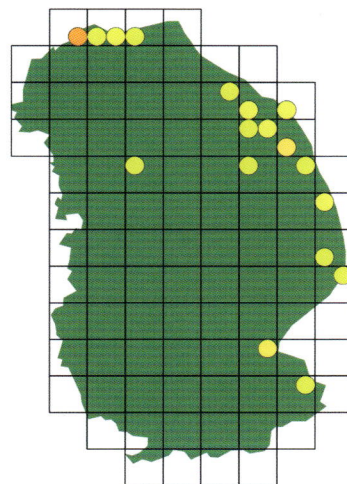

Site Name	First Date	Last Date	Count	Notes
North Cotes	20/09/1906		1	Shot
Witham Mouth	13/09/1975		1	
Alkborough	03/09/1981	14/09/1981	1	
Saltfleetby-Theddlethorpe	11/08/1982	16/08/1982	1	Adult
Cleethorpes	31/05/1989		1	
Huttoft	11/09/1995	13/09/1995	1	
Huttoft	26/09/1995	27/09/1995	1	
Nene Mouth	12/08/1999		1	
Gibraltar Point	13/09/2002		1	
Alkborough	09/09/2006	12/09/2006	1	Juvenile
Donna Nook	16/09/2006	17/09/2006	1	
Covenham Reservoir	10/10/2006	12/10/2006	1	Juvenile
Louth	22/10/2006		1	Juvenile
Barton Pits	30/07/2007		1	Adult
Reads Island	31/07/2007		1	Same
Toft next Newton	04/09/2007	05/09/2007	1	Juvenile
Alkborough	22/09/2007		1	Juvenile
Tetney	02/10/2007		1	Juvenile
Alkborough	14/07/2008		1	
Alkborough	18/07/2008	19/07/2008	1	Same
Alkborough	03/08/2008		1	Same
Alkborough	24/05/2009		1	Adult
Winteringham	08/10/2009		1	
Reads Island	28/08/2010		1	Adult
Marshchapel	14/08/2011		2	
Donna Nook	13/09/2011		1	Juvenile
Saltfleet	14/09/2011		1	Juvenile
Saltfleet	19/09/2011	22/09/2011	2	One same
Alkborough	02/10/2011		2	Juvenile
Saltfleet	12/06/2012		1	Adult
Frampton Marsh	29/09/2013	09/10/2013	1	
Middlemarsh Farm	02/06/2015		1	
Frampton Marsh	15/08/2019	28/08/2019	1	

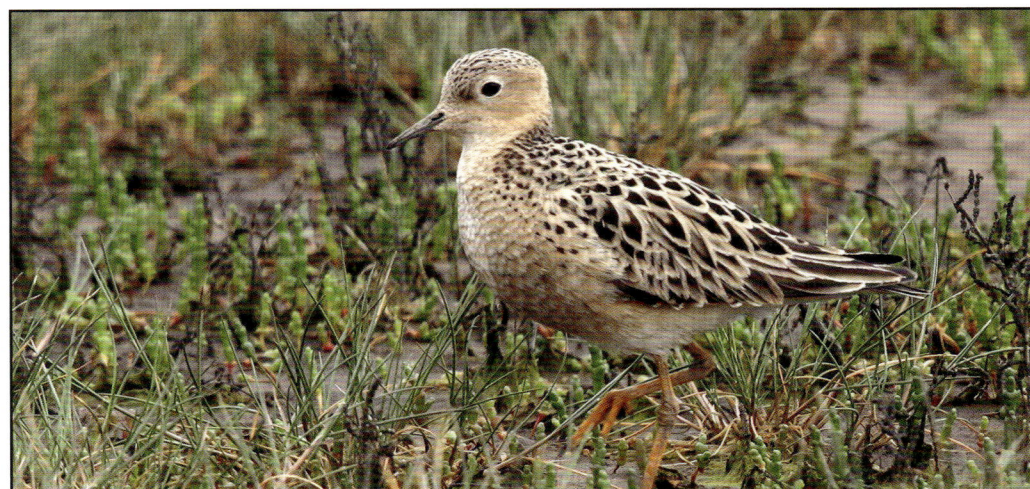

Pectoral Sandpiper *Calidris melanotos*

Very scarce autumn passage migrant and rare in spring.

White and Kehoe (2020) reported an average of 141 birds per year in Britain during 2010-18 with an increase since 1990. In Lincolnshire the *LBC Atlas* reported 80 records over the previous 50 years. LBR reports for the 10 years to 2018 reported a total of 74 "Pecs" which at seven to eight a year is a large increase over earlier times. However, with 43 in in the five years to 2013 and 31 in the five years to 2018 there has been a slight fall in the most recent period. The record year was 12 in 2011, probably the best year ever in Lincolnshire, and the poorest years were three in 2018 and one in 2019. The records over the 10 years were split between five in spring and 70 in autumn.

Semipalmated Sandpiper *Calidris pusilla*

Vagrant. North America.

A 1CY bird was trapped at Wisbech Sewage Farm in Nov 1966; one observer at the time watched it feeding along the deep furrows of an adjacent ploughed field as well as on the muddy sludge beds. Some 44 years later an adult bird turned up at Alkborough Flats in Aug 2010, just reward for the many hours spent by the finder searching through waders at this site.

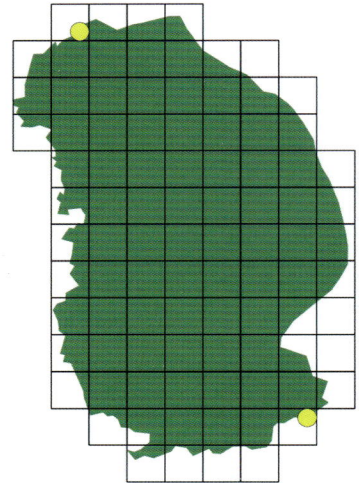

The identification of this 'peep' remains a challenge and in the early years claims of this species came under great scrutiny. Wallace (1979) published a review of 12 published British records as well as claims of Red-necked Stint *C. ruficollis* between 1953-74, and found only five to be acceptable, including the Lincolnshire bird of 1966 which was trapped, after all. Later, Jonsson and Grant (1984) published a seminal paper on the identification of stints and peeps. Including the five accepted records published in the Wallace review, there have now been 156, 1950-2018, a veritable flood. Despite this, and with double figures in some years (17 in 2011, 11 in 2013) the species remains on the BBRC list for consideration and continues to be a challenge to find.

Site Name	First Date	Last Date	Count	Notes
Wisbech STW	12/11/1966	29/12/1966	1	Trapped 13/11/1966
Alkborough	18/08/2010	24/08/2010	1	Adult

Long-billed Dowitcher *Limnodromus scolopaceus*

Rare. North America, north-eastern Russia.

The first county record was found at Wisbech Sewage Farm in Sep 1963 and accepted as either a Long-billed or Short-billed Dowitcher *L. griseus*. This record was reviewed much later and confirmed as a Long-billed Dowitcher by the BBRC (Rogers *et al* 1980). Two more followed in 1971 and 1986 before regular appearances began in the 2000s bringing the total to 12. Of the 10 birds which were aged from 2002-19, six were 1CY birds turning up in the autumn. Might some of these be coming from north-eastern Russia rather than north America?

Scolopaceus and *griseus* were formerly considered conspecific but a thorough review of British Dowitcher museum specimens and sight records by Nisbet (1961) established the acceptable historic records. Since then it has become recognised that they are most easily identified in juvenile plumage. It is more difficult with birds in breeding plumage owing to the variability seen in *griseus*, though calls are diagnostic.

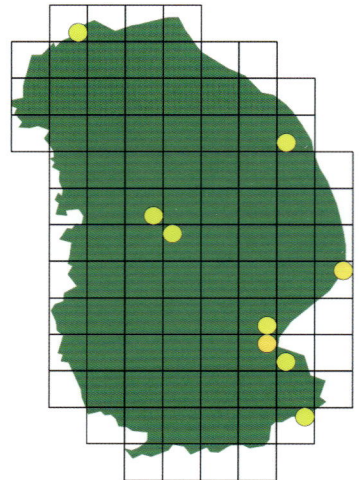

Site Name	First Date	Last Date	Count	Notes
Wisbech STW	28/09/1963		1	
Bardney	28/09/1971	10/10/1971	1	
Holbeach Marsh	19/07/1986	14/08/1986	1	Adult
Saltfleet	01/11/2002	17/11/2002	1	Juvenile
Gibraltar Point	05/07/2004	29/09/2004	1	Adult
Gibraltar Point	22/07/2006	03/10/2006	1	Adult
Branston Island	24/09/2007	14/10/2007	1	
Freiston Shore	26/09/2011	04/10/2011	1	Juvenile
Frampton Marsh	04/10/2011		1	Same
Alkborough	04/10/2012	27/04/2013	1	Juvenile
Gibraltar Point	07/09/2016		1	
Frampton Marsh	01/10/2016	08/11/2016	1	First calendar year
Freiston Shore	03/10/2016		1	
Saltfleet	16/09/2017	06/11/2017	1	First calendar year
Frampton Marsh	24/08/2018	01/05/2019	1	Adult summer
Frampton Marsh	19/07/2019	05/09/2019	1	Same adult

Woodcock *Scolopax rusticola*

Fairly common resident in woodland, passage migrant and common winter visitor.

Woodcock is a truly beautiful bird close up, but their crepuscular habits can make it difficult to get a really good view, especially when flushed in woodland. Those habits make them difficult to monitor and even full-scale atlas surveys do not always do the job.

The *LBC Atlas* found it difficult to estimate the breeding population based on roding males and put it anywhere between 250 and 600 birds in the late 1980s and it was thought there had been no change between then and the late 1990s. The BTO did a full-scale survey of roding males in 2003 and repeated it in 2013-14. It showed a UK-wide decline of some 23%. The *BTO Atlas 2007-11* showed a pronounced decline in the east of the county from where they have largely disappeared as a breeding bird. It may be a coincidence but many of Lincolnshire's woods are heavily shot over in late winter. This reveals the north-eastern European origin of the birds wintering here. In Dec 2017 a bird ringed as a nestling in Finland earlier in the year was shot at Barrowby. In some winters large "bags" are taken. Do these take out resident breeders as well as eastern migrants? The largest single count at Gibraltar Point in the 10 years to 2018 was a result of the "Beast from the East" that hit Britain in late Feb 2018 and caused thousands of dead Woodcocks to be picked up on beaches across the east of Britain. Fewer were found dead in the county but the counts of live birds at Gibraltar Point show the scale of the arrival. On Feb 28th there were 29 birds, Mar 1st there were 62, Mar 2nd there were 87 and Mar 12th a count of 119, the highest count since the last hard winter of 2009-10 when there had been 70 in Jan 2010. In the intervening years the average annual peak monthly high count was 27.

1968-1972	1980-1989	2008-2011	2016-2019	Incidence	Change
				0%	*

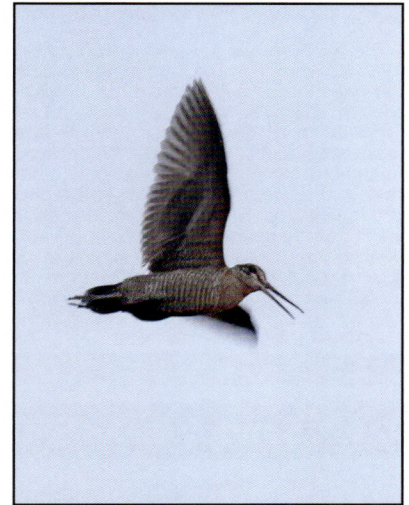

Jack Snipe *Lymnocryptes minimus*

Scarce passage migrant and winter visitor. Probably much under-recorded.

You don't usually see 'em unless you flush 'em. This makes it difficult to get any real sense of how accurate the numbers are and where they are going, but Jack Snipe can be reliably found in the same passage and wintering sites year in, year out. The *BTO Atlas 2007-11* found them wintering in 35 10km squares across Lincolnshire so it's reasonable to say they are fairly local with numbers greatest on the coast between Saltfleetby-Theddlethorpe and Humberston Fitties and in the Trent Valley. In the six years to 2017 the peak monthly count each year reported in LBR ranged between Jan, Mar, Oct and Nov and from eight in 2012 to 28 in 2013 and 2014.

WeBS online indicates that the Humber estuary is currently the ninth best location in the UK for Jack Snipe with a five-year mean winter count of seven birds. To put that in context, the top site, Chat Moss west of Manchester has a mean of 22. Its nearest breeding grounds are Sweden and Finland but it ranges across the boreal zone of Russia and Siberia.

Only one bird has ever been ringed in Lincolnshire according to BTO's online Ringing and Nest Recording Report, so it is not surprising that the movements of this cryptic wader are poorly understood.

Great Snipe *Gallinago media*

Rare. North and eastern Europe.

The first 15 county records, 1843-1952 were all of birds shot mainly Sep-Oct but with two others in Jan and May. Given that it is, disgracefully, still legal to shoot Common Snipe *G. Gallinago*, and Woodcock *Scolopax rustica*, it would be no surprise to find that this still occasionally happens. With just another four records 1978-84 the county is long overdue another. It is worth noting that on migration it can turn up in diverse habitats including short grass or sedges on lake edges or flooded fields, montane bogs, airfields and tracks in wooded areas; overall though it usually prefers drier habitat than that preferred by *G. gallinago* – witness the 1984 record where the bird was often found in sandy dunes with marram grass.

British records 1950-2018 total 168 and are stable at about three per year.

Site Name	First Date	Last Date	Count	Notes
Lincolnshire	01/05/1843		1	Male
Marshchapel	01/01/1850		1	
Tetney	01/09/1852		1	
Scunthorpe	01/09/1868		1	
Crowland	17/09/1875		1	
Stickney	04/10/1882		1	Male
Lincolnshire	02/10/1883		1	
Cowbit	01/01/1896		1	
Spalding	01/09/1896		1	
North Cotes	03/10/1899		1	
North Cotes	06/09/1901		1	
Marston	12/09/1938		1	
Tetney	20/09/1939		1	Male
Rothwell	01/09/1947		1	
Rothwell	03/10/1952		1	
Wisbech STW	21/08/1969	23/08/1969	1	
Saltfleetby-Theddlethorpe	08/10/1978	15/10/1978	1	Immature
Barton upon Humber	23/03/1983	24/03/1983	1	
Saltfleetby-Theddlethorpe	24/08/1984		1	First winter
Saltfleetby-Theddlethorpe	15/09/1984	18/09/1984	1	Same

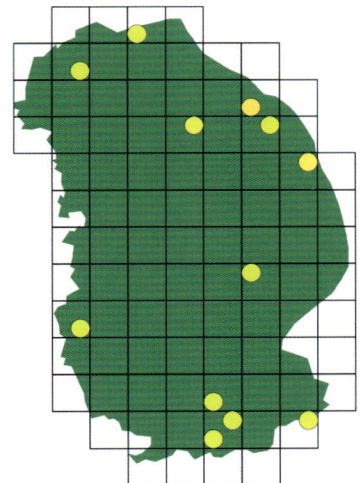

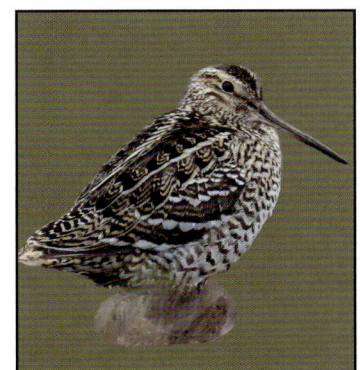

Snipe *Gallinago gallinago*

Scarce breeder, fairly common/common passage migrant and winter visitor.

Snipe has suffered a shocking decline as a breeding bird in Lincolnshire since the first *BTO Atlas of 1968-72* when it was a fairly widespread breeder in wet grassy areas throughout the county. As drainage intensified through the 1980s it went into a rapid decline such that by the end of the 1980s the *LBC Atlas* reported there were an estimated 270 pairs and the *LBC Atlas* map showed confirmed breeding in 19 10km squares. By the late 1990s the *LBC Atlas* noted there were fewer than 15 drumming males. By the last *BTO Atlas 2007-11* there was not a single instance of confirmed breeding in any Lincolnshire square and only a handful of squares had probable breeding. It is salutary to note that in the five years to 2018, LBR reported no instances of confirmed breeding and evidence of drumming from only four sites, Alkborough Flats, Blackmoor Bridge, Crowle Moors and Frampton Marsh. Snipe is less choosy about its wintering and migration habitat and being the most widespread wintering wader in Britain makes it difficult to get a handle on how many are passing through. In the five years to 2018 Alkborough Flats held all the highest peak season monthly counts as reported in LBR as follows: winter 118 Nov 2017, spring 110 Apr 2015, autumn 264 Oct 2016.

1968-1972	1980-1989	2008-2011	2016-2019	Incidence	Change
				1.5%	*

Terek Sandpiper *Xenus cinereus*

BBRC

Vagrant. North-eastern Europe

A relatively recent addition to the county list with the first record a one-day bird at Gibraltar Point in Jul 2005 located feeding near the mouth of a creek at the south end of the shore. It appeared to be moved on by a heavy rain shower and presumably moved off into The Wash with the Dunlin with which it was associating. The second record was at the same site in Jun 2009 and remained long enough to be seen by a wider audience. It was presumed to be the bird seen the day before on Teesside. The third record, at Covenham Reservoir, in May 2014 obliged by staying two days and what was considered a different individual was photographed at Far Ings NR, Barton on Humber in Aug and seems likely to have been the individual seen five days previously at Easington, Yorkshire.

The first British record was not until May 1951 in Sussex and there have been 89 in total to 2018 averaging about two per year but with four to five in recent good years (1986, 1995, 1998, 1999, 2005, 2014, 2015, 2018).

Site Name	First Date	Last Date	Count	Notes
Gibraltar Point	11/07/2005		1	
Gibraltar Point	17/06/2009		1	
Covenham Reservoir	19/05/2014	20/05/2014	1	
Barton Pits	12/08/2014		1	

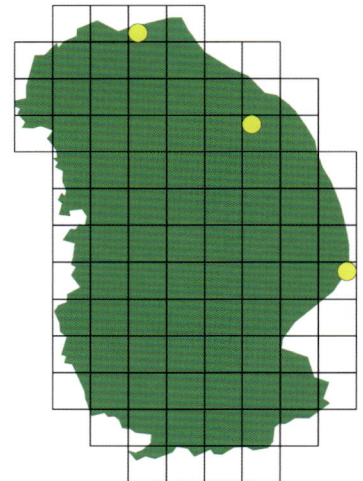

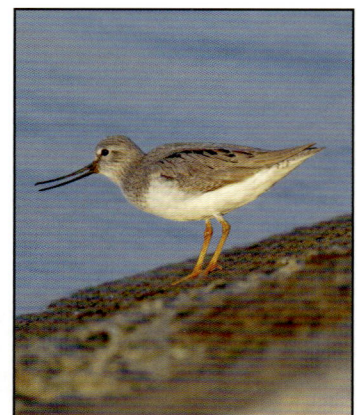

Wilson's Phalarope *Phalaropus tricolor*

BBRC

Rare. North America.

The early records all came from Wisbech Sewage Farm with two in 1967, one 1975 and two in 1979. Since then there have been eight further records most of them unsexed 1CY birds with the exception of a male of unspecified age in Jun 1975 and an adult female in Jun 1987. Ten of the 12 records were between Sep 8th and Oct 3rd, with the two spring records on 7th and 8th Jun.

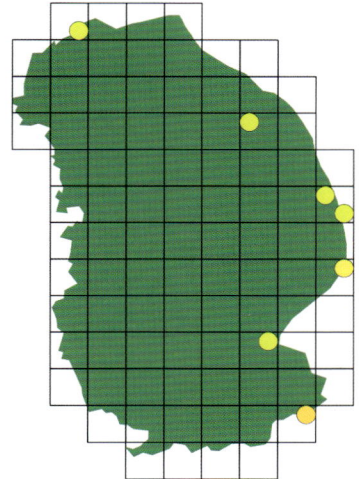

BBRC statistics show 247 records 1950-2018 with around three to four per year and probably becoming rarer. Like most American waders there is a peak from late Aug to early Oct involving mainly 1CY birds.

Site Name	First Date	Last Date	Count	Notes
Wisbech STW	28/09/1967	02/11/1967	1	
Wisbech STW	29/09/1967	15/10/1967	1	
Wisbech STW	07/06/1975	11/06/1975	1	Male
Wisbech STW	29/08/1979		2	
Gibraltar Point	11/09/1984	05/10/1984	1	
Anderby	03/10/1984		1	
Covenham Reservoir	19/08/1985	22/08/1985	1	
Gibraltar Point	08/06/1987		1	Adult female
Chapel St Leonards	26/09/1987		1	First winter
Alkborough	16/09/2008	21/09/2008	1	
Gibraltar Point	13/09/2010	15/09/2010	1	First winter
Frampton Marsh	08/09/2015		1	First calendar year

Red-necked Phalarope *Phalaropus lobatus*

Very scarce passage migrant, more in autumn than spring.

This attractive wader has become more frequent in the last 10 years but remains very scarce. LBR 2015 published a useful histogram of records from 1979 to 2016 and dates of occurrence. Using this and more recent reports from LBR we can say that the number of records by decade was as follows: 1980-89 (14), 1990-99 (13), 2000-09 (10), 2010-19 (41). The last 10 years had peaks of seven in 2015 and 2018, and eight in 2017. These 41 records occurred at the following sites: Frampton Marsh (24), Gibraltar Point (four), Middlemarsh Farm (three), Covenham Reservoir (two), Saltfleetby-Theddlethorpe (two), West Ashby (two). Four other sites had singles. Records occurred from May 3rd to Nov 17th and showed virtual parity between spring (19) and autumn (22). The European breeding population is thought to be stable so the increase in the last decade is probably down to more attractive stopover habitat and more eyes at Frampton Marsh. The recent discovery made by tagging three members of the small breeding population on Shetland with geolocators was that this population, like that of North America, winters in the North Humboldt current of the Pacific Ocean (Smith *et al* 2014). The Scandinavian population winters in the Arabian Sea which raises thought-provoking questions to which there are no definitive answers at the moment. Are passage birds in Lincolnshire coming from or going to the Pacific Ocean or Arabian Sea? Scotland's breeding population is currently around 64 males, while that of Scandinavia is 80,000 plus so the odds are on the latter but who knows?

Grey Phalarope *Phalaropus fulicarius*

Very scarce passage migrant, mainly autumn, rare in winter.

The *LBC Atlas* indicated that Grey Phalarope was rare with an average of two birds a year in the 1980s and one a year in the 1990s. Over the 10 years to 2018 LBR reflects a slight increase with annual records generally in the range from one in 2017 to six in 2013, around three to four a year. However, 2014 bucked the trend and produced the largest movement of Grey Phalarope ever seen in Lincolnshire with a total in the region of 39 birds. The largest previous annual total had been 12 in 2007. The events of 2014 took place in Oct around The Wash and started with three seen from a boat off Tabs Head on Oct 16th. Then 15 birds were seen moving south off Gibraltar Point on Oct 18th, with another 12 the following day, three on 21st and three more on 23rd. The winds through this period were predominantly south-westerly and strong. Summer records are very rare and in Aug 2015 Alkborough Flats had a moulting adult, only the third Aug record in Lincolnshire and the first since 1985. There were also three winter records during the 10-year period: Gibraltar Point in Jan 2010, Frampton Marsh in Dec 2015 and Tetney Marshes in Jan 2016, the latter reported as only the 8th Lincolnshire winter record.

Common Sandpiper *Actitis hypoleucos*

Fairly common autumn passage migrant, scarce in spring and also as a regular but very scarce winter visitor. Bred in 1979.

Common Sandpiper can turn up on any small wet edge throughout the county in spring and autumn and it is difficult to get any clear picture of the total numbers passing through the county as a whole. In the five years to 2018 LBR shows the peak spring count from any one site was 20 at Cleethorpes in May 2018, an exceptional count. The next highest was 10, and the peak autumn count was 48 at Covenham Reservoir in Aug 2015. There were five instances of birds seen in winter, four at Frampton Marsh in Dec 2015, Feb 2016, Jan 2017 and Dec 2018 and one at Gainsborough STW in Dec 2015 and Jan 2016. The sole instance of breeding reported in Lincolnshire in the historic record was a pair seen with downy young at Scunthorpe in Jun 1979.

1968-1972	1980-1989	2008-2011	2016-2019	Incidence	Change

Spotted Sandpiper *Actitis macularius*

BBRC

Vagrant. North America.

First recorded at Wisbech Sewage Farm in Nov 1970; the second was also seen there on several dates from Jun-Aug 1971, presumed to involve the same individual. The third, an adult in summer plumage was found on The Wash in 1999, present on and off for two weeks. The fourth and fifth records were also birds in summer plumage, both one day birds in May 2007 and Jun 2011. There have been very occasional reports of pairing between Spotted and Common Sandpiper *A. hypoleucos* in Britain and one report of a breeding pair of Spotted Sandpipers in the Highland Region in 1975. Another mixed pair produced three full-grown young in Jul 1991 in Yorkshire and may have been the same pair seen in that county in earlier years.

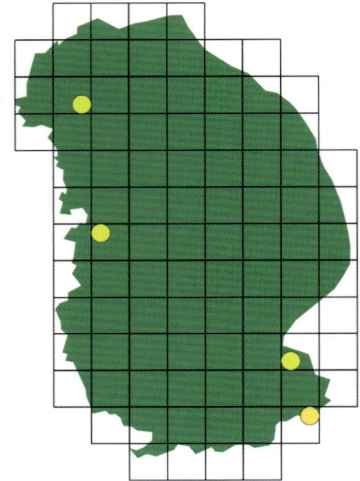

There have been more than 200 British records since 1950 (only one before then); there are around five per year and show an increasing trend.

Site Name	First Date	Last Date	Count	Notes
Wisbech STW	09/11/1970	19/12/1970	1	
Wisbech STW	13/06/1971		1	
Wisbech STW	29/07/1971		1	Same
Wisbech STW	30/08/1971		1	Same
Holbeach Marsh	15/08/1999	28/08/1999	1	
Messingham	31/05/2007		1	Adult
Whisby	17/06/2011		1	Adult summer

Green Sandpiper *Tringa ochropus*

Fairly common autumn passage migrant, scarce in spring and also as a regular winter visitor.

Green Sandpiper winters widely at low density across Lincolnshire especially in the Marsh, Fens and Trent Valley and estimating the population is all but impossible without detailed survey work. The *BTO Atlas 2007-11* found it was fairly widespread, wintering in 47 10km squares. This wintering population is probably in the order of 100-200 birds making it difficult to detect spring migration. In the autumn the position is much clearer cut. Peak passage usually takes place in Jul-Aug and the LBR reports for the five years to 2018 show a range in peak monthly counts across the county from 62 in Jul 2018 to 135 in Aug 2015. Remarkably for such a small site, Manby Wetlands has had the peak site count in three out of five years with the maximum being 25 in Aug 2015; Frampton Marsh held 26 in Jul 2014.

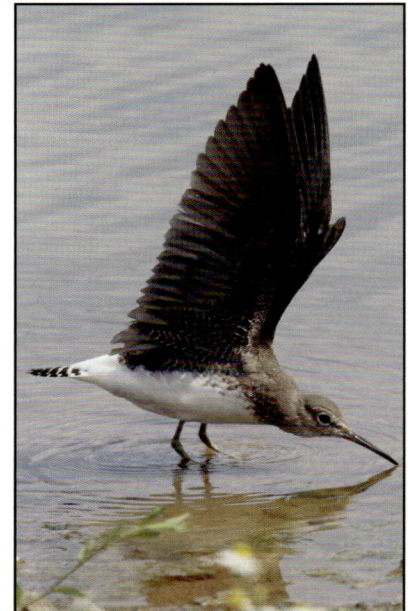

Solitary Sandpiper *Tringa solitaria*

BBRC

Vagrant. North America.

One county record of a bird found at Bardney Sugar Beet Factory Settling Ponds in Aug 1963. Very detailed notes in the LBC archive clearly show how well this bird was observed by the finders and included some quite detailed field sketches clearly showing all of the key features. It remained by a small stagnant pool to which it always returned after it had been flushed. It was noted to 'bob' in the manner of a Redshank *T. totanus*, running into cover on one occasion when a Grey Heron *Ardea cinerea*, flew over. When feeding it moved in the manner of a Green Sandpiper *T. ochropus*, and was only once seen feeding with another wader, a Redshank. It took insects from the surface of the water, but occasionally pulled a worm out of the mud in the manner of a Snipe *Gallinago gallinago*. It was heard to call several times both on the ground and in flight. On Aug 12th an unsuccessful trapping attempt unsettled it and it was last seen that evening, calling continuously as it flew away in semi-darkness.

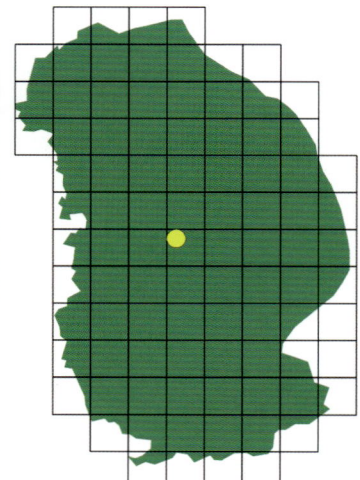

The revised list of accepted British records show that this was the seventh ever and only the second since 1950. There have now been 36 British records up to 2018 with about one record every other year. It remains difficult to catch up with as it is something of an island speciality.

Site Name	First Date	Last Date	Count	Notes
Bardney	10/08/1963	12/08/1963	1	

Lesser Yellowlegs *Tringa flavipes*

Rare. North America.

The first bird recorded in the county was shot at Tetney in Sep 1932, with 16 in all up to 2019. There were seven records from 1966-90, none 1991-2001, then 11 between 2002-19, probably involving only eight birds. There have been three records in May and Jun, two in Jul, three in Sep, six in Oct and singles in Nov-Jan thanks to the long stayer at Bagmoor Floods and Alkborough Flats Oct 2011-Jan 2012. This bird was a moulting juvenile when first found in Oct 2011 and it was presumed to be the same bird which later returned to Alkborough Flats in Oct 2012. They appear to favour sites offering largely freshwater or saline pools. The later period after 2000 coincided with the inception of more wetland sites in the county on The Humber and The Wash and elsewhere inland, all very welcome developments.

BBRC statistics show that during 1950-2018 there were 387 records with a significant minority referring to birds returning to favoured sites. They average about nine records per year but with some double-figure totals recently peaking at 17 in 2011 and 20 in 2016.

Site Name	First Date	Last Date	Count	Notes
Tetney	15/09/1932		1	
Wisbech STW	17/09/1966		1	
Killingholme	03/10/1970	30/10/1970	1	
Wisbech STW	11/10/1973		1	
Huttoft	25/07/1976	05/08/1976	1	
Cleethorpes	24/09/1978		1	
Covenham Reservoir	20/05/1984		1	
Gibraltar Point	15/06/1990		1	
Freiston Shore	31/05/2002		1	Adult summer
Killingholme	16/10/2005	22/11/2005	1	Juvenile
Freiston Shore	07/06/2006	09/06/2006	1	Adult
Gibraltar Point	09/06/2006	23/09/2006	1	Same adult
Frampton Marsh	05/05/2010		1	
Alkborough	30/10/2011	08/01/2012	1	Same
Bagmoor	30/10/2011	29/01/2012	1	Juvenile
Alkborough	28/10/2012	10/12/2012	1	Same second calendar year
Frampton Marsh	11/07/2014	26/07/2014	1	
West Butterwick	06/10/2016	09/10/2016	1	
Middlemarsh Farm	28/06/2019		1	Adult

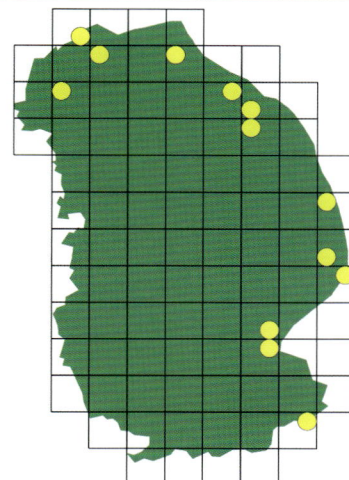

Redshank *Tringa totanus*

Nominate British and continental form a common passage migrant and winter visitor and fairly common breeding species of coastal marshes. Scarce/very scarce inland. Icelandic form (*robusta*) a common passage migrant and winter visitor.

The "Sentinel of the Marsh" breeds widely on Lincolnshire saltmarshes. The *LBC Atlas* estimated in the late 1980s that there were 650 pairs on The Wash, 30-40 pairs on the upper Humber, 75 pairs at the Humber mouth and perhaps 30 pairs inland, a total population of nearly 800 pairs. The winter population was 3,000-4,000 birds. The *BTO Atlas 2007-11* showed some small net gains inland. There has been no full survey done recently so the current position is unknown. RSPB cover three key breeding areas at Tetney Marshes, Frampton Marsh and Freiston Shore. Over six years the total breeding pairs across these sites fell from 326 in 2014 to 207 in 2019. The wintering population over the five years to 2018 based on peak monthly counts reported in LBR ranged from 3,700 in 2014 to 2,000 in 2018. The mean was around 3,000; perhaps a little down over the 30 years since 1989. The long term 25-year trend to 2018 for wintering Redshank reported by WeBS is down 17%.

1968-1972	1980-1989	2008-2011	2016-2019	Incidence	Change
				0.9%	*

Marsh Sandpiper *Tringa stagnatilis*

Vagrant. Finland and the Baltic east through Russia, Siberia to north-east China.

First recorded in Aug 1954 at Welland Mouth on The Wash, the second was 33 years later in virtually the same place, this time at the Witham Mouth in Aug 1987 on exactly the same date. Since then they have remained a vagrant with further records in May 1992, Jul 2011 and Jul 2017. The individual in 2011 was seen twice at Alkborough Flats and presumed the same at Boultham Mere in Aug 2011, having spent the intervening time at Blacktoft Sands, Yorkshire. The 2011 bird was a moulting adult, that in 2017 was a juvenile.

Between 1950-2018 there were 141 British records averaging about three per year and possibly decreasing (although there were six in 2018).

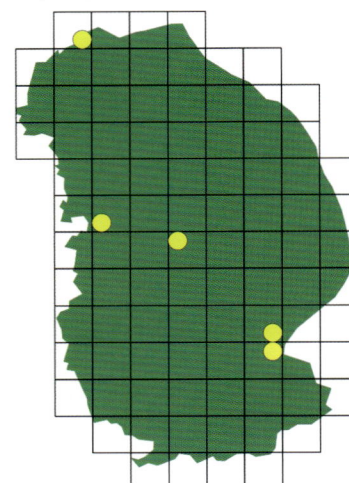

Site Name	First Date	Last Date	Count	Notes
Welland Mouth	08/08/1954		1	
Witham Mouth	08/08/1987	15/08/1987	1	
Bardney	18/05/1992	19/05/1992	1	
Alkborough	11/07/2011		1	Adult
Alkborough	21/07/2011		1	Same adult
Boultham Mere	04/08/2011		1	Same adult
Freiston Shore	25/07/2017		1	First calendar year

Wood Sandpiper *Tringa glareola*

Scarce passage migrant, in spring and autumn.

Wood Sandpiper migration varies from year to year. Birds turn up from Apr, though May is usually the peak spring month. LBR reports show that in the five years to 2018 peak May counts ranged from a low of two in 2017 to a high of 20 in 2016 with an average May peak of around 13. Autumn migration usually peaks in Jul-Aug and ranged from a low of 13 in 2017 to a high of 32 in 2015 with an average autumn monthly peak of around 18. The autumn of 2019 was exceptional with a peak count of 172 birds in Jul. The *LBC Atlas* suggested that in the 20 years to the late 1990s numbers had averaged 25 per year, so recent years have certainly seen an increase. Of the 10 top sites in 2019, six new wetlands did not exist 25 years ago when the *LBC Atlas* was being written. These sites (numbers in brackets are the 2019 peak autumn count): Freiston Shore (41), Tennyson Marsh, Gibraltar Point (35) Middlemarsh Farm, nr Skegness (19), Manby Wetlands (18), Frampton Marsh (16), and Alkborough Flats (two), are all excellent for waders and seem to be favoured by Wood Sandpipers.

Spotted Redshank *Tringa erythropus*

Passage migrant, scarce in spring but fairly common in autumn. Very scarce but regular in winter.

One of our most attractive waders and handsome in all plumages. LBR reports for the five years to 2018 show that Alkborough Flats is now the best place to see "Spotshank". It tends to top the charts for numbers of birds in all seasons. Spring passage has always been light and is at its peak in Apr. In the five years to 2018 it ranged from peak monthly counts of 17 in 2014 to 22 in 2015. Autumn passage usually peaks in Sep and ranged from 45 in 2015 to 104 in 2018, with an average of 63. Wintering occurred every year with around 13 birds per year. Interestingly the *LBC Atlas* reported occasional flocks of over 50 on The Wash. There has been nothing like that in recent years with the largest flocks at Alkborough Flats on the Humber every year, with a max of 34 there in Sep 2018. The largest count from The Wash was 29 at Frampton Marsh also in Sep 2018, the best recent season.

Adult Spotted Redshanks moult in staging areas before undertaking a trans-Saharan migration to wintering grounds in the Sahel and northern savanna zones in Mali, Nigeria and Chad. They are present in the tropics mainly Oct-Apr, although the first birds reach West Africa in Aug and the last leave Ethiopia in mid-May. They migrate north along the west European coast late Apr to mid-May, arriving in Finland from first week of May. An adult bird ringed in The Netherlands on Sep 10th 1964 was found dead at Tetney on Jan 1st 1969. A second bird colour ringed as a 1CY bird on Aug 4th 2019 in Norway was seen alive in Norfolk on Aug 13th 2019, and then twice at Frampton Marsh on Aug 20th and Sep 22nd 2019. The value of the Lincolnshire coastal and estuarine areas as part of a key stage of the East Atlantic Flyway cannot be overstated.

Greenshank *Tringa nebularia*

Passage migrant, scarce in spring, fairly common in autumn. Very scarce in winter.

Greenshank occurs widely in small numbers across Lincolnshire in spring and its "chu chu chu" call is always a welcome sound. Reports from LBR for the five years to 2018 indicate that peak spring migration is usually in May, though in 2018 it was in Apr. Peak counts across the county ranged from 16 in 2015 to 58 in 2016 with an average of 28. Autumn migration is much more prolonged and can peak anywhere from Jul-Sep and extends into Oct. The peak monthly count ranged from 113 in Jul 2018 to 285 in Jul 2014 with an average of 185. The largest counts are always on The Wash with Freiston Shore boasting 108 in Jul 2014. Wintering has become more obvious in recent years. The *LBC Atlas* reported that "recently one or two birds have occasionally wintered on the coast". In the five years to 2018 wintering occurred annually with a range of one in 2016 to seven in 2017, an average of four per year.

The Wadden Sea is used by many Fennoscandian birds as a stopover and moulting site from late Apr to mid-May, and most Palaearctic Greenshanks are trans-Saharan migrants wintering right through Africa to its southernmost countries. Two interesting recoveries of birds ringed in Lincolnshire are of interest in this context. DD15102 was ringed as a 1CY bird on The Wash on Aug 4th 2007 and retrapped on May 3rd 2009 at Castricum on The Netherlands coast. A second bird, DE41539, had been colour ringed as an adult Aug 31st 2015, also on The Wash and was seen alive on its breeding grounds in Norway on Jun 21st 2016 west of Tromsø and again on May 9th 2017 south-west of Trondheim, possibly still on its way further north.

Greater Yellowlegs *Tringa melanoleuca*

BBRC

Vagrant. North America.

One of the rarer North American waders to occur in Britain, the first county record wasn't until Apr 2007 at Freiston Shore then in May 2007 it turned up at Gibraltar Point, presumably having 'hidden' somewhere on The Wash between times. The second at Baston GP in late Sep 2007 was considered to have been a different bird – it was quite short billed compared to the bird at Gibraltar Point, which was particularly long billed, and it occurred at the height of autumn wader passage. It came in from the north-west, stopped for 20 mins and flew off south and was in Hampshire the following day.

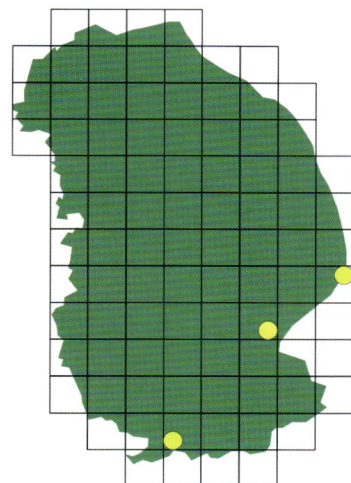

Identifying Greater Yellowlegs has been well-covered in the literature but can still be tricky on lone individuals where the differences in size and structure are more difficult to evaluate. In breeding plumage Greater shows heavier black barring on the underparts than Lesser Yellowlegs *T. flavipes*, and sometimes barring extending across the belly, something admirably described in the finders account at Gibraltar Point in *LBC Rare and Scarce 2003-07*. Between 1950-2018 there were 27 British records although in some years the mobility of the bird(s) involved made assessment difficult; averages about one every two years.

Site Name	First Date	Last Date	Count	Notes
Freiston Shore	09/04/2007		1	Adult
Freiston Shore	19/05/2007		1	Same adult
Gibraltar Point	30/05/2007	31/05/2007	1	Same adult
Baston	25/09/2007		1	

Cream-coloured Courser *Cursorius cursor*

BBRC

Vagrant. North Africa.

An exhausted bird was caught on the coast near Marshchapel in 1840. Most British records of this vagrant are historic with 30 documented prior to 1949 and only eight since 1950. The most recent record was in 2012, itself the first since one on the Isles of Scilly in 2004. Records elsewhere in northern Europe reflect those in Britain, with an absence of spring birds and the majority of autumn vagrants in Oct.

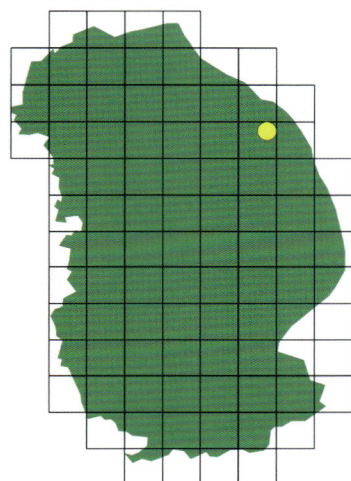

The breeding range of Cream-coloured Coursers is wide and extends from the Atlantic archipelagos of Cape Verde and the Canary Islands to North Africa. They breed patchily eastwards along the Sahel zone to the Middle East, through to Central Asia and north-west India. Recently a few pairs have bred in southern Spain in spring (2001).

The prospects of another county record appear extremely slim, to say the least.

Site Name	First Date	Last Date	Count	Notes
Marshchapel	01/01/1840		1	Caught in about 1840

Realignment Freiston Shore RSPB

Collared Pratincole *Glareola pratincola*

BBRC

Vagrant. Southern Europe.

The first county record was of a bird shot at Branston Hall near Lincoln in 1827, with the second in May 1973, 146 years later, at Gibraltar Point. Three further records between 1981-2011 with none since, confirms its status as a vagrant in the county. That in 1981 was found at Donna Nook on Jul 11th and the one at Frampton Marsh Aug 8th-9th 2009 proved to be quite elusive disappearing off south-eastwards on its first day. It reappeared for three hours the next day, hawking insects over the scrapes and reed beds before flying off to the north. It was noted to be missing some primaries in its left wing. The 2011 bird was found on Apr 27th during a breeding bird survey at Killingholme Marshes. It frequented pools just inland of there amidst the industrial installations of the Humber bank oil industry and remained until May 7th. It was the only British record that year.

With 77 British records between 1950-2018 and several pending 2019-20, the species remains a rarity with about one record a year.

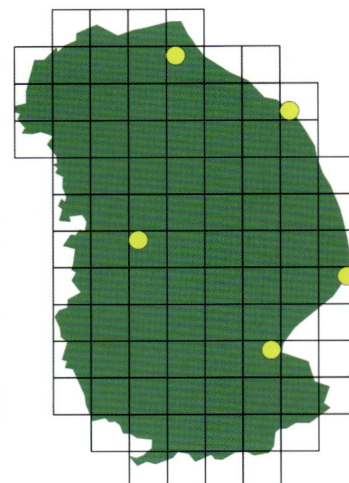

Site Name	First Date	Last Date	Count	Notes
Branston	15/08/1827		1	
Gibraltar Point	21/05/1973		1	
Donna Nook	11/07/1981		1	
Frampton Marsh	08/08/2009	09/08/2009	1	
Killingholme	27/04/2011	07/05/2011	1	

Oriental Pratincole *Glareola maldivarum*

BBRC

Vagrant. East Asia.

One at Frampton Marsh in May 2010 was a county first and was the seventh and most recent British record. This follows the sixth record in Kent and Sussex in May-Jun 2009 and may conceivably have been the same returning bird. The species was first recorded in Britain in Jun 1981 (Suffolk and later Essex) but only accepted after lengthy debate and eventually published in the BBRC report of 1988, it was also a first for the Western Palaearctic. The bird in question was a 2CY bird which later in its stay suffered at the hands of some severe rainfall which obscured some of its main plumage features. It was also seen later in Essex. The Frampton bird presented no such problems being a pristine adult in near perfect plumage. It also followed an established pattern of turning up in the spring in the south-eastern quarter of England.

Oriental Pratincoles breed from northern Pakistan and Kashmir into China, wintering in south-east Asia and Australia. North-eastern populations are long-distance migrants and return north to breed in East Asia May-Aug and are common migrants in central and north-eastern China. The mechanism of the species vagrancy to Britain is speculative but is presumed to relate to wandering juveniles finding their way to Africa in their first autumn and subsequently migrating north with Collared Pratincoles *G. pratincola*. Might it also be possible that some birds from China undertake the long migration into Africa similar to that seen in Amur Falcons *Falco amurensis*? Whatever the truth, vagrant Pratincoles are always a very welcome find.

Recent estimates put the wintering population of Oriental Pratincoles in south-east Asia to Australia at 2,880,000 birds with up to 1,000,000 wintering in the Indian sub-continent where they are largely a short-distance migrant.

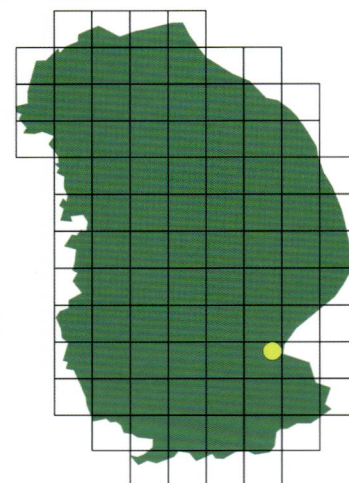

Site Name	First Date	Last Date	Count	Notes
Frampton Marsh	09/05/2010	19/05/2010	1	Adult

Black-winged Pratincole *Glareola nordmanni*

BBRC

Vagrant. South-east Europe through Ukraine to SW Russia and N Kazakhstan.

Two records of this Pratincole in Jul 2014 and Jun 2019, the latter completing a very nice hat-trick for Frampton Marsh after Collared Pratincole *G. pratincola*, in 2009 and Oriental Pratincole *G. maldivarum* in 2010. The first record at Gibraltar Point dropped into Tennyson's Marsh early morning and proved to be an immaculate adult, an excellent find. It showed well for a fortunate small band of observers before disappearing off to the south a few hours after it was found. The BBRC report on rare birds in 2014 reports that there were records from seven recording areas between Sussex and Northumberland but given that Pratincoles have a track record of ranging widely it was assumed that these referred to just one individual. This bird disappointed many, so a repeat at Frampton Marsh in Jun 2019 was more than welcome, especially as it stayed for four days.

This was the 38th British record, 1950-2019, and the only one in 2019. European breeders are confined to the northern Black Sea in Romania and Ukraine where they are rare and declining. To the east, they are more numerous across the steppes of S Russia to E Kazakhstan. In Britain, they remain a major rarity turning up every one to two years.

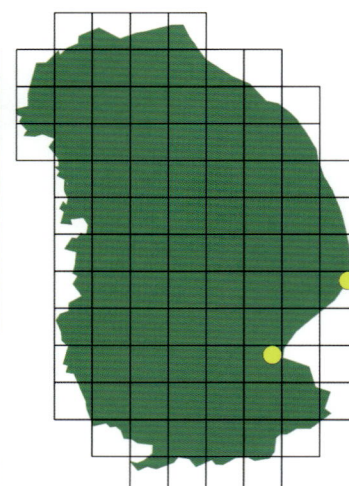

Site Name	First Date	Last Date	Count	Notes
Gibraltar Point	14/07/2014		1	Adult
Frampton Marsh	18/06/2019	21/06/2019	1	Adult

Kittiwake *Rissa tridactyla*

Fairly common but declining passage migrant and winter visitor with small numbers in summer. Scarce inland

Kittiwake has been declining for some time. The JNCC Report; *Seabird Population Trends and Causes of Change: 1986-2018* indicates that the UK population fell 50% between 2000-18. It's not surprising then that apart from notable passage days with strong northerly winds, Kittiwakes are getting harder to see in Lincolnshire. The nearest breeding population is Flamborough Head/Bempton Cliffs, about 40 miles north of the Humber Mouth. That population has actually increased by 7% during the period of overall fall. How many of these relatively "local" birds are actually seen off Lincolnshire is open to conjecture. The following information is based on LBR reports for the five years to 2018. Records usually occur in all months of the year with the quietest months being Apr, Jul and Aug. Of 15,443 birds over five years, 93% occurred on just 10 days in the five-year period. The biggest movement was on Nov 21st 2016 when, in a strong northerly, 1,800 were recorded heading south off Sutton on Sea and 1,517 off Gibraltar Point. Altogether during the five years, Nov had 69% of records Oct 24% and Sep 1%. With peaks coming several months after breeding dispersal birds seen in Lincolnshire could be coming from anywhere in the North Sea and further north.

Sabine's Gull *Xema sabini*

Very scarce coastal migrant in autumn, mainly Aug-Oct. Exceptional inland.

Breeds in Greenland, the North American arctic and north-east Siberia and winters off west and southern Africa. An adult Sabine's Gull in breeding plumage makes any autumn seawatch special, and it remains a very scarce bird in Lincolnshire, possibly because, as one of the more oceanic members of the gull family, its main migration route lies in the Atlantic to the west of the British Isles.

The *LBC Atlas* reported that there had been 30 records in the 20 years up to 1997. Over the period 2010 to 2020 there have been slightly more with an average about three per year; in 2012 and 2014 there were none, while 2018 had seven and 2020 had nine; all were seen between Aug 9th and Oct 30th. The only inland bird, and also the longest stayer, was a juvenile at Leadenham Tip from Sep 16th-19th 2011. The most seen at one site on one day were three adult (2CY+) birds at Witham Mouth on Aug 30th 2020 and there were two adults in early moult on the beach between Horseshoe Point and Tetney Marshes on Aug 29th 2018.

Bonaparte's Gull *Chroicocephalus philadelphia*

Vagrant. North America.

One county record of this small American gull: a 2CY bird found on fields at Chowder Ness, Barton-upon-Humber in Jun 2010 among several hundred small gulls roosting there.

They are widespread in North America, mainly in Canada and Alaska in the taiga zone, where they primarily breed in spruce and larch trees and only exceptionally on the ground. There has been a steady increase in British records during recent decades, with double-figure totals in seven of the years since 2000. This may reflect increasing observer awareness of Bonaparte's Gull, multiple recording of itinerant individuals and an increasing population in North America. In fact in Jun 2017 a pair were discovered nesting in Iceland within a colony of Black-headed Gulls *C. ridibundus*.

There have been 247 British records 1950-2018 with about seven per year and showing an increasing trend. They can be encountered in any month of the year.

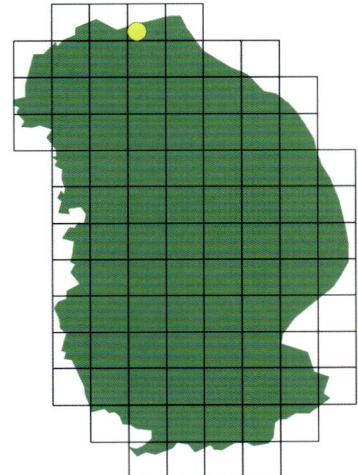

Are they under-recorded in the county amidst the teeming masses of Black-headed Gulls?

Site Name	First Date	Last Date	Count	Notes
Barton Pits	28/06/2010		1	Second summer

Black-headed Gull *Chroicocephalus ridibundus*

Very common resident, passage migrant and winter visitor.

Our most frequently recorded gull all over the county. In the late 1980s, the *LBC Atlas* estimated the breeding population at some 3,000 to 4,000 pairs with a winter population approaching 100,000 birds. In the ensuing 30 years all the large colonies that formerly bred on saltmarshes like Grainthorpe Marsh, Frampton Marsh and Holbeach Marsh appear to have been abandoned for islands in former gravel pits like Whisby and lagoons such as those created at Frampton Marsh and Gibraltar Point. There has been no formal, recent estimate of the size of our seabird populations and the latest JNCC Seabird Census is currently ongoing. Probably the largest colony at present is at Whisby NR which has held around 1,000 pairs in recent years. The wintering population is difficult to assess as WeBS counts tend not to be held during evening roosts. The average five-year rolling mean count to 2018/19 is around 45,000 which is likely to be a large underestimate. Numbers tend to peak in late summer/early autumn during peak agricultural ploughing. The origin of our wintering birds has been illuminated by colour ringing and the site faithfulness shown by some birds is remarkable. JHX6 turned up on Aug 2nd 2020 for its third winter stay at Cleethorpes boating lake, its summer home is at Stavanger, Norway. Its countryman JGN8 from Oslo beat it to Cleethorpes by four days, for its eighth consecutive winter. At Skegness HROR originally ringed at Sopron, Hungary as an adult in Feb 1999 visited from Jul-Oct 2014 then switched to Switzerland from Dec 2014-Feb 2015 after which it was seen in Poland in Jul 2015 before returning to Gibraltar Point in Aug 2017. It is easy to see why observing and reporting colour ringed wildfowl, gulls and waders is an increasingly popular birding activity.

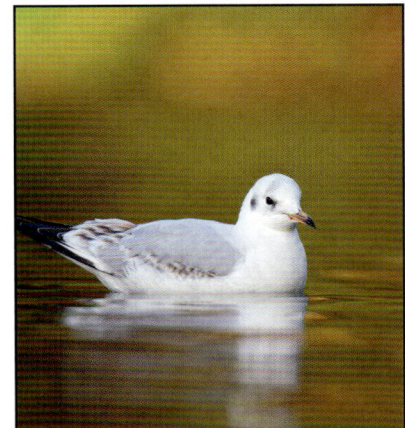

1968-1972	1980-1989	2008-2011	2016-2019	Incidence	Change
				39%	*

Little Gull *Hydrocoloeus minutus*

Fairly common passage migrant in spring and autumn, scarce/very scarce in summer and winter.

Considering that thousands are seen in the Hornsea area every year, less than 20 miles north of the Humber from Jul onwards, Lincolnshire doesn't do nearly as well for Little Gulls. Most records are of coastal birds although it is a well-known overland migrant and small numbers also occur at inland water bodies too. There is a clear passage period in Mar-May, usually peaking in Apr. Apr 2018 was exceptional with over 200 birds; in the four earlier years there were around 15-20 a year. There have been variable numbers summering in Jun, but Jul is much quieter. There is an extended and variable period of autumn passage from Jul to Nov which tends to peak in Nov with an average of around 40 birds in the month. The largest single flock reported in the period was 80 at Witham Mouth in Aug 2015. The winter months are usually very quiet with few at all in Dec-Feb, but some very variable occurrences can be seen. For example, in Jan 2014 and 2018 there were no records at all yet there were 51 in Jan 2016. Outside the breeding season Little Gulls seem to be at least semi-pelagic, feeding and roosting at sea in coastal waters of the NE Atlantic from the Irish Sea south to Iberia and Morocco as well as in the Mediterranean, Black and Caspian Seas. Winter appearances seem most likely to be associated with severe weather movements.

Laughing Gull *Leucophaeus atricilla*

Vagrant. North America.

The first two county records were both in 1979. A 3CY bird was seen on Feb 24th at Donna Nook and the second was a 1CY bird seen on Oct 6th at Huttoft. Four further records probably involving just three birds have followed with the last, now rather distantly, in 1998. This last individual had been a long-stayer in Norfolk, seen displaying in the Black-headed Gull *Chroicocephalus ridibundus* colony at Titchwell and later coming regularly to bread in Hunstanton town centre. Amazingly this bird followed the Franklin's Gull from Norfolk, pitching up at Kirkby GP eleven days later.

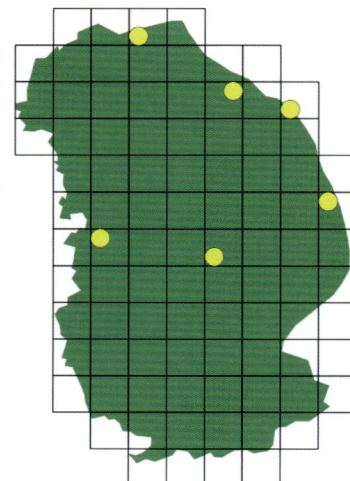

Thanks to an 'invasion' year in 2005 when there were 58 records, spilling over into 2006 (22), there have now been more than 200 British records of Laughing Gulls. These two years distort the picture somewhat though and Britain can expect around five per year on average.

Site Name	First Date	Last Date	Count	Notes
Donna Nook	24/02/1979		1	Second winter
Huttoft	06/10/1979		1	First winter
Thorpe on the Hill	23/05/1984		1	First summer
Barton upon Humber	28/12/1984		1	Same second winter
Cleethorpes	21/10/1997		1	Adult
Kirkby on Bain	24/05/1998		1	Adult

Franklin's Gull *Leucophaeus pipixcan*

Vagrant. North America.

One county record of a 2CY bird at Kirkby GP in May 1998 present for a single evening only. This bird was presumed to have been the individual seen three days earlier at Titchwell, Norfolk and then two days after leaving Kirkby GP it was found at Hillwell, Shetland.

A remarkable record all round and indicative of the birds urge to travel northwards in spring, even if not on its native continent.

These small, black-hooded gull nest in small to very large colonies in the marshes of interior North America. The first British and Western Palaearctic record was as recently as 1970 in Hampshire, and with a total of 80 British records up to 2018 this remains one of the rarer north American gulls with two to three per year.

How many of these delightful gulls track through the county unnoticed?

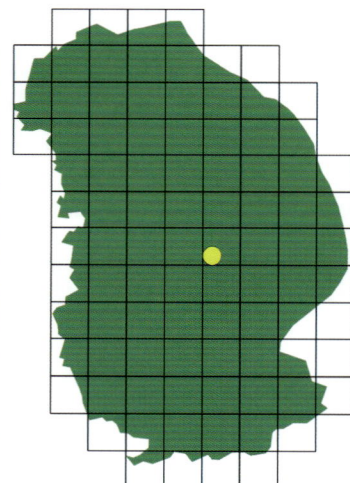

Site Name	First Date	Last Date	Count	Notes
Kirkby on Bain	13/05/1998		1	First summer

Kirkby Pits

Audouin's Gull *Ichthyaetus audouinii*

BBRC

Vagrant. Mediterranean.

One county record of a near-adult bird on Aug 15th 2008 at Huttoft Bank, seen just briefly there before flying off. Thankfully, it was relocated in the Chapel Point area Aug 17th-23rd. It showed intermittently there during its stay but would disappear out to sea for extended periods. One example of this from the many messages on the LBC Forum: "Firstly (it showed) at around 15:30 when it was resting on the beach - it flew off out to sea and was lost from view. A growing group of birders gathered throughout the rest of the day at Chapel Point and at about 20:15 in the gathering dusk it was spotted some distance away on the sea where it bobbed about giving only brief views".

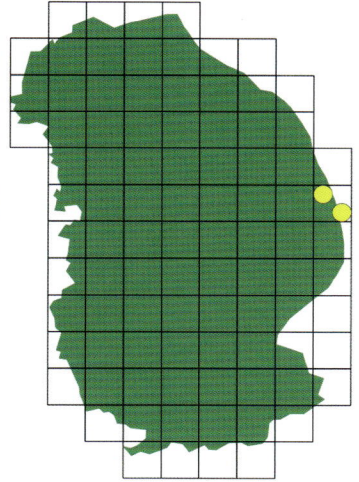

The first record for Britain wasn't until 2003 but this was the fifth individual since then and there have now been eight British records in all up to 2018. Formerly a rare and localised species, it has undergone significant expansion in the Mediterranean basin largely due, it is thought, to the availability of waste from fisheries. Numbers have risen from around 1,000 pairs in the 1960s and 1970s to a global population of more than 21,000 pairs in Europe, as well as a few small colonies in North Africa. The majority winter along the Atlantic seaboard from Morocco to Senegal and Gambia.

Site Name	First Date	Last Date	Count	Notes
Huttoft	15/08/2008		1	Sub-adult
Chapel St Leonards	17/08/2008	23/08/2008	1	Same sub-adult

Mediterranean Gull *Ichthyaetus melanocephalus*

Scarce, but increasing visitor throughout the year, most frequent in late summer/early autumn.

With a British breeding population estimated at 1,200 pairs between 2013-17 and a hybrid pair with a Black-headed Gull at Messingham SQ in the early 1990s, Mediterranean Gull has been a Cinderella species in Lincolnshire for nearly 30 years. By the time of the *BTO Atlas 2007-11*, Lincolnshire had been leapfrogged and birds were successfully breeding in Yorkshire and Northumberland. Pairs have summered in Lincolnshire in each of the nine years to 2020 but the first confirmed breeding record finally came in 2020 at a flooded gravel pit site near Lincoln. Two pairs fledged two and one young respectively and a third pair that included a 2CY bird, failed. Back in the 1990s the *LBC Atlas* suggested that there were around 25 birds a year in Lincolnshire. Between 2014-18 annual peak numbers reported in LBR tables fluctuated from 69 (2017) to 135 (2014, 2016) and the peak month varied between Apr, Jul and Aug. The top monthly count was 24 in Apr 2018.

1968-1972	1980-1989	2008-2011	2016-2019	Incidence	Change
				0%	*

Common Gull *Larus canus*

Very common passage migrant and winter visitor.

Large numbers of Common Gulls arrive from the continent from Jul-Aug onwards to feed on the plough after harvest. They spend the winter feeding on the fields being especially common on the Wolds but also in towns and the few remaining domestic refuse tips. Interestingly Common Gull is also the most frequently encountered gull offshore from the Humber mouth in the winter. Counting them is difficult. They frequently leave roost sites on the coast and inland ones like Covenham Reservoir and the former gravel pits around Lincoln before or around dawn and fly back in at dusk. WeBS does not accurately monitor the species as most counts are done on waterbodies when the birds are not there. Very few accurate roost counts are undertaken. The wintering population is well over 10,000 birds per annum, but it would be conjecture to suggest by how much. They usually depart back to the breeding grounds by the end of Mar and those few that summer are usually 2CY birds.

Ring-billed Gull *Larus delawarensis*

Vagrant. North America.

Despite the number of British records standing at nearly 2,000, 1958-2018, this remains a vagrant in the county. The first record was a 2CY bird in Jul-Aug 1988 and there have been just four more, most recently at Trent Port Wetlands, Marton in 2019. Although 'ex-BBRC 1988' it remains a tricky species to identify for some and it may be overlooked if occurring with large congregations of gulls on the coast or in the Wolds.

A common species across North America, it prefers fresh water on migration and occurs inland more often than other species of gull, but its overwintering habitat includes coastal sites as well as inland reservoirs, lakes, landfill sites and conurbations. White and Kehoe (2019) showed that there has been a decadal decline in British records since 1990 with annual mean totals of 77 (1990-99), 54 (2000-09) and 17 (2010-17). However, there is huge variability between years and in 2019 27 birds were reported with 15 thought to be returning birds and 12 being new arrivals, one of these being the one-day 2CY bird at Trent Port Wetlands.

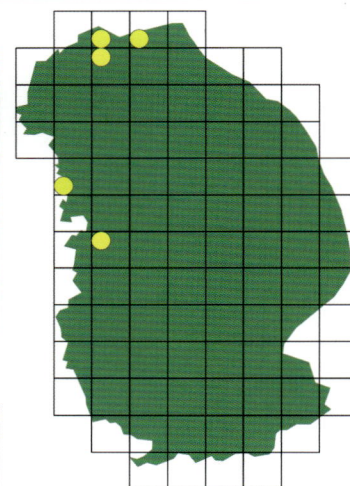

Site Name	First Date	Last Date	Count	Notes
Bagmoor	24/07/1988		1	Second calendar year
Bagmoor	03/08/1988		1	Same second calendar year
Whisby	27/04/1990	29/04/1990	1	Adult summer
Whisby	17/04/1992	19/04/1992	1	First summer
Barton Pits	28/06/2010	30/06/2010	1	Second calendar year
Marton	05/02/2019		1	Second calendar year

Great Black-backed Gull *Larus marinus*

Very common passage migrant and winter visitor with some immatures remaining in summer.

The British population is thought to be declining but no long-term trend information is available from the JNCC Seabird Monitoring Programme. Birds from Scotland, Norway and Denmark winter in Lincolnshire, based on ringing data. The relative abundance map in winter for this species in *BTO Atlas 2007-11* shows that Lincolnshire is one of the most important wintering sites for Great Black-backed Gull in Britain. The top site for this large gull on the coast is Gibraltar Point. During the five years to 2018, LBR reports reflect that the largest peak site counts were here in four out of five years. The peaks were 500 in Oct 2014, Aug 2015 and Oct 2017. Barton Pits also had a peak of 500 in Sep 2014. The closure of domestic refuse tips across the county over the last 15 years has had an impact on the volume of wintering gulls in general. One of the few remaining open tips in the county, at Kirkby on Bain, had a peak count of 250 in Dec 2016 and a lowest peak count of 75 in Jan 2017. The actual passage and wintering populations are very difficult to estimate.

Glaucous Gull *Larus hyperboreus*

Scarce passage migrant and winter visitor, rare in summer.

The entire world population of the Glaucous Gull breeds in the circumpolar Arctic. It is an annual winter visitor to the county, a bird with a catholic diet, liable to be found on the coast feeding on vertebrate carcasses, at rubbish tips or coming to bread with other large gulls at seaside promenades. Most records are of 1CY or 2CY birds and the *LBC Atlas* reported in the 1980s that there were around 20 records per year, which was around half the number in the 1990s. An analysis of records in LBR for the five years to 2018 indicates that there was a minimum of 68 records ranging from six in 2015 to 30 in 2017, averaging about 14 per year. Most records (75%) occurred in Jan-Apr with Jan the busiest month. Summer tends to be very quiet with one each in Jun 2018 and Jul 2017. There were no records in Aug-Sep. Autumn passage tends to be very late with no birds before the end of Oct. Most birds seen between Oct and Dec are coastal fly pasts.

Iceland Gull *Larus glaucoides*

Scarce passage migrant and winter visitor Aug-May. The subspecies known as Kumlien's Gull *L. g. kumlieni* is a vagrant (LBRC). Thayer's Gull *L. g. thayeri* is a vagrant (BBRC).

IOC currently recognises three subspecies – nominate *glaucoides* ('Iceland Gull') breeding in Greenland, a regular winter visitor; *kumlieni* ('Kumlien's Gull'), breeding mainly on Baffin Island, vagrant; and *thayeri* ('Thayer's Gull') breeding in the Canadian high arctic, also a vagrant.

Nominate *glaucoides* is a regular winter visitor to Lincolnshire, generally in small numbers but with occasional larger influxes. There are earlier claims but one at Gibraltar Point in Oct 1962 has the best credentials as a county first. This was a rare winter visitor 1962-97 with only nine records but thereafter became more regular. From 2000-10 there were five to six records per year with peaks of nine in 2005 and 11 in 2006. An unprecedented influx into north and north-western Britain occurred in 2012 with perhaps as many as 20 birds in Lincolnshire, mostly 2CY birds. Double figure records were seen in four other years and numbers averaged eight to nine overall, though movement between sites makes accurate estimates almost impossible.

L. g. kumlieni is a vagrant and the first accepted county record was Dec 18th 1992 at Apex Pit, North Hykeham when the taxon was still considered by BBRC; indeed, it was the only record falling under BBRC's jurisdiction before being devolved to county records committees from Jan 1st 1999. Since then there have been a further 16 records involving about 12 birds. Adults were at Apex Pit, North Hykeham in 1999 and Thurlby Pits in 2009, the remainder being 2CY birds.

There is one county record of Thayer's Gull in Apr 2012 expertly pinned down with a large gull flock feeding on a field treated with liquid slurry. Identification of *thayeri* is potentially problematic, but claims are welcomed if accompanied by detailed notes and good photographs. Details of a ringed or marked bird would of course provide additional evidence.

Kumlien's Gull

Thayer's Gull

Herring Gull *Larus argentatus*

British form *L. a. argenteus* a Red listed common passage migrant and winter visitor; fairly common and increasing breeding species. Nominate northern European form *L. l. argentatus* a green listed fairly common winter visitor.

There was no evidence of Herring Gull breeding during the *LBC Atlas* period from 1980 to 1989 but by the time of the *BTO Atlas of 2007-11* it had colonised at least seven 10km squares in Lincolnshire principally in Grimsby, Skegness and Boston. That spread has continued and it is now breeding in inland towns like Louth. Flat-roofed factory and warehouse buildings and old houses with tall chimney stacks and pots are preferred nesting locations. For birders, as a red data species it may be a positive addition to the day list but for many householders the raucous pre-dawn calls are less than welcome and some chimney stacks now sport gull repellent netting. Ironically, prior to its recent spread it had started to become a rare bird at many inland sites because of the closure of open domestic refuse tips. There is no current accurate estimate of the breeding population, but it is well above 100 pairs. The wintering population is also poorly counted and the current WeBS Online figures of some 8,500 are likely to be an underestimate.

1968-1972	1980-1989	2008-2011	2016-2019	Incidence	Change
				13%	*

Caspian Gull *Larus cachinnans*

Very scarce all-year round visitor. Eastern Europe to Asia.

The taxonomy of the Herring Gull group of species is continually evolving and the BOURC recognised Caspian Gull as a full species separate from Herring Gull *L. argentatus*, and Yellow-legged Gull *L. michahellis*, Oct 2007. Its identification still poses a big challenge to many birders. Probably as a result of the fluctuating taxonomic position, the first Caspian Gull was not reported in the county until Dec 16th 2001. Birds are mobile moving between feeding and roosting sites often over large distances so some duplication in the records is unavoidable, but there were probably three to four other individuals in the county that year.

Since then they have been recorded in variable numbers in every year to 2019 with 10-20 individuals in good years. They have been seen in every month of the year with non-breeding birds arriving from Jul onwards and Jul-Aug are the only months when fresh plumaged juveniles may be seen, peaking mid-Sep with significant numbers remaining into Nov. Most have left by Dec but return migration begins Feb-Mar. A returning, one-legged individual was seen in the Gainsborough and Lincoln areas in most years between Dec 2004-Jun 2014. Colour-ringed individuals from Polish and German colonies are also regularly recorded.

Broad black tail band with spotted uppertail-coverts

Long rear-end

Gleaming white head/neck and small eye

long slender bill, base just begining to pale up

Fine black streaking to neck

Scapulars showing variable markings

Extremely long neck when alert

1st winter Caspian Gull

Kirkby Pits

2/2/02

A perfect individual —

whitish tips to median and greater coverts forming a double wing-bar

Long thin legs, Pale pink

Yellow-legged Gull *Larus michahellis*

Scarce visitor, mainly in late summer/autumn. Azorean Yellow-legged Gull *L. m. atlantis* is a vagrant.

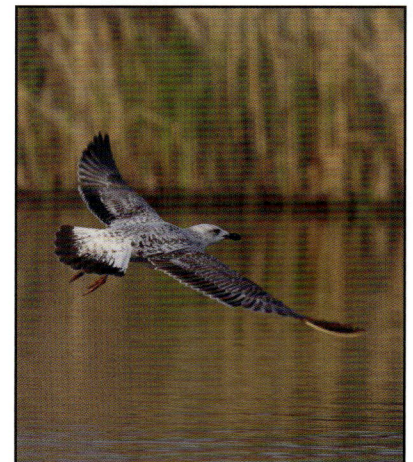

Records of this species in the UK increased during the 1960s and 1970s and Yellow-legged Gull was formally recognised as a full species by the BOU in 2005 (Sangster *et al*), although keen gull watchers had been recording it as such for some years previously. Ironically the latest DNA research suggests it is most closely related to Great Black-backed Gull *L. marinus*. It is polytypic with at least two sub-species, *L. m. michahellis* and *L. m. atlantis*. It is fair to say that the story is not yet over.

The widespread nominate Mediterranean race was said by the *LBC Atlas* to occur in Lincolnshire at a rate of around 15 birds per year in the 1990s. Before that less attention was paid to it. An analysis in LBR 2012 showed that in the years 2010-12 it was occurring at a rate of 160 birds per year with birds all year round but a pronounced peak in Jul. In the seven years from 2013 to 2019 the total annual monthly peaked counts ranged from 37 in 2018 to 172 in 2015 with an average of 140 per year. Interestingly the peak month of occurrence appears to have shifted from Jul to Aug from 2014. For the three years to 2019 the annual rate fell to 50 birds a year. Whether this is a longer-term shift remains to be seen. In the seven-year period the largest counts were 38 in the Crowle/Eastoft area in Aug 2014 and 23 there in Aug 2015 and 29 at Norton Disney Pits in Jul 2016.

The first and only record of Azorean Yellow-legged Gull *L. m. atlantis*, in Lincolnshire occurred at Marston on the morning of Oct 26th 2015 before flying across the border into Nottinghamshire. It was thought to be the same bird as originally found in Oxfordshire in 2009. It was added to the British list in 2016 based on a bird in Cornwall in 2008.

Lesser Black-backed Gull *Larus fuscus*

Present all year: western European form *L. f. graellsii* common on spring and autumn passage, scarce but increasing in winter and as breeding species. Danish and southern Scandinavian form *L. f. intermedius* is a fairly common passage migrant. Birds showing characters of the northern Scandinavian form *fuscus* (Baltic Gull) have been claimed and one was recently accepted.

Lesser Black-backed Gull is a widespread spring migrant with passage peaking in Apr-May. Like its congener the Herring Gull *L. argentatus*, it didn't breed in the *LBC Atlas* period but has shown a similar pattern of spread to that species over the same period in similar nesting sites and often nesting in mixed colonies with it. As a breeder it is still fairly local. There is no accurate estimate of the current breeding population, but it is probably greater than 100 pairs currently. The wintering population at the top 10 sites for this species during Dec-Jan from recent LBR data is around 100 birds and the true wintering total is likely to be higher but not massively so.

In most years birds belonging to the northern race *L. l. fuscus* Baltic Gull are seen but BBRC acceptance criteria are very strict and there are only a handful of accepted British records to date. BBRC published the first acceptable record for the county in Jun 2020. This was a 2016 submission concerning an adult at least six years old which had been colour-ringed at a colony in Norway in 2010 (Black/white J2LL). It was seen by two astute 'gullers' at Norton Disney quarry Jul 22nd 2016. Recent updated information from the BBRC advise that claims of unringed second calendar year birds based on detailed notes (including moult analysis) and good photographs result in more accepted records.

1968-1972	1980-1989	2008-2011	2016-2019	Incidence	Change
				26%	*

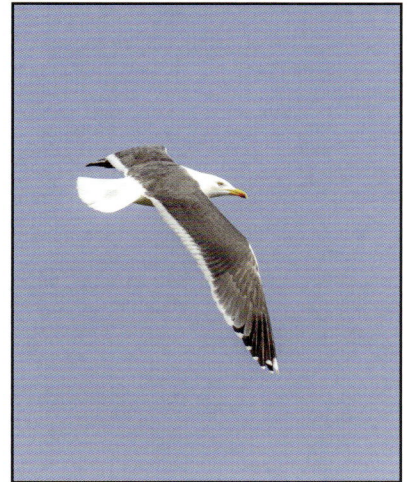

Gull-billed Tern *Gelochelidon nilotica*

BBRC

Vagrant. Western Europe.

The first Gull-billed Tern record was in Aug 1967 at Gibraltar Point. Four more followed 1969-78, with the immature bird in 1972 staying for nearly four weeks, unusual for this notorious wanderer. The description of the 1972 bird to BBRC noted head as adult winter with a black 'mask' and a darkish band extending from the base of the primaries along the centre of the wing to the outer secondaries presumed to indicate a 2CY bird. There followed a very long wait until a sixth in Jun 2013 present for one afternoon until dusk but gone by the next morning. A well-watched and photographed bird at Gibraltar Point in Aug 2020 awaits ratification by BBRC.

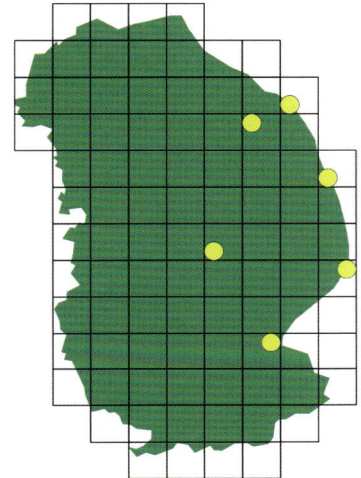

With 318 British records 1950-2018 the average number of records per year is around three to four per year, but hardly ever an easy bird to pin down.

Site Name	First Date	Last Date	Count	Notes
Gibraltar Point	30/08/1967		1	
Donna Nook	04/08/1969		1	Adult
Covenham Reservoir	19/09/1972	14/10/1972	1	
Witham Mouth	07/09/1975		1	Adult
Trusthorpe	03/09/1978		1	Adult
Kirkby on Bain	18/06/2013		1	Adult

Goxhill Mudflats, the Humber Esturay

Caspian Tern *Hydroprogne caspia*

BBRC

Vagrant. Western Europe.

There is one historic record of a bird predictably shot at Caythorpe, Grantham on May 17th 1853. The next was not until Sep 5th 1971 at Gibraltar Point and since then there have been another 17 records involving about 12 birds; the two birds in Jul 1979 are regarded as separate individuals. In Jun 2016 the two birds at Gibraltar Point were the first and only time there have been two together. One of these two carried a metal ring on the right leg although no details could be read. There are five recoveries of foreign-ringed Caspian Terns in Britain, one from America (1939), three from Sweden (1976, two in 2019) and one from Finland (1972). The two in 2019 were colour-ringed and had been ringed as nestlings, one in 2015 and one in 2016 but both seen together in Devon.

Clearly Scandinavian colonies are most likely to be the source of Lincolnshire records.

Site Name	First Date	Last Date	Count	Notes
Caythorpe	17/05/1853		1	Shot
Gibraltar Point	05/09/1971		1	
Messingham	04/07/1979		1	
Donna Nook	14/07/1979		1	
Messingham	14/06/1981		1	
South Ferriby	15/06/1981	16/06/1981	1	Same
Spalding	05/05/1988		1	Adult
North Hykeham	08/05/1988		1	Same
Torksey	06/06/1989		1	
Messingham	23/06/1992		1	
Boultham Mere	10/08/2002		1	Adult
North Hykeham	10/08/2002		1	Same
Freiston Shore	22/05/2009		1	Adult
Gibraltar Point	30/05/2015		1	Third calendar year
Middlemarsh Farm	13/06/2016	15/06/2016	1	Third calendar year
Gibraltar Point	14/06/2016	18/06/2016	2	One same
Gibraltar Point	19/06/2016		1	Same
Alkborough	24/07/2016		1	
Baston	15/07/2017	16/07/2017	1	Same third calendar year

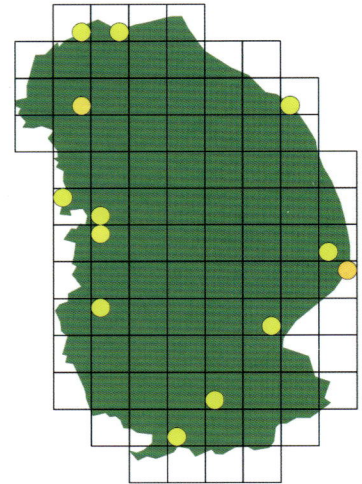

Lesser Crested Tern *Thalasseus bengalensis*

BBRC

Vagrant. Oceanic.

One record on Jun 20th 1993 at Saltfleetby-Theddlethorpe; this bird had flown south from Spurn Point on the evening of Jun 18th suggesting that it might have visited Lincolnshire. One astute observer had a look at the shoreline at North Cotes on the evening of Jun 19th without success. The next day, Jun 20th, and after a report the bird had departed south from Beacon Ponds at Spurn again around mid-day, the same observer decided to visit Rimac at Saltfleetby-Theddlethorpe. He located the Lesser Crested Tern with two Sandwich Terns *T. sandvicensis*, some way to the south of the mouth of Saltfleet Haven. Closer observation showed that it was paired with a Sandwich Tern colour-ringed on the left leg and BTO-ringed on the right leg. This was the same bird that it had been with at Beacon Ponds and, apparently, the same Sandwich Tern it had been with on the Farne Islands for the two previous years.

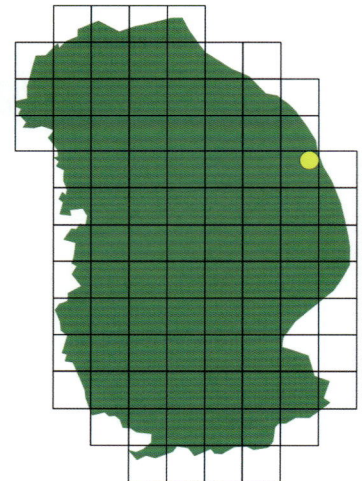

This famous individual was a female, affectionately christened 'Elsie' by the birding fraternity. She turned up every year during 1983-97 in the Sandwich Tern colony on the Farne Islands. She was first noted brooding an egg in 1985 and produced single hybrid young in 1989, 1992, 1996 and 1997. To complicate matters, one of the hybrid young bred with a Sandwich Tern in 1994 also fledging a chick (RBBP reports, www.rbbp.org.uk). 'Elsie's' long stay perhaps encouraged observers to take this species almost for granted as an annual visitor but In fact, the last new arrival was in 1998 and the last British record was in 2005.

Site Name	First Date	Last Date	Count	Notes
Saltfleetby-Theddlethorpe	20/06/1993		1	

Sandwich Tern *Thalasseus sandvicensis*

Passage migrant, scarce in spring but common in autumn. Rare inland. Occasional pairs in summer.

Barely qualifies as an irregular Lincolnshire breeder on the basis of six pairs that nested in a Common Tern *Sterna hirundo* colony at Friskney in 1950 but failed to raise any young. Although they breed close by in large colonies on the Norfolk coast at Scolt Head Island and Blakeney Point, the Lincolnshire coast appears to be too disturbed for them. However, the large numbers that use the coast from North Cotes to Gibraltar Point in Jul-Aug, bringing their recently fledged young to feed here, suggests there is no problem with food availability. Peak counts reported in LBR 2014-18 were 800 in the north at Saltfleet/Mablethorpe in Aug 2016 and in the south 5,500 at Gibraltar Point in Jul 2015.

Colour ringed birds from the Blakeney Point colony are regularly recorded at Gibraltar Point.

1968-1972	1980-1989	2008-2011	2016-2019	Incidence	Change
					*

Little Tern *Sternula albifrons*

Scarce summer visitor and fairly common passage migrant. Occasional inland.

Little Tern is for many the signature species of the Lincolnshire coast and its status has been causing concern, on and off, for decades. The *LBC Atlas* states "throughout the 1980s colony counts indicate that 100-200 pairs attempted to breed in most years, but numbers declined at many sites in the 1990s". RBBP records show that an average of 20 pairs per year bred during the period 2013-17 (range 18-30) but in the years 1990 to 2012 for which RBBP published data the numbers fluctuated between zero (1997, 1998, 2000) and 92 in 1992. Breeding success depends on shingle ridges that don't flood at high tide, that aren't accessible to foxes and do not experience human disturbance. A difficult cocktail to manage on the heavily visited Lincolnshire coast. At present although terns have used at least four sites in the last 30 years, only the one at Gibraltar Point is currently tenable. That is partly because of funds made available under an EU Life project that has now expired, and the active hard work that is put in by the management team there. Nesting and fledging are usually completed by Aug and birds have usually gone by mid-Sep. There is a variable passage of birds moving south from the numerous colonies further north on the east coast from Jul onwards.

1968-1972	1980-1989	2008-2011	2016-2019	Incidence	Change
				0%	*

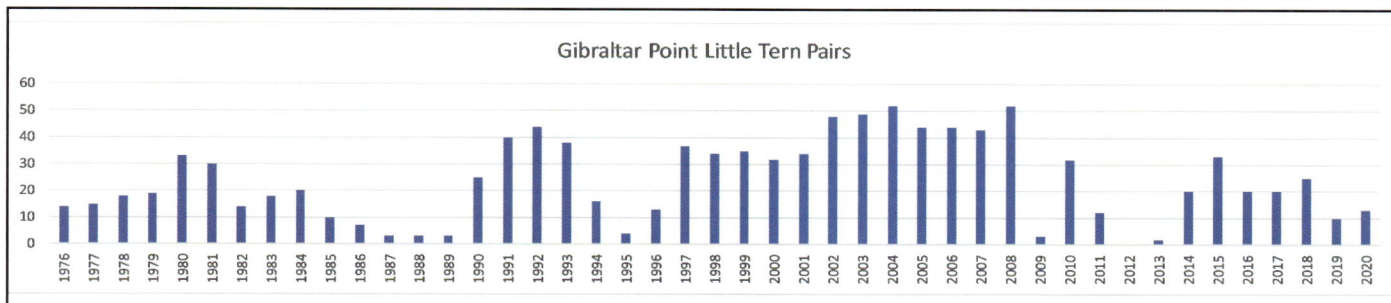

Gibraltar Point Little Tern Pairs

Roseate Tern *Sterna dougallii*

Very scarce passage migrant, May-Oct.

Two in the spring and four in the autumn of 1964 at Gibraltar Point were the first confirmed county records. After 1964 they were recorded in 14 of the 36 years to 2000, mostly just one to three birds, but with an exceptional 15 in Sep 1968 roosting on the beach at Gibraltar Point. From 2001-19 they were recorded in every year except for 2002, 2005 and 2013 mostly single birds except for 2008 (four), 2009 (five) and 2009 (eight). Roseate Terns have been recorded in every month Apr-Oct, peaking in mid-Aug to mid-Sep. The earliest in spring was Apr 27th 1979 at Gibraltar Point and the latest in autumn Oct 22nd 2006 at Freiston Shore. There has been one ringing recovery in the county: SX96930 was ringed as a nestling at Rockabill (Dublin) on Jul 7th 1999 and was found (long dead) at Gibraltar Point on Jan 9th 2000, some 423 km east.

The fortunes of Roseate Tern in Britain show a significant downward 25-year trend of -47% to 2017. From a peak of 172-177 prs in 1989 a nadir was reached in 2012 when only 73 pairs bred although there has been a recent upturn 2015-17 with confirmed breeding by 113, 106 and 112 pairs respectively. The best locations to look for them in the county are the very large tern roosts in the autumn at Horseshoe Point or Gibraltar Point, and coastal sites having breeding colonies of Common Terns *S. hirundo*, such as Freiston Shore.

Common Tern *Sterna hirundo*

Fairly common summer visitor and common passage migrant. Most breeding colonies are now inland.

Based on the 1973 Operation Seafarer Survey which established a population of 120 breeding pairs in the county the *LBC Atlas* concluded that the breeding population remained around the same but breeding colonies had shifted from the coast to suitable sites inland. They usually nest on islands in disused gravel pits or islands in man-made lagoons. The high level of disturbance on the coast means there are little or no natural nesting sites. There is no current accurate assessment of the population, but it is not lower than 120 pairs and may well be higher. Coastal passage can be spectacular with evening roosts at North Cotes Point during Aug of anywhere between 5,000 and 10,000 birds in recent years. Most birds pass through by the end of Sep, but records occur through to Nov.

1968-1972	1980-1989	2008-2011	2016-2019	Incidence	Change
				6%	29%

Arctic Tern *Sterna paradisaea*

Fairly common passage migrant, mainly coastal but there is a regular inland passage in spring. Has bred.

Arctic Tern was thought to have bred sporadically in Lincolnshire in the past and in the *LBC Atlas* period there was one instance of confirmed breeding at Frampton Marsh in 1980. There has been no evidence of breeding since then. The *BTO Atlas 2007-11* shows it breeding in low numbers in tern colonies on the north Norfolk coast, but the next nearest county where it breeds in any numbers is Northumberland. Flocks of up to 100 are reported widely at inland waterbodies from mid-Apr through to late May. Autumn passage from mid-Jul through to Oct is mainly coastal.

1968-1972	1980-1989	2008-2011	2016-2019	Incidence	Change

Whiskered Tern *Chlidonias hybrida*

BBRC

Vagrant. Southern Europe.

One at Covenham Reservoir in Jun 1987, thought to be a 2CY bird, was the first record. Although present for six days it frequently flew off and there were several long absences. There have been a further six records 2004-16, all adults in summer plumage. Four were at coastal sites and three at inland water bodies, although all were found over freshwater ponds and lagoons as befits a 'marsh' tern. The Fiskerton Fen individual was also thought to have been seen in Derbyshire and may have been one of the incredible flock of 11 Whiskered Terns in Derbyshire in late Apr 2009 which wandered widely thereafter.

There have been 202 British records, 1950-2018, averaging about five per year and including occasional large influxes as occurred in 2008 (12), 2009 (25) and 2016 (11).

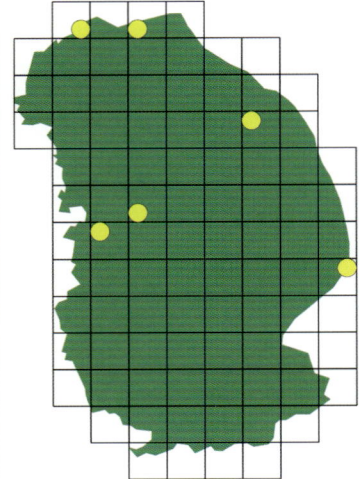

Site Name	First Date	Last Date	Count	Notes
Covenham Reservoir	10/06/1987	15/06/1987	1	First summer
Gibraltar Point	08/08/2004		1	Adult
Gibraltar Point	20/06/2006	21/06/2006	1	Adult
Barton Pits	07/06/2008		1	Adult
Alkborough	27/06/2009		1	Adult
Fiskerton	02/08/2009		1	Adult summer
North Hykeham	13/05/2016		1	

Low tide mud flats in the Wash

White-winged Black Tern *Chlidonias leucopterus*

BBRC

Rare/very scarce migrant. Central and eastern Europe.

First recorded at Marston STW in Aug 1957, there has been a total of 45 records, all of single birds apart from two at Gibraltar Point in May 1973. Records have spanned seven months May-Nov with the earliest May 14th and the latest Nov 18th. They have predominantly turned up in Aug-Sep (23) with fewer in May-Jun (11) but with Jul records over the period 5th-26th (seven) suggestive of wandering birds summering in Lincolnshire and adjacent counties. This pattern closely mirrors that for Britain as a whole summarised in White and Kehoe (2020). The majority of the records have occurred at inland water bodies with most at Covenham Reservoir (10).

The species was ex-BBRC from Jan 1st 2006 by which time there had been 800 British records. Unlike its congener, Black Tern *C. niger,* there have been no records of attempted breeding despite occasional large influxes into Britain such as those in 1970 (37) and 1992 (50).

Site Name	First Date	Last Date	Count	Notes
Marston	17/08/1957	21/08/1957	1	
Killingholme	24/08/1958	26/08/1958	1	Immature
Wisbech STW	21/09/1958	07/10/1958	1	
Marston	03/06/1959		1	Adult
Nene Mouth	13/09/1963		1	Immature
Huttoft	28/09/1963		1	Adult
Huttoft	02/08/1964		1	Adult
Tetney	05/08/1964		1	Same adult
Huttoft	07/08/1964		1	Same adult
Bardney	23/08/1966	07/09/1966	1	Immature
Donna Nook	05/08/1969		1	Adult
Wisbech STW	14/08/1969	27/08/1969	1	Immature
Anderby	15/05/1970	17/05/1970	1	
Bardney	06/08/1970	08/08/1970	1	Immature
Wisbech STW	06/08/1970	09/08/1970	1	Adult
Wisbech STW	08/08/1970		1	Adult
Gibraltar Point	24/05/1973		2	
Gibraltar Point	01/09/1973		1	Juvenile
Covenham Reservoir	08/08/1974		1	Immature
Covenham Reservoir	30/08/1974	10/09/1974	1	Immature
Covenham Reservoir	30/08/1977	11/09/1977	1	
Covenham Reservoir	31/05/1992		1	
Covenham Reservoir	23/08/1992	02/09/1992	1	Juvenile
Saltfleetby-Theddlethorpe	21/06/1993		1	
Barton upon Humber	16/07/1994		1	Adult
Messingham	25/07/1994		1	Juvenile
Covenham Reservoir	02/06/1996		1	Second summer
North Cotes	12/09/1999		1	Juvenile
Kirkby on Bain	27/09/2001	04/10/2001	1	Juvenile
Greetwell (Scunthorpe)	18/11/2002		1	Adult
Deeping St James	12/07/2003		1	Adult
Covenham Reservoir	01/10/2004	08/10/2004	1	Juvenile
Barton Pits	23/05/2006	25/05/2006	1	Adult
Worlaby (Brigg)	30/06/2006		1	Adult summer male
Covenham Reservoir	04/08/2006		1	Juvenile
Covenham Reservoir	20/08/2008	21/08/2008	1	Juvenile
Covenham Reservoir	14/05/2009		1	Adult summer
Frampton Marsh	25/07/2012		1	Adult
Kirkby on Bain	31/05/2013		1	Adult summer
Frampton Marsh	03/09/2014		1	Juvenile
Frampton Marsh	05/07/2015	07/07/2015	1	Adult summer
Gibraltar Point	09/09/2015		1	Juvenile
Barton Pits	26/07/2017	05/08/2017	1	Juvenile
Covenham Reservoir	21/05/2018		1	Adult
Frampton Marsh	26/07/2018		1	Adult

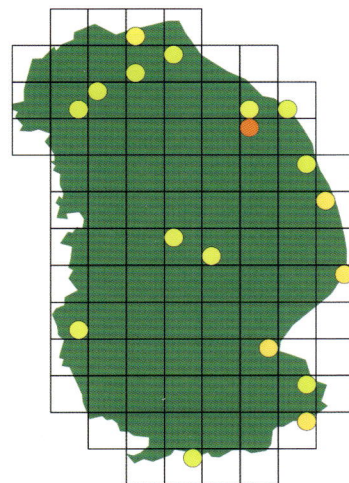

Black Tern *Chlidonias niger*

Scarce/fairly common passage migrant in spring and autumn. American Black Tern *C. n. surinamensis* is a vagrant (BBRC).

A familiar passage migrant recorded more frequently in autumn than in spring and in very variable numbers. It is regularly seen on inland waters as well as on the coast. The *LBC Atlas* notes that large numbers used to nest in the Fens before drainage and they last bred in Crowland Wash in the 1840s. In more recent times a pair built a nest at Bardney Ponds in 1961. Peak numbers are seen in May and Aug and while most larger flocks are of the order of 20-40, some as large as 100 or more have been recorded. Most have moved through by the end of Sep with stragglers into Oct or, exceptionally, Nov.

An American Black Tern was present at Covenham Reservoir Sep 17th-Oct 7th 2011, the first record for the county. This sub-species has been treated as a separate species by some authorities on account of "plumage, moult, structural and other differences" but further research needs to be done and it has not been accepted as a species in its own right by IOC.

Images below are of the American Black Tern Covenham Reservoir in 2011

Great Skua *Stercorarius skua*

Fairly common autumn passage migrant, mainly Aug to Nov and very scarce in winter and spring. Rare inland.

In contrast to Arctic Skua *S. parasiticus*, the British breeding population of "Bonxie" has been booming though the Seabird Monitoring Programme has not published definitive trend data for the period from 2000; more than half of the world's population breed along the north-western seaboard of Scotland and Ireland. The figures that follow are illustrative from LBR. They refer to annual totals of site peak monthly counts. These ranged from 150 in 2015 to 400 in 2014 with an average of around 250 per year. Jan usually sees a few birds but then it is very quiet until Aug migration begins. Unusually, in Apr 2012 a bird colour ringed as a nestling on Bear Island, Norway in 2011 was seen inland at Kirkby GP. The autumn migration as a proportion of total annual records spreads between Aug (5%), Sep (35%), Oct (45%) and Nov (13%). The largest single flock reported was 180 at Sandilands on Oct 13th 2014. From Sep 14th-15th 2017 170 birds were counted at Huttoft and a further 65 off Gibraltar Point. It is thought migrating birds use the rivers heading south-west from The Wash as a short cut to the Severn estuary and 24 birds were seen circling high at Gedney on Sep 17th 2016 seemingly heading inland.

Pomarine Skua *Stercorarius pomarinus*

Generally scarce autumn passage migrant, although occasional larger movements, mainly Sep to Nov. Rare in winter and spring and exceptional inland.

The following illustrative figures from LBR refer to annual totals of site peak monthly counts. These ranged from 50 in 2017 to 500 in 2018 with an average of around 200 per year. Jan-Feb usually sees a few birds but from Mar to Jul records are rare. It seems to be our latest migrating skua and migration doesn't really get under way until Sep. The autumn migration as a proportion of total annual records spreads between Aug (1%), Sep (5%), Oct (54%), Nov (13%) and Dec (14%). The largest single flock reported was 188 at Gibraltar Point on Oct 27th 2018 and there were at least 90 elsewhere on the coast that day. Two days later on Oct 29th 100 were reported off Witham Mouth and 90 along the coast. Another big movement occurred on Dec 2nd 2014 when 81 were reported off Huttoft and a further 40 off Gibraltar Point. Previous to these two events the largest day count of "Poms" in Lincolnshire had been 91 off Huttoft on Sep 29th 1988. It is interesting to note that 58% of all peak counts occurred on just three days of the five-year period.

Arctic Skua *Stercorarius parasiticus*

Fairly common, but decreasing, late summer and autumn passage migrant, mainly Jul to Oct. Very scarce in spring and winter and exceptional inland.

According to the Seabird Monitoring Programme UK breeding Arctic Skuas have crashed by 70% over the period 2000-18. Certainly, migrants off the Lincolnshire coast have declined too. The *LBC Atlas* referred to numbers of several hundred per day in the 1980s-90s but in the five-year period to 2018 nothing like that has been observed. The figures that follow are illustrative from LBR. They refer to annual totals of site peak monthly counts. These ranged from 130 in 2017 to 210 in 2014 with an average of around 180 per year. Autumn migration starts mid-Jul and tends to peak in late Aug to late Sep but continues through Oct into Nov. The migration as a proportion of total annual records spreads between Jul (5%), Aug (31%), Sep (43%), Oct (16%) and Nov (2%). The largest single day counts were both from Gibraltar Point with 92 on Aug 26th 2014 and 76 on Sep 5th 2017.

Long-tailed Skua *Stercorarius longicaudus*

Very scarce/scarce autumn passage migrant, mainly Aug-Oct and exceptional inland.

Long-tailed Skua was considered as a BBRC rarity up until 1979. The nearest breeding grounds are Arctic Scandinavia and spring passage primarily occurs off the west coast of Britain and especially the Outer Hebrides. The *LBC Atlas* reported two particularly good autumn passage years in Lincolnshire in 1988 and 1991 when more than 60 birds were recorded. Interestingly, compared to the other skuas it has a narrow window of occurrence in Lincolnshire and in the 10 years to 2019 no records at all were received for the period Dec-Jul. A spring plumaged adult would be a welcome sight in the county. Extreme dates were from Aug 13th to Nov 21st with the bulk of passage occurring in Aug-Oct. The numbers of birds per annum over 10 years to 2019 ranged from a low of five in 2019 to 47 in 2013, with an average of about 16 birds per annum. The highest day counts were of seven at both Gibraltar Point and Huttoft Bank on Sep 10th 2013. Counts of six at Gibraltar Point on Sep 13th 2017 and Witham Mouth on Oct 13th 2013 and five at Sandilands on Aug 30th 2013. Very recently on Aug 29th 2020 in a big Long-tailed Skua passage, Gibraltar Point had a day count of 19 birds, smashing all recent records.

Little Auk *Alle alle*

Scarce passage migrant and winter visitor, with occasional larger influxes.

Little Auks are tiny and susceptible to wrecks and being eaten by large gulls, especially Great Black-backed Gulls *Larus marinus*, so they usually winter well out at sea to avoid these perils. They tend to be driven towards the coast by strong north and north-easterly winds in late Oct-Nov. The *LBC Atlas* reported a wreck of over 200 birds picked up dead in Lincolnshire in Feb 1983 and a day count of over 1,000 birds on Nov 2nd 1986. An even bigger influx occurred from Oct 29th-Nov 4th 1995 when over 2,000 birds were counted along the coast. The mother of all Little Auk passages occurred on Nov 9th 2007 when 2,200 birds were counted moving south along the coast and the month total was 3,400. An interesting table in LBR 2013 reports annual totals from 1979 to 2013. Numbers reported in LBR in the five years to 2018 were more modest, ranging from a sum of monthly peak counts of 13 in 2017 to 222 in 2014. The best single day was Nov 4th 2014 when 94 were counted at Gibraltar Point with 98 more along the coast. There were few reports from Dec-Feb and none at all from Mar-Aug. In summary, over the 40-year period to 2018, three years mentioned above had more than 1,000 birds, three years had more than 500 birds (1983, 1991 and 2006) and in 33 years there were less than 250 birds.

Guillemot *Uria aalge*

Fairly common passage migrant and winter visitor but can occur in any month. Rare inland.

Guillemots breed 40 miles to the north of the Humber Mouth in the Flamborough/Bempton area and some 85,000 birds were counted there in 2017. They tend to feed and winter well offshore and most of the year surprisingly few make it to Lincolnshire inshore waters where they can be seen from the coast. Fortunately, oiling incidents and wrecks have not been a feature in recent years since the *LBC Atlas* reported that over 1,000 were picked up dead in in Feb 1983. They are seen in small numbers off the coast, particularly in Jan and Feb and most months of the year but larger movements tend to be in Oct. LBR does not cover this species in great detail, so the following comments refer to peak one day counts from the five years to 2018. The poorest year was 2014 when a max of only five was reported from Huttoft Car Terrace in Oct, the location that often gets most birds. The best day was Oct 29th 2017 when 290 birds were reported at Sutton on Sea and a further 154 from sites further south, 444 in total. It should be noted that auks can often be seen passing too far out to be identified. The peak day count of auk sp. was 165 on Oct 14th 2015. Since Guillemots apparently outnumber Razorbill *Alca torda* some 10:1 in the North Sea, they are most likely of the former species.

A rough day at Pyewipe, Grimsby

94

Razorbill *Alca torda*

Western European form *islandica* scarce passage migrant and winter visitor but can occur in any month. Exceptional inland.

The *LBC Atlas* reported that the wreck of Feb 1983 included 2,500 Razorbills, an astonishingly high figure, given the relative scarcity of this species compared to Guillemot *Uria aalge*, in the North Sea. LBR reports for the five years to 2018 showed that the yearly total of peak counts ranged from a low of 15 in 2015 to a high of 80 in 2016 with an average of 40 birds per year. The largest count each year was from Gibraltar Point with an overall high of 43 on Jun 2nd 2014. On Jun 1st 2016 Gibraltar Point had 23 and on the same day 18 were counted moving north at Saltfleetby-Theddlethorpe.

Interestingly from locations like Rimac in the 1980s and 1990s, warm onshore breezes in Jul would often reveal Razorbills with chicks, a phenomenon unseen in the five years to 2018. This is surprising given that Razorbills breed 40 miles to the north of the Humber Mouth in the Flamborough/Bempton area. Some 28,000 birds were counted there in 2017 and that site holds the fastest growing population in the UK according to the Seabird Monitoring Programme.

Black Guillemot *Cepphus grylle*

A rare passage migrant and winter visitor. Amber List.

One shot at Washingborough in Jan 1899 was the first county record and there have been very few confirmed records since, all of which have appeared between Aug 30th and Feb 13th. Apart from the first record all have occurred off the coast or in the Wash.

An exciting recent record was the discovery of a very white-looking Black Guillemot in the Wash off the Witham Mouth, Dec 7th-11th 2017. After some research it was tentatively identified as being one of the high arctic races *C. g. mandtii* or *C. g. ultimus*. BOURC accepted the record as Mandt's Black Guillemot *C. g. mandtii*, the first occurrence of this taxon in Britain.

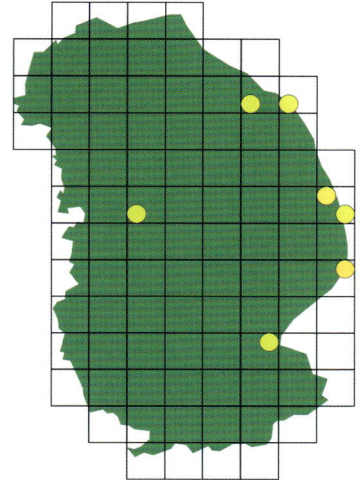

Site Name	First Date	Last Date	Count	Notes
Washingborough	21/01/1899		1	
Chapel St Leonards	21/09/1963		1	
Gibraltar Point	13/11/1965		2	
Gibraltar Point	15/01/1966		1	
Donna Nook	30/08/1967		1	
Gibraltar Point	16/11/1968		1	Immature
Witham Mouth	29/12/1977	03/01/1978	1	
Donna Nook	05/11/1980		1	
North Cotes	23/11/1980		1	
Huttoft	13/02/1983		1	
Huttoft	02/11/1985		1	
Huttoft	05/11/2006		1	
Witham Mouth	07/12/2017	11/12/2017	1	Adult winter *C. g. Mandii*
Chapel St Leonards	30/10/2019		1	

Images Mandt's Black Guillemot (SG)

Puffin *Fratercula arctica*

Very scarce passage migrant but can occur in any month. Exceptional inland.

The Puffin's rarity in Lincolnshire may partly be because the Flamborough/Bempton colony is small and the Coquet and Farne Island colonies that hold 95% of the English Puffin population are some 200 miles north of the Humber mouth. Up until the 1980s the *LBC Atlas* described Puffin as fairly common and it became rare off Lincolnshire in the 1990s. Interestingly, the British population was thought to be growing at that time though it has declined some 21% since 2000. The big wreck of Feb 1983 produced 300 dead Puffins along our coast. Analysis of LBR reports shows that in the 10 years to 2019 an average of around seven birds a year was reported and accepted, ranging from one in 2012 to 19 in 2017. Peak passage occurs in Sep-Oct and the only month with no records in the 10-year period was Dec. The largest flocks reported were 10 north at Gibraltar Point on Sep 18th 2017, with two at Crook Bank, Saltfleetby-Theddlethorpe the same day. The second largest was eight at Donna Nook on Aug 11th 2010. Also at Donna Nook, six birds were picked up dead from Mar 29th to Apr 1st 2013.

Another factor that probably limits their presence off the Lincolnshire coast is their extreme pelagic tendency, demonstrated by geolocator tracking, which sees them many thousands of kilometres away in oceanic waters during the winter, though some remain in the North Sea.

Red-throated Diver *Gavia stellata*

Fairly common offshore passage migrant and winter visitor Aug-May with peaks in mid-winter. Rare inland and in summer.

Henderson and Wilson in LBR 2015 provided a summary and discussion of Red-throated Divers wintering off Lincolnshire over the period 1978-79 to 2014-15. The general trend was of an increase in the wintering population and they highlighted the value and limitations of shore-based sea watching for detailed assessment of such changes. The largest count by far in the entire period was 1,308 seen off Gibraltar Point on 20th Feb 1999 and the mean peak for the period 2006-7 to 2014-15 was around 250. Their analysis showed peak use of Lincolnshire waters In Jan-Mar. The Greater Wash Special Protection Area, which extends from Bridlington to Great Yarmouth covering all the North Sea inshore waters off Lincolnshire, holds over 8% of the 22,000 UK wintering population, second only to the outer Thames estuary. It is now known from satellite radio tracking that birds breeding as far east as the Ob River Delta in Arctic Russia winter off Lincolnshire following a cycle of movements through the Baltic and North Sea through the winter. Since 2015 the peak Lincolnshire count was 506 off Crook Bank, Saltfleetby-Theddlethorpe on Jan 21st 2018.

Red-throated Divers are particularly averse to marine windfarm developments and it is to be hoped that as such developments proceed, they will be sited so as to minimise displacement of divers from prime feeding habitat.

Black-throated Diver *Gavia arctica*

Scarce winter visitor to coastal and inland waters. Mainly Sep-Apr, rare in summer.

The *LBC Atlas* reported that in the 1980s and 1990s around 10 birds were seen a year from Sep-Apr. Black-throated Diver remains scarce but the numbers have increased with sea watching effort and the average in the five years to 2018 based on LBR reports is around 28 per year ranging from 26-30 so remarkably consistent from year to year. Numbers are therefore less than 5% of their better studied cousin Red-throated Diver *G. stellata*.

Identification especially at distance at sea from the shore remains an issue and although the species is no longer considered by LBRC records are monitored and checked as required. Most records are of singles but a flock of six was reported heading south at Chapel Point on Sep 23rd 2017 and Sep-Oct is peak passage time. Five were seen at sea off Witham Mouth on Jan 5th 2014. Jan-Feb is the winter peak period with numbers around half the autumn passage. Over the five years there were four inland records,three at Covenham Reservoir and one at Cleethorpes CP which stayed 33 days from Jan 3rd-Feb 5th, 2014 and was by far the longest stayer. Summer birds between May and late Aug are exceptional but an adult in full breeding plumage was seen off Crook Bank Saltfleetby-Theddlethorpe on Jun 3rd, 2016.

Great Northern Diver *Gavia immer*

Scarce winter visitor Aug-May. Mainly coastal and exceptional in summer.

The *LBC Atlas* reported fewer than five birds per year were seen during the 1980s and 1990s. Great Northern Divers wintering around Britain are thought to breed in Iceland and Greenland. Their numbers show a long-term increase in British waters but the numbers inshore in the Greater Wash from Bridlington to Great Yarmouth remain low with WeBS data reporting a running five-year mean to 2018-19 of around seven birds per annum in that massive area. Of course, more birds could be wintering further offshore only coming closer in on occasions. During the five years to 2018 LBR reported around 28 birds a year, the same as Black-throated Diver *G. arctica*. Numbers ranged from 21 in 2015 to 36 in 2016 with peak months from Oct-Jan. The largest day counts were five at Gibraltar Point on Nov 4th 2016 and three birds at each of Freiston Shore May 2nd 2016, Witham Mouth Dec 11th 2017 and Gibraltar Point Oct 27th and Dec 11th 2018. There were five birds on inland waters in the period with four in Jan and one in Feb.

White-billed Diver *Gavia adamsii*

Vagrant. Arctic

The first county record of this spectacular High Arctic diver unfortunately involved an adult bird found dead in the Wash at Friskney Marsh on Mar 26th 1976. Twenty years later however, Lincolnshire birders got the chance to catch up with a live White-billed Diver in the County in a most remarkable occurrence and in the most unexpected of locations. On Feb 29th 1996 a detour by one local birder was made along the road which runs alongside the River Witham near Tattershall Bridge. At around midday, about a mile before reaching Tattershall Bridge, a large diver was noticed underneath the nearside bank at a range of around 10 metres. The bird was identified as a White-billed Diver, an amazing sighting for what is an exceptionally rare bird in Britain away from Scotland. Unfortunately, on Mar 2nd the bird took an angler's pike bait. The bird was brought to the bank and was subsequently taken into care, but it had swallowed the hooks and did not survive. This was only the second British record of a White-billed Diver away from the coast. Strangely, the third record was on the same stretch of the River Witham 21 years later in 2017, this time surviving and staying almost two weeks. The fourth record was of a freshly dead bird found at Freiston Shore in May 2018.

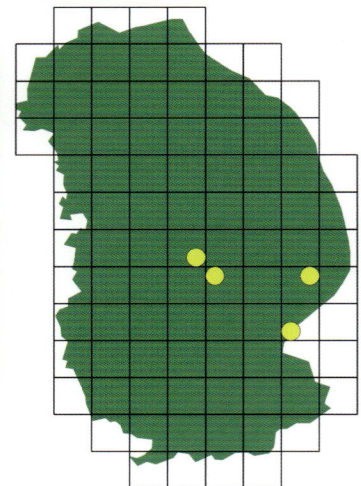

Site Name	First Date	Last Date	Count	Notes
Friskney	26/03/1976		1	Adult found dead
Tattershall	29/02/1996	02/03/1996	1	Died in care
Kirkstead Bridge	20/01/2017	01/02/2017	1	Second calendar year
Freiston Shore	22/05/2018		1	Found dead

Yellow-nosed Albatross *Thalassarche chlororhynchos*

Vagrant. South Atlantic.

One extraordinary county record of an immature bird found and photographed by an angler on the Diawa Manton fishing lakes just outside Scunthorpe in Jul 2007. It was found late in the day on Jul 2nd settled on the water and being mobbed by various gulls, *Larus sp*. The angler in question was fishing overnight and awoke the next morning, Jul 3rd, to discover that the bird was still present and just a few metres away from his swim. It attempted to take a cast carp bait on two occasions but fortunately dropped it. Pestered by Mute Swans *Cygnus olor*, among others, it drifted away out of view. It was then seen again for the last time around midday when in flight; it circled the lake for a few minutes and disappeared away to the south unseen by a single birder. This bird had been taken into care in Brean, Somerset Jun 29th-30th before being released and later found in Lincolnshire.

More emerged about other records of this species in 2007 beginning with one found in a cow pasture in Maine, USA in Apr and taken into care and released about three weeks later. There was a subadult off the Norwegian coast the day before the Somerset record seen following a fishing boat, with further records in the same general area over the next few days. Then, after the British records another, maybe the same bird, was seen off the west coast of Sweden towards Malmö before apparently heading inland. In all there were clearly at least two individuals in northwest Europe, and maybe more.

Site Name	First Date	Last Date	Count	Notes
Messingham	02/07/2007	03/07/2007	1	Immature

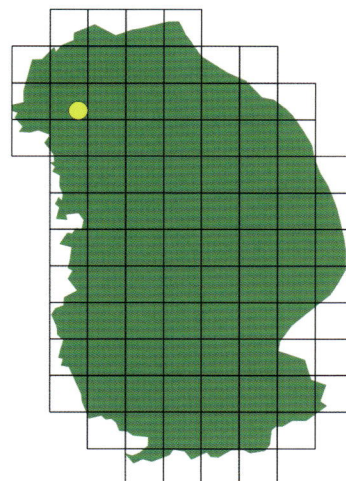

Storm Petrel *Hydrobates pelagicus*

Very scarce offshore migrant Jul-Nov averaging more than one per year but variable between years; exceptional inland and in other months.

Lorand and Atkin (1989) noted some historic records of a small number of storm-blown birds being picked up on the coast and at inland localities in the autumns between 1869-1889. Unusual numbers along the whole of the east coast in Nov 1883 resulted in at least four Lincolnshire records but there are no other details.

In the modern era, Storm Petrel remains a very scarce offshore migrant with just 45 records involving 63 birds, albeit with some duplicate sightings as birds moved north or south past coastal seawatching sites. The most in any one year was three. Some nocturnal ringing activity 1989-2001 using a tape lure resulted in a small number of captures and one recapture but gave some very valuable insight into Storm Petrel movements off our coast. Six birds were ringed, and two of these birds were later controlled: one ringed on Aug 29th 1989 was recaptured at Tarbat Ness, Highland on Jul 19th 1991; the second was ringed on Aug 7th 1992 and recaptured at Kråkenes Vågsoy, an island off southwestern Norway on Aug 25th 1993. One caught on Jul 15th 1994 had been ringed at Tynemouth two days earlier and was subsequently recaptured in Jul 1994 and 1995 on the Northumberland coast.

Leach's Petrel *Oceanodroma leucorhoa*

Scarce offshore migrant generally Sep-Nov, but also May and Dec-Jan. Exceptional inland.

Historically, there are a good number of reports of birds found well inland and also on The Wash particularly at Friskney where several were netted by bird-catchers up to 1881. During a large wreck in 1952 the following records were summarised by Boyd (1954): One picked up alive between Scunthorpe and Appleby on Sept 19th was kept and fed for two days then allowed to fly away; one shot on the Brigg side of Scunthorpe about end of Sep; one found dead, Brocklesby, Nov 2nd and another Nov 10th or 11th; one found dead, south of Grantham, before Nov 5th; one long dead, Tetney Marsh, Nov 16th. That year in England Boyd (1954) estimated that at least 6,700 birds perished and that birds were found in every English county except Rutland.

Leach's Petrel records are usually in single figures most years, but occasionally weather conditions result in many more. The weekend of Sep 9th-10th 1989 (see Catley, LBR 1989) was one such occasion with the kind of seabird passage more usually associated with northern and western coasts. Leach's Petrels in particular figured prominently and the total was a minimum of 183 birds with most seen at Gibraltar Point with at least 32 birds on Sep 9th and 35 in four hours on Sep 11th. There had been just 38 records 1960-88. This remarkable event has not been replicated since and in the nine years 2011-19 there was an average of three records/four birds per year but with huge variation and five blank years. The average is greatly skewed by the 16 records of 27 birds in 2013.

Fulmar *Fulmarus glacialis*

Offshore visitor, declined from fairly common to scarce, mainly Mar-Nov. Rare in winter and very infrequent inland.

The Seabird Monitoring Programme reports that Fulmar has declined by 38% from 2000-18. In Lincolnshire the decline feels much worse. The *LBC Atlas* reported a one-day count of 2,000 past Chapel Point in Sep 1989. The largest one-day count in the period 2003-18 was 200 at Gibraltar Point on Feb 29th 2004. The mean peak one day count reported in LBR has fallen as follows: five years to 2008, 115 birds, five years to 2013, 50 birds, five years to 2018, 20 birds. In the latest five-year period, the peak count ranged from eight in Oct 2018 to a high of 63 at Gibraltar Point on Feb 8th 2016. The decline has been attributed to several different causes. These include the reduction in the North Sea white-fish fishery with resulting declines in offal discharge, and the impact of climate change on sand eel populations, an important natural food source. Also, mortality as bycatch in the longline and gill net fisheries in Norwegian waters may be a factor. It should be emphasised that birds are still seen regularly; on 45 dates between Jan-Oct 2018 but in much smaller numbers than hitherto. The absence of birds in Nov-Dec is attributed to Fulmars wintering in distant waters.

Cory's Shearwater *Calonectris borealis*

Vagrant. Oceanic.

Still a great county rarity, there have only been six records involving five birds. The first flew south at Saltfleetby-Theddlethorpe in Aug 1985; the next two in 1988 and 1995 were also on very similar dates in Aug. Another was picked up during a seawatch at Witham Mouth in Sep 2009; it drifted slowly north past Freiston Shore and away up the west side of The Wash, briefly landing on the sea. The final record to date flew south at Trusthorpe in Oct 2011. The rarity of this species in the county is most likely related to geography with no peninsulas or prominent headlands. For example, the species is regularly recorded in small numbers on the Yorkshire coast, usually heading south and with occasional larger influxes, one of which occurred in 2001 with 41 records. Although good numbers of Sooty Shearwaters *Ardenna grisea*, were seen off the Lincolnshire coast on that date, no Cory's Shearwaters were reported. There are well known late-summer movements in the South-western approaches, with occasional huge influxes such as the 17,250 recorded in 1980 including an astonishing 10,940 past Cape Clear Island, Co. Cork on Aug 16th. Unsurprisingly, with such huge numbers arriving in some years, the species ceased to be considered by BBRC in 1983.

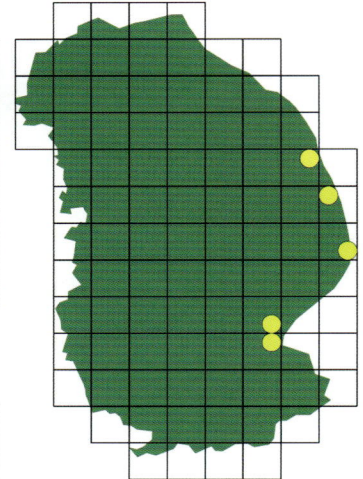

Site Name	First Date	Last Date	Count	Notes
Saltfleetby-Theddlethorpe	14/08/1985		1	
Huttoft	20/08/1988		1	
Skegness	22/08/1995		1	
Freiston Shore	13/09/2009		1	
Witham Mouth	13/09/2009		1	Same
Trusthorpe	10/10/2011		1	

Sooty Shearwater *Ardenna grisea*

Scarce offshore passage migrant Jul-Nov, mostly Sep. A few winter records; Dec-Jan.

The migration of Sooty Shearwater around the South and North Atlantic Ocean to their breeding grounds in the very southern Atlantic Ocean sees birds forced into the North Sea by north-west winds. Most birds seen off Lincolnshire are heading north, presumably returning to the Atlantic, and they are seen off the coast in small numbers most years with bigger numbers in some. The *LBC Atlas* mentions big years of 100 records in 1977 and 200 In 1996. The all-time record year was in 2005 when 358 birds were recorded off Gibraltar Point in a fierce north-easterly blow on Sep 16th.

During the five years 2014-18 annual numbers ranged from 12 in 2017 to 268 in 2016, a big year. Dates of occurrence ranged from Jul 18th-Nov 30th, with Sep-Oct by far and away the peak passage period. The biggest counts came in 2016 with Gibraltar Point counting 62 on Sep 17th with a further 48 along the coast to the north that day, a total of 110. A couple of weeks later, on Oct 2nd, a total of 152 were recorded at five sites from Donna Nook to Gibraltar Point. Including the 2016 total the average was around 80 per year, without it the average was around 30.

Chapel Point Observatory

Great Shearwater *Ardenna gravis*

Vagrant. South Atlantic.

Even rarer than Cory's Shearwater *Calonectris borealis* in the modern era, there have only been eight records involving nine birds. As ever, a significant proportion of the records involved birds which were wrecked (1882), shot (1893 and 1902) or found dead (1970). The records of birds seen alive and well were in mid-Sep 1968 when two were seen off Anderby Creek, followed by singles in 1974, 1975 and 1977 all in mid-Sep.

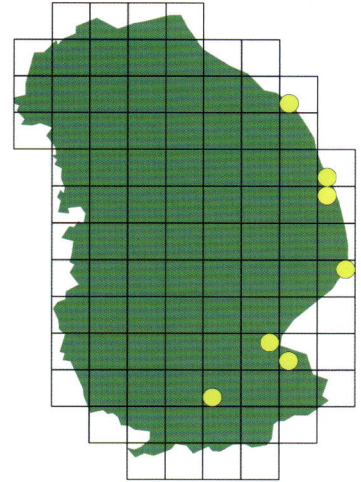

Great Shearwaters are trans-equatorial migrants which breed on Tristan da Cunha, Gough Island and Kidney Island (Falklands) in the South Atlantic. Around 6,000,000 pairs breed on the first two islands. They move clockwise around the North Atlantic reaching British waters from Jul to Oct, and as with Cory's Shearwaters the bulk of British records have been in the South-western Approaches. With a gap of 43 years since the last one, the Lincolnshire coast looks unlikely to improve on this meagre return but there is hope. Nearby Spurn has had records in five years between 1992 and 2013 with 2007 delivering six different birds between Sep 3rd-15th.

Site Name	First Date	Last Date	Count	Notes
Spalding	01/11/1882		1	Captured then sent to London Zoo
Holbeach Marsh	15/12/1893		1	Shot mid-Dec
Welland Mouth	27/11/1902		1	
Anderby	15/09/1968		2	
Gedney Drove End	28/02/1970		1	Adult, probably dead for 3 weeks
Gibraltar Point	15/09/1974		1	
Trusthorpe	29/09/1975		1	
Donna Nook	17/09/1977		1	

Manx Shearwater *Puffinus puffinus*

Scarce/fairly common offshore visitor/passage migrant, Apr-Nov with most Sep. Occasional inland records after south-westerly gales.

Most of the world population of 400,000 Manx Shearwaters breed around Britain and 90% of those are on the Scottish Island of Rum and islands off the Pembrokeshire coast. Very few breed in the Scottish North Sea and none in the English part. Remarkably few are seen off Lincolnshire. The *LBC Atlas* reported regular yearly totals of less than 200 with a spectacular movement in 1989 when 1,200 were recorded. In the five years to 2018, LBR reports showed an annual total of monthly peak counts of around 110 with a range from 29 in 2015 to a high of 225 in 2014. Records occur in small numbers from Apr-Jul with peak passage in Aug-Sep. The largest day counts in the five years to 2018 were both at Gibraltar Point with 182 north on Aug 18th 2014 and 76 on Sep 23rd 2018. Passage falls away in Oct and is usually over by early Nov. The British population winters in the South Atlantic off South America and can wander anywhere in the southern oceans from Dec-Mar.

Anderby Creek

Balearic Shearwater *Puffinus mauretanicus*

LBRC

Rare migrant. Mediterranean.

Balearic Shearwater was split from Manx Shearwater *P. puffinus* in 1991 when it became known as the western race of Mediterranean Shearwater *P. yelkouan*. A decade later, a further split led to recognition of Balearic Shearwater *P. mauretanicus* and Yelkouan Shearwater *P. yelkouan* as separate species (Sangster *et al* 2002).

Despite this taxonomic progression, *mauretanicus* is still a recognisable taxon in the field and the first county record was of a bird seen off Gibraltar Point in Oct 1963. It has remained a rare migrant in the county with 21 more records to 2019. Three were seen off Gibraltar Point in Sep 1997 and two in Sep 2007 and Sep 2019. The extreme dates were Aug 1st to Nov 10th.

Balearic Shearwaters are one of the world's most endangered seabirds, with an estimated breeding population of 2,000-2,400 breeding pairs and 10,000 individuals in all (Wynn and Yésou 2007). They breed on islands and coastal cliffs in the Balearic Islands. It is a regular visitor in varying numbers to west European (Atlantic) inshore waters during the summer and autumn, most commonly seen off Iberia, France and the south-western coasts of Britain and Ireland during its post-breeding dispersal.

A huge count of 6,500 birds in the Bay of Biscay in 2003 represented 65% of the world population. In Britain, flocks of several 100s are regularly seen off the English south and south-west coasts, and the peak total in British and Irish waters was 3,474 in 2001; the vast majority are seen Jul-Oct.

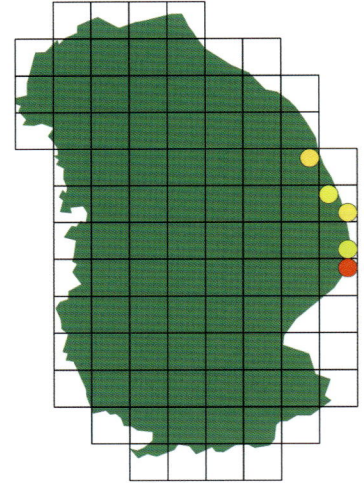

Site Name	First Date	Last Date	Count	Notes
Gibraltar Point	05/10/1963		1	
Gibraltar Point	01/08/1971		1	
Gibraltar Point	15/09/1976		1	
Saltfleetby-Theddlethorpe	28/07/1984		1	
Saltfleetby-Theddlethorpe	28/08/1984		1	
Saltfleetby-Theddlethorpe	09/09/1989		1	
Gibraltar Point	27/10/1989		1	
Gibraltar Point	09/09/1997		3	
Huttoft	13/09/1998		1	
Gibraltar Point	10/11/1999		1	
Huttoft	09/09/2001		1	
Gibraltar Point	21/09/2004		1	
Chapel St Leonards	17/09/2005		1	
Chapel St Leonards	18/09/2007		2	
Gibraltar Point	18/09/2007		1	
Gibraltar Point	22/09/2007		1	
Gibraltar Point	26/09/2007		1	
Skegness	09/11/2010		1	
Gibraltar Point	15/09/2011		1	
Gibraltar Point	16/09/2011		1	
Huttoft	01/10/2018		1	
Gibraltar Point	17/09/2019		2	

Chapel Point Observatory

Black Stork *Ciconia nigra*

BBRC
British Birds Rarities Committee

Rare. Western Europe. First record 1984, recent records reflect the presence of a growing breeding population in northern France.

Only seen for the first time at Freiston Shore in Apr 1984, there was a gap of 22 years before the second in 2006 since when they have been recorded in 2014, 2015 and 2017. The 2015 records involved at least three wandering individuals seen at many coastal sites, one of which carried colour ring F05R. This bird had been ringed as one of a brood of three in the Ardennes, north-east France. It was the bird last seen Aug 11th and was sadly found dead just north of Madrid, Spain on Oct 20th 2015. One of its siblings F05P was in north-east Scotland Aug 6th-24th.

There has been something of a recovery in the northern European population. In France there were just 13 pairs in 2002, but by 2019 they had increased to 60-80 pairs. The Austrian population of just three pairs in 1960 numbered 200-300 pairs by 2002 and breeding birds have returned to Belgium, where there were 31-41 pairs by 2002 and now there are between 100-150 pairs (European Environment Agency Report Mar 2020).

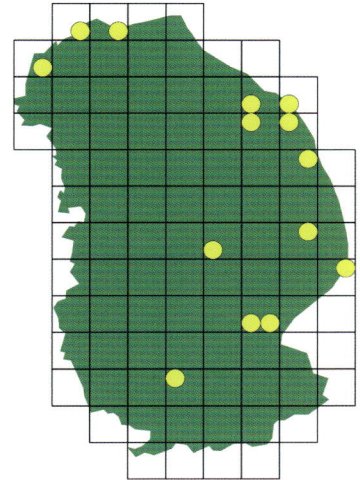

Site Name	First Date	Last Date	Count	Notes
Freiston Shore	25/04/1984		1	
Welton le Marsh	20/06/2006	21/06/2006	1	Adult
Crowle	26/05/2014		1	
Boston	22/05/2015		1	
Kirkby on Bain	22/05/2015		1	
Alkborough	03/08/2015		1	Juvenile
Tetney	03/08/2015		1	Same
Conisholme	04/08/2015		1	Same
Donna Nook	04/08/2015		1	Same
Reads Island	04/08/2015		1	
Saltfleetby-Theddlethorpe	04/08/2015		1	Same
Gibraltar Point	05/08/2015	08/08/2015	1	Same
Covenham Reservoir	11/08/2015		1	
Dunsby	24/07/2017	25/07/2017	1	First calendar year

White Stork *Ciconia ciconia*

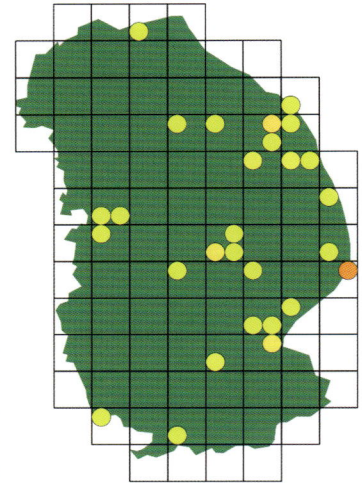

Very scarce migrant. Status complicated by re-introduction schemes in Britain and Europe.

One was shot at Marshchapel in Jun 1832 since when subsequent records have been more fortunate. An influx in 1967 involving about 11 birds (Fraser 2013) and described in the rarities report for that year as "perhaps the most striking influx ever recorded" resulted in the first modern record with three seen at Marshchapel on May 21st that year. This was a BBRC species until 1983 and there were another three records in 1968, 1971 and 1976 before the species was devolved to county record committees. Since then, taking into account multiple records of wandering birds and excluding probable escapes and birds from reintroduction schemes, there may have been as few as 27 individuals involved. Suspect individuals identified by coloured plastic rings or non-standard metal rings were noted in 2007, 2011, 2015, 2016 and 2018. The overall total is thus a conservative one.

Since the first record of three birds there have only been two years (2008, 2012) where more than one bird was involved. There are records from seven months of the year with the earliest Jan 29th 2012; this bird frequented Kirkby landfill site on and off to Mar 3rd of that year. Another long-stayer was in the Stickney area Apr-Jun 1971. Most records were in Apr -Jun (21) with singles in Aug, Oct and Nov.

Site Name	First Date	Last Date	Count	Notes
Marshchapel	21/05/1967		3	
Benington	25/11/1968	15/12/1968	1	
Stickney	29/04/1971	12/06/1971	1	
Gibraltar Point	01/06/1976		1	
Gibraltar Point	16/04/1996		1	
Boston	12/05/1996		1	
Gibraltar Point	12/04/1998		1	
Louth	07/05/1998		1	
Wold Newton	10/04/1999		1	
Frampton Marsh	05/06/1999		1	
Morkery Wood	19/08/2001		1	
Gibraltar Point	24/04/2002		1	
Lincoln	16/05/2002		1	Same
Walcott	25/05/2002		1	Same
Kirkby on Bain	28/05/2002		1	Same
Donna Nook	03/06/2002		1	Same
Gibraltar Point	10/08/2002		1	Same
Kirkby on Bain	17/05/2004		1	
Saltfleetby-Theddlethorpe	28/04/2005		1	
Marshchapel	29/04/2005		1	Same
Huttoft	11/06/2005		1	
Market Deeping	07/10/2005	13/10/2005	1	Wide ranging later found dead
Claxby (Market Rasen)	04/05/2006	05/05/2006	1	
Saltfleetby St Peter	02/06/2006		1	
Boultham Mere	29/05/2008	30/05/2008	1	
Alvingham	30/05/2008		1	Same
Louth	27/06/2008	28/06/2008	1	
Gibraltar Point	28/06/2008		2	
Freiston Shore	29/06/2008		1	Same
Conisholme	29/01/2012	01/02/2012	1	
Scrivelsby Park	05/02/2012		1	Same
Wood Enderby	07/02/2012		1	Same
Kirkby on Bain	09/02/2012	03/03/2012	1	Same
Gosberton	23/04/2012	24/04/2012	2	
Whisby	19/06/2013		1	
Frampton Marsh	03/04/2017		1	
Gibraltar Point	24/04/2017		1	Same
Barton upon Humber	03/05/2018		1	
Frampton Marsh	04/05/2018		1	Same
Carlton & Manby Washlands	12/05/2018		1	Same
Gibraltar Point	06/08/2018		1	Same
Middlemarsh Farm	06/08/2018		1	Same

Gannet *Morus bassanus*

Regular offshore visitor/passage migrant most of the year. Fairly common/common Jul-Nov, but scarce at other times and rare inland.

The British Isles are home to 55% of the world population of Gannets and the nearest colony, Bempton Cliffs, just over 40 miles north of the Humber mouth grew from 3,940 pairs in 2003 to 13,392 in 2015 and is probably still growing, so there are lots of Gannets breeding on our doorstep. As they take five years to reach maturity and they move large distances on foraging flights, this is one sea bird we see plenty of. In the five years to 2018 LBR reported total monthly peak counts by site averaging at 3,700 per year ranging from 2,000 in 2014 to 4,800 in 2016. It occurs in every month of the year but 96% of the peak counts come in the main passage period from Aug-Nov. The largest counts included a massive 1500 at Gibraltar Point on Oct 2nd 2016 with over 900 there on Oct 2nd and 3rd 2018 and 500 at Witham Mouth on Sep 5th 2015. Apart from the coast and The Wash, birds are also regularly recorded up the Humber past the Humber Bridge and sometimes inland. Four were recorded in different Wolds villages in Oct 2015.

Cormorant *Phalacrocorax carbo*

Common winter visitor, mainly coastal but increasing numbers inland and in summer.

Previously only a winter visitor, an inland breeding colony was established at Deeping Lakes in 1992 which rapidly increased to 120 pairs by 1995. The colony has continued but had declined to around 26 nests by 2018. The Grey Heron *Ardea cinerea* heronry that shares the trees on the same island has also declined but whether this decline of both species is due to competition for nest sites, reduced fish availability or persecution from angling interests is not known. There are a handful of other growing colonies in the county, the largest one currently being at the LWT reserve, Woodhall Spa Airfield, that holds over 30 pairs. The wintering population over the five years to 2018-19 has been around 1,400 birds with around 600 inland, 500 on The Wash, especially Gibraltar Point and 300 on the Humber. Summering and wintering birds are widespread in the county and can be encountered on most waterbodies from time to time.

1968-1972	1980-1989	2008-2011	2016-2019	Incidence	Change
				8%	*

Alkborough Flats

Shag *Phalacrocorax aristotelis*

Usually very scarce/scarce winter visitor Sep-May to the coast and inland, with occasional larger influxes. Exceptional in summer.

Shags breed widely around Britain on rocky coasts from Bempton Cliffs northwards. The largest influx recorded was 149 birds in 1993, which included 132 flying south at Mablethorpe on Jan 25th. In the five years to 2018 LBR reported an average of 16-17 birds per year with a range from seven in 2013 to 40 in 2016. Birds were recorded from Sep-May with none in Jun-Aug. The largest single flock was of 13 birds seen at Donna Nook on Jan 1st 2016.

Of 11 birds picked up dead or having their colour rings read in Lincolnshire from 2007 to 2018, seven were ringed as chicks on the Isle of May and four as chicks elsewhere in the Firth of Forth. The eldest was a bird of over 25 years of age ringed as a nestling at Craigleith, Firth of Forth in Jun 1991 and picked up freshly dead at Cleethorpes on Feb 7th 2017. These records give a good indication of where birds seen in Lincolnshire are coming from.

Glossy Ibis *Plegadis falcinellus*

Rare but increasing passage migrant; breeding attempt in 2014.

The first for Lincolnshire was one shot on Read's Island in autumn 1869. There were two more in 1881 and a fourth in 1923. Much later there were records in 1975 and 1976 followed by a 32-year gap before the next, and a breeding attempt in 2014. The breeding attempt came out of the blue and was the first such recorded event in the UK. A bird arrived at Frampton Marsh on Jun 14th 2014 and was joined by a second on Jun 26th. In the period to Jun 29th, display, mating and nest platform building was observed. One bird disappeared and the remaining one stayed until Dec 7th. There has been no repeat performance.

1968-1972	1980-1989	2008-2011	2016-2019	Incidence	Change

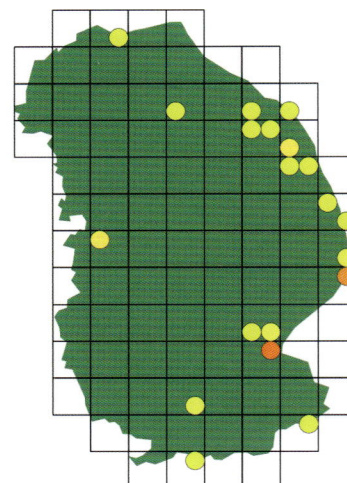

Site Name	First Date	Last Date	Count	Notes
Reads Island	15/09/1869		1	
Skegness	09/09/1881		1	Immature male
Skegness	27/10/1881		1	
Tetney	29/09/1923		1	
Saltfleetby-Theddlethorpe	05/11/1975		1	
Gibraltar Point	16/05/1976		1	
Wisbech STW	16/05/1976	18/05/1976	1	
Donna Nook	30/01/2008	19/03/2008	1	Second winter
Grainthorpe	21/09/2009	27/09/2009	1	
Covenham Reservoir	23/09/2009	26/09/2009	1	
Gibraltar Point	09/03/2012		3	
Saltfleet	09/03/2012	10/03/2012	3	
Caistor	17/03/2012		1	
Gibraltar Point	27/10/2013		2	
Gibraltar Point	02/12/2013		1	
Frampton Marsh	17/12/2013		1	
Deeping St James	31/12/2013	07/01/2014	1	
Chapel St Leonards	10/02/2014	17/02/2014	1	
North Hykeham	12/02/2014		1	
Willow Tree Fen	15/05/2014	25/05/2014	1	
Frampton Marsh	03/06/2014	04/06/2014	1	Same
Willow Tree Fen	07/06/2014		1	Same
Frampton Marsh	12/06/2014	25/06/2014	1	Same
Gibraltar Point	13/06/2014		1	Same
Freiston Shore	14/06/2014		1	Same
Frampton Marsh	26/06/2014	29/06/2014	2	One same
Frampton Marsh	01/07/2014	07/12/2014	1	Same
Whisby	23/08/2014		1	
Frampton Marsh	18/01/2015	19/01/2015	1	Same
Witham Mouth	18/01/2015	19/01/2015	1	Same
Boston	28/01/2016		1	
Carlton & Manby Washlands	12/05/2016		1	
Frampton Marsh	24/10/2016	05/11/2016	1	
Frampton Marsh	15/02/2018		1	
Anderby	22/05/2018	23/05/2018	1	
Freiston Shore	24/05/2018		1	
Gibraltar Point	24/05/2018		1	
Frampton Marsh	26/05/2018	28/05/2018	1	
Whisby	23/07/2018		1	

Spoonbill *Platalea leucorodia*

Scarce but increasing visitor, mainly coastal in Apr-Sep, very scarce in winter.

Nested in the Fens around Crowland in the mid 17th century and probably later. Became extinct as a breeder and was subsequently a rare visitor. Lorand and Atkin (1989) noted eight records in the 19th century of which six were shot. There were 11 birds in the first half of the 20th century up to 1948 and then 80 up to the late 1980s. Since then, it has become an increasingly regular visitor to the coast with birds summering from Apr to Sep. More birds occur in Jul to Sep, related to the post breeding dispersal of breeding colonies in Norfolk, Yorkshire and The Netherlands as demonstrated by the presence of birds colour ringed as nestlings. The late summer population is now approaching 60 birds with Alkborough Flats, Frampton Marsh and Gibraltar Point the most favoured sites. In Apr 2018 surveyors counting an inland mixed Grey Heron *Ardea cinerea* and Little Egret *Egretta garzetta* colony were surprised when at least two Spoonbills flew out in a melee of herons. They hadn't been there in Mar and weren't seen again. No Spoonbills had been seen in the adjacent potential feeding areas. This experience suggests prospective breeding Spoonbills may be stealthier than one might imagine for such a big white bird. Hopefully they will return as a breeder in the next decade.

Winter records were formerly exceptional but from 2013 to 2018 there were eight winter records in six years at four different sites indicating an increasing propensity for Spoonbills to winter.

1968-1972	1980-1989	2008-2011	2016-2019	Incidence	Change

Bittern *Botaurus stellaris*

Very scarce breeder and winter visitor. Bred to early 19th century, from 1940s-79 and in recent years.

Eliminated as a breeding bird by drainage by 1850, it was a rare bird in Lincolnshire until the late 1940s when it began breeding in the Humber Clay Pits. There were up to six booming males until 1979 by which time the reed beds had deteriorated and breeding stopped. A great deal of management work by LWT saw them return to the Humber in 1999 and they have appeared every year since, though fledging success has been poor until very recently.

RBBP records show an average of three boomers per year during the period 2013-17 and fledging has been reported at Far Ings and Alkborough Flats in 2018 and 2019. It remains a very scarce winter visitor and bad winters like 2010 can knock the breeding population back. In the last 10 years peak monthly counts have tended to fall between Nov and Feb and have ranged from six in Jan 2011 to 15 in Jan 2013 with an average of nine.

1968-1972	1980-1989	2008-2011	2016-2019	Incidence	Change

Little Bittern *Ixobrychus minutus*

BBRC

Rare. Western Europe.

One shot near Gainsborough in May 1870 was the first country record and another, also shot in the county "sometime before 1872" was the second. Since then, a further 11 have occurred 1953-2019, in Apr (one), May (four), Jun (four), and Aug (two). The earliest in spring was Apr 19th, the latest Jun 26th. Eight of the records have been of adult males and, unsurprisingly, the autumn birds were an 'immature' (unaged) and a juvenile. Interestingly, the *LNU Transactions* (vol 3, Dec 1954) recorded that the immature bird at Goxhill in Aug 1953 was first flushed from a reed-bed and eventually, while doing its 'freezing' act, was caught, ringed and photographed (although the last has not survived).

This colourful small heron has almost as many British records before 1949 (247) as after, up to 2018 (271). Presumably like many of our more colourful rare birds it offered an irresistible target to the gunmen of the day. It averages around three to four records per year and has occasionally bred. The first occasion was in 1984 in Yorkshire followed by a series of records from the south-west, the first in 2010 when breeding was confirmed at the Ham Wall reserve in Somerset.

The latest publication on rare breeding birds in Britain in 2017 (Holling *et al* 2019) reported one to six pairs at four sites, three in the south-west and one in northern England.

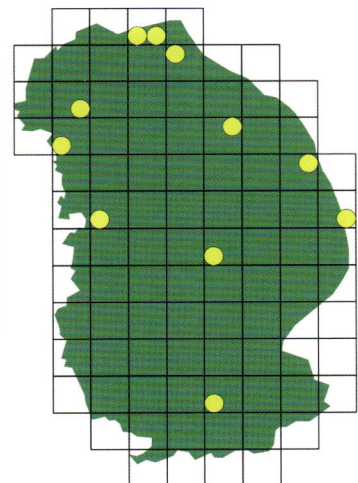

Site Name	First Date	Last Date	Count	Notes
Gainsborough	20/05/1870		1	
Lincolnshire	01/01/1872		1	
Goxhill	12/08/1953	16/08/1953	1	Immature trapped
Chapel St Leonards	21/05/1964	26/05/1964	1	Male
Burton Pits	13/05/1970		1	
Birchwood	26/08/1976	27/08/1976	1	Juvenile
Barton upon Humber	23/06/1977		1	Male
East Halton	19/04/1978	29/04/1978	1	Male
Saltfleetby-Theddlethorpe	11/05/1994	13/05/1994	1	Adult male
Spalding	12/05/2003	13/05/2003	1	Male
Messingham	01/06/2004	02/06/2004	1	Adult male
Kirkby on Bain	28/06/2015		1	Male
North Thoresby	04/06/2019		1	Adult male

Night-heron *Nycticorax nycticorax*

Rare. Western Europe.

An immature bird shot on the foreshore at Tetney in Nov 1888 was the first record, but it was nearly a century later that the second, also an immature, was found along a waterway in Boston in Nov 1973. The third record, another immature, was found at an old brick pit in Skegness in Dec 1979, then later found dead, shot, in Jan 1980. It had been ringed as a nestling near Belyayevka on the Black Sea coast of Ukraine in Jun 1979, some 2,250 km to the east. The total number of county records now stands at 21 involving 19 individuals. Of those that were aged, 10 were juvenile or immature birds, seven were adults. The adults were seen between Mar 25th-Jun 3rd, the immatures were seen Apr-May in the spring and Aug-Jan in the autumn/winter, extreme dates Aug 29th-Dec 30th. Over the years occasional free-flying birds of captive origin have been seen around Britain and one such individual was found roosting in Hawthorns *Crataegus monogyna*, along the Maud Foster Drain in Boston. It had three plastic rings identifying its location as Edinburgh Zoo. This is not included in the county record.

There have been some extraordinary influxes into Britain some years, with the 53 in 1987 and 61 in 1990. The species was ex-BBRC in 2002 by which time there had been 449 records, 1950-2002.

Site Name	First Date	Last Date	Count	Notes
Tetney	26/11/1888		1	Immature
Boston	13/11/1973	19/11/1973	1	Immature
Skegness	30/12/1979	04/01/1980	1	Immature found shot
Saltfleetby-Theddlethorpe	14/05/1983		1	
Goxhill	05/10/1986	12/10/1986	1	Second winter
Gibraltar Point	03/05/1987	04/05/1987	2	Immature
Barrow Haven	03/06/1987		1	Adult
Wainfleet	01/06/1988		1	Adult
Frampton Marsh	25/03/1990	04/04/1990	1	
Dunsby	29/08/2002		1	
Buslingthorpe	13/11/2002		1	First calendar year
North Hykeham	15/04/2006	30/04/2006	1	
Gibraltar Point	19/05/2007		1	Adult
Gibraltar Point	22/05/2007		1	Same adult
Gibraltar Point	23/05/2007		1	Same adult
Gibraltar Point	21/09/2007	17/10/2007	1	Second winter
Barton Pits	05/04/2008		1	
Barton Pits	19/04/2008		1	Same
Saltfleetby-Theddlethorpe	30/05/2009		1	
Holbeach St Marks	29/05/2012		1	
Orby	10/05/2015		1	

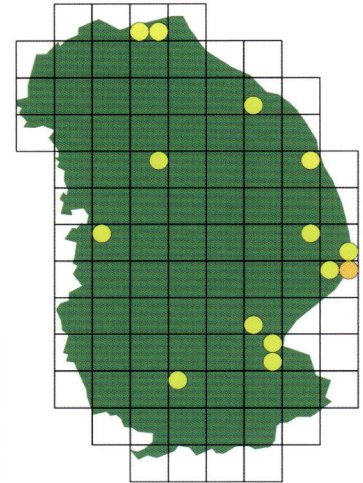

Green Heron *Butorides virescens*

Vagrant. North America.

The discovery of a Green Heron at Messingham SQ was the ornithological event of 2001, present from Sep 24th-Oct 2nd, the only county record, and often very elusive. The bird was discovered when feeding in the open in an area of cut reeds and fortunately it stayed for nine days enabling many local birders and those from further afield to see it.

This was the first live Green Heron in Britain since 1982; a dead bird, probably killed by a fox was discovered in Scotland in 1987.

There have been just eight British records in all to 2018, seven of them since 1950. Records are split evenly between east and west coast and the most recent record turned up in Pembrokeshire on a garden pond in Apr 2018, the first spring record. Whether this had turned up the previous autumn and over-wintered is purely speculative. It stayed for 10 days, not quite as long as the previous record in Cornwall in 2010 present for 57 days after turning up on Oct 6th.

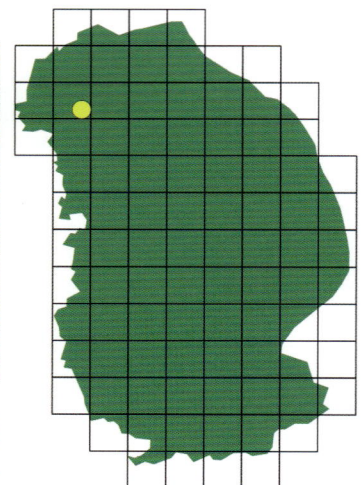

Site Name	First Date	Last Date	Count	Notes
Messingham	24/09/2001	02/10/2001	1	Immature

Saltfleet Haven

Squacco Heron *Ardeola ralloides*

Vagrant. Western Europe.

The rarest of the county's herons after Green Heron *Butorides virescens*; one shot at Fillingham Lake near Gainsborough in Jun 1861 was the first record, with the second, an immature, also shot on the Humber Bank at Great Coates in Sep 1910. There was a large hiatus before the third record of a summer-plumaged adult in Jun 1999, since when three followed in quick succession in 2015 (2CY+), 2016 (2CY) and 2019 (Adult). They have occurred between Jun 1st-Sep 29th with no particular pattern, other than they are a late arriving vagrant in spring.

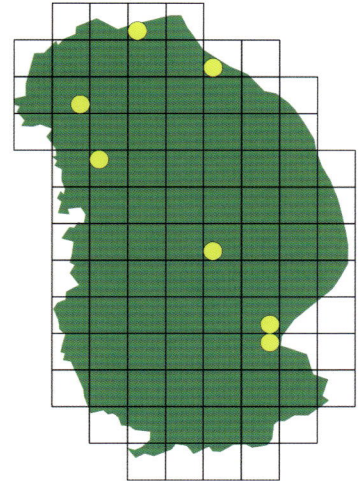

There were 68 British records before 1949 and another 106 in the modern era, 1950-2018. Eight in 2007 and 10 in 2015 are the most recorded in any year. There are about three records per year and the trend is of a slight increase. The bias in records is to the south-west (42%) and the south-east (20%) maybe suggesting an Iberian origin for British vagrants.

Site Name	First Date	Last Date	Count	Notes
Fillingham	01/06/1861		1	
Great Coates	29/09/1910		1	Immature
Messingham	03/06/1999	11/06/1999	1	Adult summer
Kirkby on Bain	27/06/2015		1	
Barton Pits	04/08/2016	09/08/2016	1	Second calendar year
Frampton Marsh	05/07/2019		1	Adult
Freiston Shore	06/07/2019		1	Same adult

Farming in the Wolds, Louth

Cattle Egret *Bubulcus ibis*

Rare/very scarce and increasing. Southern England and Western Europe.

The first record of this globally expanding species was not until 1986 when one was found, appropriately enough, in a cattle field at North Hykeham; it roosted at nearby Apex Pit and stayed for two weeks. The single at Messingham in May 1992 and flock of four at Mablethorpe at the same time were associated with a marked movement into the near continent and southern Britain at the time. Eighteen birds were recorded in Britain that year. Strangely, given the dynamic nature of the Cattle Egret as a species, there were no more until 2008 since when their appearances have accelerated and there have been records in eight of the years 2008-19. They have become much more familiar in the county during this time and with some long stayers in 2016-19 especially.

They became ex-BBRC in 2009 and a huge influx into Britain in winter 2007-08 saw 160 records in 2008 and led to the first successful pair breeding, in Somerset. A similarly huge influx occurred in 2016-17 with nearly 200 in Britain, with 84 in Cornwall alone in Mar 2017. The colonisation of Britain has seemingly begun and in 2019 they bred in three counties for the first time: Essex, Hampshire and Northamptonshire. Given their record of expansionism across the globe, it may only be a matter of time before they are found breeding in Lincolnshire.

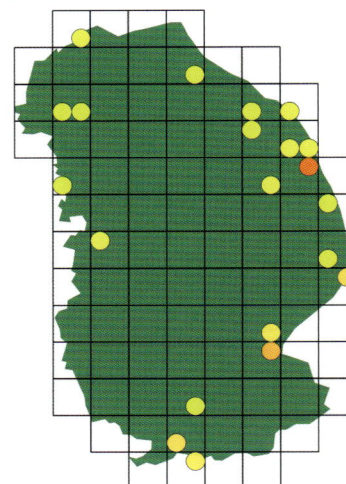

Site Name	First Date	Last Date	Count	Notes
North Hykeham	12/10/1986	25/10/1986	1	
Messingham	06/05/1992	10/05/1992	1	
Mablethorpe	07/05/1992		4	
Saltfleetby-Theddlethorpe	30/01/2008	31/01/2008	1	
Little Cawthorpe	03/02/2008		1	
Legbourne	04/02/2008	18/02/2008	1	Same
Donna Nook	10/10/2010	18/11/2010	1	
Tetney	07/09/2013	27/09/2013	1	
Gibraltar Point	27/09/2013	27/09/2013	1	Same
Susworth	03/08/2014	11/08/2014	1	
Willow Tree Fen	28/11/2014	29/11/2014	1	
Deeping St James	12/01/2016	14/01/2016	1	
Saltfleetby-Theddlethorpe	17/04/2016		1	
Immingham	04/11/2016		1	
Covenham Reservoir	08/11/2016		1	
Skidbrooke	13/11/2016	21/12/2016	2	
Marton	21/02/2017		1	
Saltfleetby-Theddlethorpe	01/03/2017	15/03/2017	1	
Alkborough	19/03/2017	24/03/2017	1	
Gibraltar Point	19/03/2017	20/03/2017	1	
Deeping St James	03/04/2017	16/05/2017	1	
Anderby	16/04/2017		1	
Frampton Marsh	30/05/2017	10/07/2017	2	Adult summer
Baston	24/08/2017		3	
Theddlethorpe St Helen	09/09/2017	13/09/2017	1	
Saltfleetby St Clement	14/09/2017	18/09/2017	2	One same
Saltfleetby-Theddlethorpe	23/09/2017	05/11/2017	3	One same
Freiston Shore	03/09/2018	15/09/2018	1	
Frampton Marsh	08/09/2018		1	Same
Freiston Shore	20/09/2018	23/09/2018	1	Same
Gibraltar Point	22/09/2018	23/09/2018	1	
Frampton Marsh	23/09/2018	09/10/2018	1	
Alkborough	27/09/2018		1	
Baston	12/10/2018	14/10/2018	1	
Frampton Marsh	14/10/2018		1	
Gibraltar Point	26/02/2019		1	
Deeping St James	19/04/2019	19/05/2019	1	Adult
Donna Nook	15/05/2019	19/05/2019	1	Adult
Croft	28/07/2019		1	Adult
Frampton Marsh	07/09/2019		1	
Frampton Marsh	08/09/2019	23/09/2019	1	Same
Freiston Shore	11/09/2019		1	Same
North Hykeham	13/11/2019		1	

Grey Heron *Ardea cinerea*

Fairly common resident, partial migrant and winter visitor.

The status of Grey Heron in Lincolnshire is a cause for concern. LBR 2016 published a chart showing the number of nests counted in the county by the BTO Heronries Census since it began in 1928 at which point around 150 nests were counted. The peak count of more than 400 nests was in 1989-90. Since then there has been a steady decline and in 2019, 168 nests were counted. There seems to be a trend towards more, much smaller Heronries and that may mean some get missed by observers. The chart below shows the Lincolnshire index of Grey Herons counted on BBS. It is not an accurate reflection of their breeding population, but it shows the same worrying long-term decline in birds counted. There are no definite identified causes but adult mortality through persecution is a possibility as is reduced chick productivity caused by food shortages or the impact of poisons in the freshwater food chain. It should be noted the decline is general across England but especially pronounced in Lincolnshire. Ironic, as it is the emblematic bird of the Lincolnshire Bird Club.

1968-1972	1980-1989	2008-2011	2016-2019	Incidence	Change
				20%	-30%

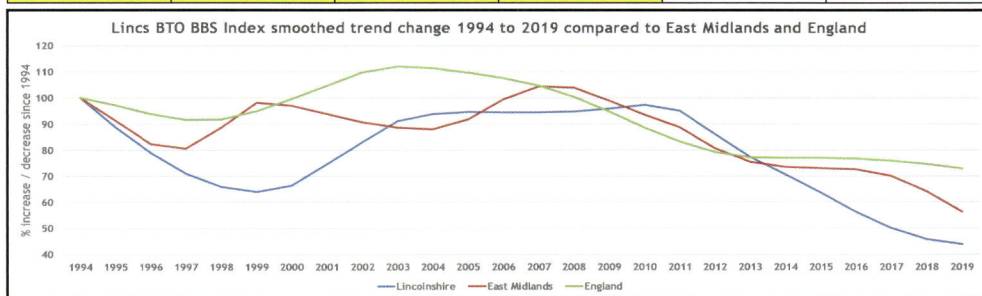

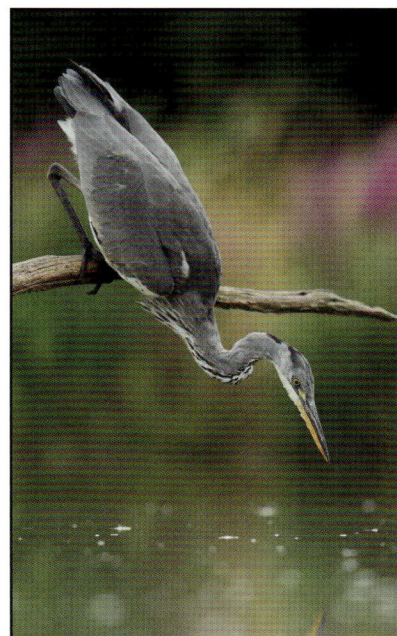

Lincs BTO BBS Index smoothed trend change 1994 to 2019 compared to East Midlands and England

Purple Heron *Ardea purpurea*

Rare. Western Europe.

Two Purple Herons were discovered at a lake on the Brocklesby Estate in Jul 1952 and at least one was reportedly still present Sep 1st. There has been a total of 31 county records in all, with 17 during 2000-19 involving around 15 birds. The earliest spring bird arrived Apr 12th, but the vast majority have turned up May 6th-May 30th (14). There are five Jun records and four in Jul. They are practically unknown in autumn, the sole record being of a juvenile which turned up on Nov 6th 1986 which was later found dead at the end of the month. Most remarkably two juvenile birds were discovered together at Manby Wetlands in Jul 2019. There is no evidence that they were the product of a breeding pair in the county or elsewhere in Britain but having arrived together they may well have come out of the same nest. Occasionally summering birds have been suspected of breeding in Britain, but it wasn't until 2010 that a pair successfully bred at Dungeness in Kent. Present Apr 27th-Aug 23rd, they fledged two young.

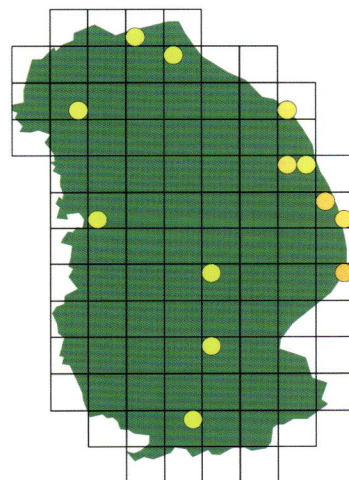

Site Name	First Date	Last Date	Count	Notes
Donna Nook	11/05/1968		1	Immature
Chapel St Leonards	07/05/1970	08/05/1970	1	Adult
Anderby	23/05/1970		1	Same
Huttoft	04/06/1975	06/06/1975	1	
Chapel St Leonards	30/05/1976		1	
Saltfleetby-Theddlethorpe	17/04/1977		1	
Tattershall	19/05/1977	22/05/1977	1	
Chapel St Leonards	12/04/1980		1	
Donna Nook	18/05/1981		1	Immature
Gibraltar Point	26/05/1981		1	
Deeping Fen	06/11/1986	30/11/1986	1	Immature
Gibraltar Point	10/05/1988		1	Sub-adult
Gibraltar Point	12/05/1999		1	Adult
Chapel St Leonards	01/06/2000	03/06/2000	1	Immature
Gibraltar Point	04/06/2000		1	Same adult
Barton Pits	01/07/2001	02/07/2001	1	
Messingham	20/05/2002	21/05/2002	1	
Huttoft	06/05/2003		1	
Gibraltar Point	20/05/2004		1	
Killingholme	08/06/2005		1	Immature
Messingham	17/06/2006		1	
Huttoft	20/06/2006		1	Same adult
Saltfleetby-Theddlethorpe	13/05/2007		1	
Huttoft	20/04/2011	24/04/2011	1	First summer
Bicker	26/05/2011		1	First summer
Boultham Mere	14/05/2017		1	Adult
Donna Nook	21/06/2019		1	Immature
Gibraltar Point	24/07/2019		1	Adult
Carlton & Manby Washlands	30/07/2019		2	Juvenile
Carlton & Manby Washlands	31/07/2019		1	Same juvenile

Great White Egret *Ardea alba*

Scarce but increasing visitor recorded in all months, scarcer Jul-Aug.

The first record was in Aug 1979 at Messingham SQ, when Lorand and Atkin (1989) described the species as 'a very rare vagrant'. Two more records followed in Nov 1983 and Jul 1993 with a lull until after the new millennium. Between 2002-19 there were records in every year; there were two to eight records 2002-10 and thereafter a slow rise in numbers accelerated into an exponential increase. Records reached double figures in 2011 (17), 2014 (14), 2015 (17) and 2016 (19), increasing in 2017 to 28, but rocketing to 54 in 2018 and 72 in 2019. It is extraordinarily difficult to know how many birds are concerned in this very mobile species moving sometimes over large distances along the north and north-east coast, in the Witham valley, the Fens and the southern rivers and water bodies, but perhaps a minimum of 15-20 individuals. It is notable that the first record of two birds together was not until 2015 and in 2018 four were found wintering on the Witham at Woodhall Spa on Mar 4th, later moving to Kirkby GP. Three were at Boultham Mere Aug-Sep 2018, peaking at five birds on Sep 29th; a group of five came in off the sea at Gibraltar Point Oct 31st and four were at Freiston Shore Nov 26th. The number of sites involved has been between 35-54, 2017-19.

Birds have been recorded in the county every month of the year and, given their catholic choice of nest sites (often trees but also reedbeds, artificial nest platforms and even on the ground), and confirmed breeding in neighbouring Norfolk, a breeding attempt may not be far away. They first bred in Britain in 2012, when two pairs fledged four young in Somerset. They have bred every year since to 2018 when 12-18 pairs bred or attempted to breed, including two pairs at one site in Norfolk that raised two young each and three possible pairs in Cambridgeshire. The changing status of Great White Egret in Britain was well covered by Holt (2013) who noted that the source of the increasing numbers in Britain is most likely the Netherlands where there is a well-established breeding population and more than 2,000 wintering.

Little Egret *Egretta garzetta*

BBRC to 1991. Formerly a rare/scarce visitor in all months, a few wintering since 1993-94. Now a fairly common breeding bird and increasingly common resident.

First confirmed record Jul 23rd 1966, the spread of Little Egret since then has been nothing short of phenomenal from a rarity to a fairly common, but fairly local, breeding bird in 25 years. This change was documented in detail by Hyde in LBR 2015. RBBP records show an average of 104 pairs per year bred during the period 2013-17 and these had increased to 118 pairs in 2019, breeding in nine widely dispersed locations both inland and coastal throughout the county. Though primarily a bird of the coast, wintering birds are even more widely dispersed inland and can turn up at unlikely spots in the middle of the Wolds. Birds colour-ringed as nestlings at one egretry in the north-east of the county have been seen in the Channel Islands, Somerset, Surrey, Shropshire, Cheshire, Northumberland, Tyne and Wear and Scotland. Peak numbers are usually noted on the Humber and The Wash in Aug-Sep as fledglings start to move.

1968-1972	1980-1989	2008-2011	2016-2019	Incidence	Change
				4%	55%

Osprey *Pandion haliaetus*

Scarce passage migrant, rare in summer.

The *LBC Atlas* reported that there was an average of eight birds a year in the 1980s, but this had doubled to 16 a year by the 1990s. The successful breeding Osprey reintroduction programme at Rutland Water started in 1996. By 2018 RBBP reports indicated that there were six confirmed pairs breeding in Leicestershire/Rutland and a further three in Northamptonshire. There is no real evidence of prospective breeding in Lincolnshire as yet but increasing numbers of immature birds are summering. In the five years from 2014 to 2018 the number of birds reported in summer (Jun-Jul) was two, one, two, five and seven in successive years. Now may be the time for a sustained programme of nesting platform erection at suitable sites, particularly in the south-west of the county. Though some already exist, they have not proved sufficiently tempting as yet. In this same period earliest spring migrants arrived between Mar 16th-Apr 17th at a rate of around 16 per year. The last autumn migrants departed between Oct 2nd -15th and averaged around 16 per year with autumn 2016 being the peak year with 35. All told, the number of birds per year ranged from 30 in 2014 to 52 in 2016 at an average of 36 per year.

Honey-buzzard *Pernis apivorus*

Very scarce passage migrant, with occasional larger autumn influxes.

Lorand and Atkin (1989) note that historically there had been around 58 records with (inevitably) many falling victim to the gun. Four were shot in Sep 1896 and there were other records in the early 20th century but they remained a rare migrant with one at Tetney in Sep 1967, the first since 1955. There were six more records to 1976 when there were four, one in May and three in Sep. Then five in 1980 were the next records, all Sep-Oct with blank years 1981-85. Since 1986 though there have been only three more blank years, 1990, 2003 and 2018. In most years there have been around five records but with influxes in some years the first of 12 birds in 1993 with the LBR noting that this was "the highest annual total of birds ever". This was utterly eclipsed by the next big year in 2000 when around 97 were recorded including 25 SSW on Sep 20th at Gibraltar Point. A static anticyclone over Scandinavia with strong easterlies along its southern edge was the reason for this and around 500 were reported across Britain. Seventeen occurred in 2006 before another big influx occurred in 2008 involving some 61 birds with just one in May and 60 during Sep 13th-21st. Since then 'normality' has resumed with three to six per year 2009-19. In spring the earliest date was Apr 17th 1973 at Gibraltar Point with most recorded during May and in the autumn dates span Aug 7th-Oct 27th with the vast majority in Sep.

The RBBP reports to date show a five-year mean of around 40 breeding pairs of Honey Buzzards in Britain though point out that the true total may be double this given their very secretive nature. There have been no records of attempted breeding in Lincolnshire up to 2020, but a nationwide survey in 2021 may prove otherwise.

Golden Eagle *Aquila chrysaetos*

Vagrant. Northern Britain/Europe.

Sadly, there is no further update since the account in Lorand and Atkin (1989) with five historic records only. That in 1834 was trapped (outcome unknown) and the second, an immature male, was shot at Normanby Park, Scunthorpe in Nov 1881. An immature bird seen at Langton, Spilsby in Nov 1920 was eventually picked up "in a dazed condition" and sent to London Zoo where its identity was confirmed. Another was seen that year in Maltby Wood, Louth in Dec. Finally, another immature bird frequented Brocklesby Woods from Dec 1927 until mid-Jan 1928, where it fed largely on Rabbits *Oryctolagus cuniculus*, until it was shot nearby at Beelsby. An account by one of the observers of this bird appears in the LNU Transactions of 1929 when he noted that "…with eight fellow naturalists, I had the unusual pleasure of watching the flight of a large Eagle, as it soared leisurely over the Earl of Yarborough's woodlands near Caistor Double Gates". No mention of the optics in use is made, but they saw it at a distance of 100 yards and described the bird as being all dark brown, apart from a large patch on each wing and the basal half of the tail, which were white, an accurate description of an immature Golden Eagle.

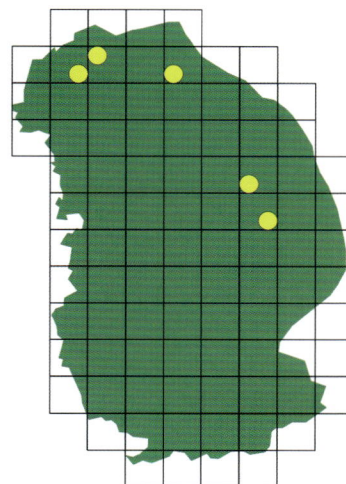

A pair famously bred at Haweswater in the Lake District from 1969 with some turnover of male and female partners and 16 young were produced 1970-96, and a second pair bred 1975-83 fledging four chicks. This micro-population gradually died out and the last remaining bird, a male, failed to appear in spring 2016; the female died in 2004. Adult Golden Eagles stay on territory throughout the year, but immatures wander, and a few have made it into England with an immature in County Durham in the spring of 2002 for example. The prospects of a repeat county record look bleak.

Site Name	First Date	Last Date	Count	Notes
Appleby	01/01/1834		1	
Scunthorpe	01/11/1881		1	
Langton (Spilsby)	10/11/1920		1	
Maltby Wood	04/12/1920		1	
Brocklesby	15/12/1927	15/01/1928	1	

Sparrowhawk *Accipiter nisus*

Common resident and passage migrant. After extinction as a breeder in 1950s, due to pesticides and persecution, recolonised rapidly from the early 1980s.

The return of the Sparrowhawk as a Lincolnshire breeding bird after almost 30 years as a scarce visitor was one of the great success stories of the late 20th and early 21st centuries. In 1987 there were still only breeding records from 16 sites yet within five years the LBR reported no numbers but a widespread breeding population. Numbers appeared to continue to increase to at least 2010 with pairs breeding within large urban areas, in large gardens and isolated copses in open farmland but there may have been a small decrease in the last ten years. An estimated 250-300 pairs were noted in the *LBC Atlas* at the end of the 1980s and the same publication suggested a doubling of this figure by 2000. It is likely that there are now at least 1,500 pairs in the county and probably 2,000. Coastal migration takes place in spring and autumn with recoveries of birds from Scandinavia recorded. Winter roosts of up to five birds have been recorded but this behaviour is seldom documented.

1968-1972	1980-1989	2008-2011	2016-2019	Incidence	Change
				8%	-15%

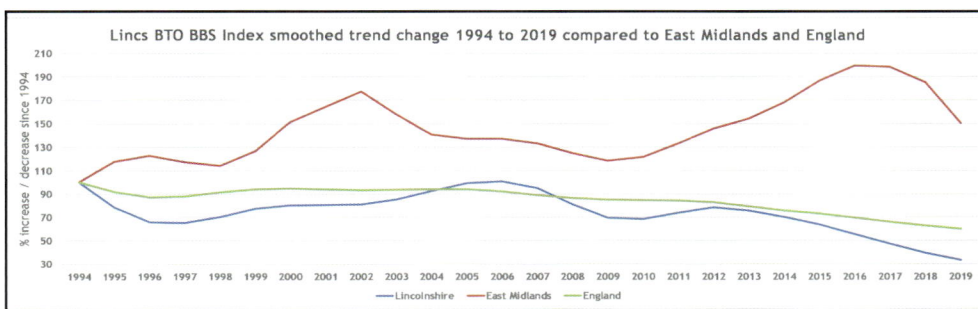

Lincs BTO BBS Index smoothed trend change 1994 to 2019 compared to East Midlands and England

Goshawk *Accipiter gentilis*

Rare visitor/passage migrant, mainly autumn to spring. Bred in 2019 first proven record since 1864

Most records are coastal and a bird ringed as a nestling in Norway on 26th Jun 1994 was trapped by a ringer at Theddlethorpe on 18th Oct 1994. Immature birds tend to range widely and can easily be missed. This was well illustrated by the occurrence of two satellite-tracked birds ringed and tagged as nestlings in Breckland Norfolk. One of these birds turned up briefly in the county in 2018 and the second in 2019, both immature females. Neither bird was seen by a birdwatcher.

With 58 confirmed pairs in adjacent counties: Norfolk (24), Nottinghamshire (12) and Yorkshire (22) in 2018, per the annual RBBP report, it is surprising that until very recently there has been no confirmed breeding in Lincolnshire. A pair was thought to have bred at Burwell Woods during 2009-11 but no nest was seen. In 2019 breeding was reported from two areas of the north-west of the county. In one area a nest was present and recently fledged young were seen but no adults were noted by known observers; the adults may have been of captive origin. Elsewhere, a pair of Goshawks fledged two young; this pair were both 2CY birds, an unusual occurrence in this species. It is to be hoped that these recent successes will pave the way for the return of this magnificent raptor to suitable woodlands across the county.

1968-1972	1980-1989	2008-2011	2016-2019	Incidence	Change
				0%	

Marsh Harrier *Circus aeruginosus*

Fairly common passage migrant and scarce summer visitor, increasing in winter.

Recolonisation took place from 1983, when a pair nested near the Wash, with birds increasing in number and nesting spreading to the Humber in 1992 and the Witham Fens in 1994. Further expansion took birds down other river corridors and into other wetland sites with the national census in 2005 locating 91 nesting females that fledged 236 young in the county. Since then 108 pairs were recorded in 2009 but there appears to have been a decline in confirmed breeding numbers thereafter, but no full census has taken place. Regular monitoring in the north of the county shows a peak in numbers in 2012 with an ongoing decline since then. Wintering numbers have increased to peak at around 50 birds in the mid 2000s with the bulk on the Humber but birds are increasingly regular on the coast and inland. A former large roost at Nocton Fen appears to have been abandoned possibly due to changes in cropping regime. Spring and autumn passage still occur with wing-tagged birds reared in Kent, Suffolk and Norfolk occurring in the county. RBBP records show an average of 45 pairs per year bred during the period 2013-17.

1968-1972	1980-1989	2008-2011	2016-2019	Incidence	Change
				7%	-24%

Hen Harrier *Circus cyaneus*

Scarce passage migrant and winter visitor, mainly Sep-May; rare in summer. Bred to 1872.

This species has become less numerous since the 1990s. Former roosting sites on The Wash used to hold double figures throughout the winter periods. The description of this species as a common breeding bird on the heaths and commons of the north of the county in the early 19th century is best illustrated by the killing of 50 birds in one season on Middle Rasen Moor in the 1820s but there is no verification of such numbers. With no breeding record after 1872 even wintering numbers appeared to be low in the years to c.1970 when a small increase occurred mainly along the coast and Wash with winter roosts holding up to 40 birds in the mid 1970s to early 2000s. Up to 18 birds were recorded from a favoured Wash roost at this time and up to eight at a coastal roost. Inland small roosts develop in some winters if a suitable productive feeding area is available, often the result of high vole numbers in what are often temporary habitats. The number of wintering birds has declined rapidly in the last 15 years due mainly to the persecution of birds on British grouse moors and the county wintering population now is probably less than 30 birds in total. A male summered and displayed at an inland locality for three summers but failed to attract a mate while odd immatures linger through to May and may occasionally summer. Ringed birds from Sweden, the Netherlands, Orkney and at least two wing-tagged birds from Langholm in SW Scotland have been found wintering in the county.

1968-1972	1980-1989	2008-2011	2016-2019	Incidence	Change

Pallid Harrier *Circus macrourus*

BBRC

Vagrant. Eastern Europe.

A 2CY bird at Gibraltar Point in May 2012 was the first county record. The second wasn't far behind, a 1CY bird at Donna Nook in Oct 2015 also seen at Gibraltar Point later in the month and into early Nov. It had been seen at Spurn before arriving at Donna Nook and after its departure it was seen in Norfolk Nov 15th-17th. Two birds were seen in 2017, the first a 2CY bird at Conisholme in May and the second another 1CY bird at Donna Nook in mid-Nov 2017. There were 17 Pallid Harriers in Britain in 2017 and BBRC noted the species' transition from a seldom-encountered vagrant to one of the more regular national rarities. Since the discovery of regular breeding Pallid Harriers in Finland in 2011, they have spread westwards in Europe with a pair breeding in the Netherlands in 2017 and another in Spain in 2019. Two 3CY+ males had extended stays in Britain in 2017, and a breeding attempt in Britain in the near future is within the bounds of possibility

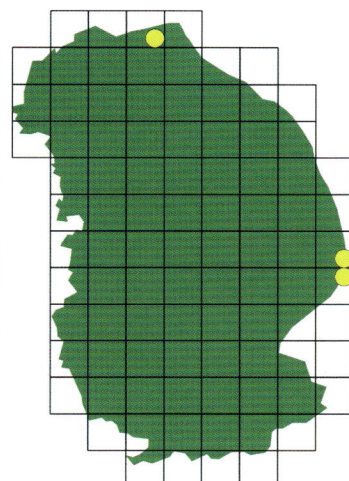

The majority of Pallid Harriers winter in Africa south of the Sahara or in southern Asia, but a few remain in North Africa and the Middle East.

Occasionally a few also overwinter in Europe including Britain: a juvenile was in Norfolk 24th Dec 2002-30th Mar 2003, and another remained in Cornwall from 4th Dec 2009-25th Jan 2010. Interestingly, a juvenile female that wintered in France 27th Nov 2014-7th Mar 2015 was identified from photographs as the same individual that had been at Steart, Somerset (600 km to the north), until 7th Nov 2014. It seems that Pallid Harriers are well able to survive in winter in the absence of locusts, their prime winter prey.

Site Name	First Date	Last Date	Count	Notes
Gibraltar Point	08/05/2012		1	Juvenile
Donna Nook	27/10/2015	28/10/2015	1	First calendar year
Gibraltar Point	29/10/2015	08/11/2015	1	Same first calendar year
North Somercotes	05/05/2016		1	Second calendar year
Donna Nook	08/11/2016	17/11/2016	1	First calendar year

Montagu's Harrier *Circus pygargus*

LBRC

Scarce passage migrant and summer visitor, though breeding irregularly.

After a spate of good years in 2000-05 when an average of five pairs per year bred, RBBP records show an average of 0.4 pairs per year bred during the period 2013-17.

The first Lincolnshire nests were found in 1951 and 1956 but no young fledged; two pairs bred in 1965 fledging two young after which single pairs bred in 1969 and 1980 before more regular breeding began from 1987 but even then nesting attempts were irregular until a small colony became established around the Wash from 2000 to 2011.

The annual maximum was five pairs during 2000-04 with four pairs again in 2010 but the most young reared in a season was only seven. Away from the Wash breeding pairs have appeared at inland sites with birds sometimes returning in the following one or two years but losses on migration mean that single pairs seldom survive at a locality for many years without replacement. The last successful nest in the county was in 2012 when an inland pair fledged four young with the last nesting pair recorded in 2014, the nest being predated soon after the chicks hatched.

The first migrants appear in mid to late Apr and most of the annual records refer to passage migrants many of which are 2CY or 3CY birds. The last records of the year are typically in Sep or early Oct, latest 11th. The number of birds occurring has declined in line with the British population which peaked in 2007. No birds bred in Britain in 2020 for the first time in 40 years.

1968-1972	1980-1989	2008-2011	2016-2019	Incidence	Change
				0%	

Red Kite *Milvus milvus*

Scarce resident and, from 2008, very scarce and now scarce breeder.

This species was not covered by RBBP after 2012, and during the period from 2009 to 2012 an average of 12 pairs per year bred. Numbers have continued to increase. A study tracking the increase in 10km square TF02 (Fleming LBR 2017) showed the scale of the increase in the south-west with a rise from zero annual occupied territories in 2009 to nine in 2013 and 14 in 2017.

The return of the Red Kite to the county as a breeding bird has been a slow process. Expansion out of the core breeding range in the south-west has taken time but odd pairs have bred away from this area in recent years and the Wolds, from where they were extirpated in the 1870s, is being prospected.

It is possible the population now is over 30 pairs. The bulk of the annual records still refer to wandering immatures, the species does not breed in its first summer, with the traditional peaks in Mar-Jun and Sep-Oct being apparent from submitted sightings. The peak coastal count to date is of 31 south at Gib Point May 2017.

Winter roosts have become established in the south-west with peaks of 25 in 2017 and in the Wolds a winter roost in 2018-19 held a peak of 14 birds.

1968-1972	1980-1989	2008-2011	2016-2019	Incidence	Change
				10%	514%

Black Kite *Milvus migrans*

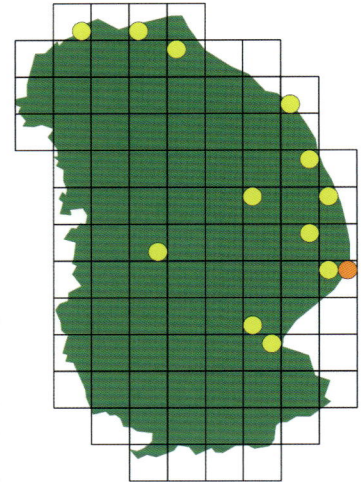

Rare. Western Europe.

The first record for the county was on May 18th 1979, since when there have been about 11 further records allowing for duplications in years when the same individual was presumed seen at several sites. It remains a county rarity with records in only four years since 2000, most recently in May 2019. A bird seen on the Humber bank in Nov 2014 was thought to be an escaped Yellow-billed Kite *M. aegyptius* and is not included in these records.

It is noteworthy that most European Black Kites of the nominate race have largely moved out of Europe by early autumn with peak crossing at Gibraltar in Aug and few in Sep, so any occurring in late autumn in Britain should prompt close examination. This species ceased to be considered by BBRC from Jan 1st 2006 by which time there had been 360 British records; the upsurge in records started 1979-80 when they began breeding in Belgium. BBRC also reminded us that more than 50% of records, 2000-04, have been rejected, female/immature Marsh Harriers *Circus aeroginosus* being the bogey bird.

Site Name	First Date	Last Date	Count	Notes
East Halton	18/05/1979		1	
Burgh le Marsh	14/04/1988		1	
Gibraltar Point	17/04/1988		1	
Croft	17/04/1988		1	Same
Gibraltar Point	25/04/1988		1	Same
Gibraltar Point	01/05/1988		1	Same
Gibraltar Point	12/05/1988	13/05/1988	1	Same
Wainfleet	12/05/1988		1	Same
Frampton Marsh	09/06/1989		1	
Gibraltar Point	30/04/1993		1	
Gibraltar Point	16/05/1993		1	
Boston	04/06/1994		1	
Barton Pits	21/05/2007		1	
Huttoft	24/05/2007		1	Same
Saltfleetby-Theddlethorpe	24/05/2007		1	Same
Nocton	16/07/2007	07/11/2007	1	Same
Farforth	23/04/2011	24/04/2011	1	
Gibraltar Point	03/05/2011		1	
Gibraltar Point	12/05/2011	13/05/2011	1	Same
Gibraltar Point	31/05/2011		1	Same
Donna Nook	15/04/2015		1	
Alkborough	03/05/2019		1	Adult

A juvenile Black Kite of an as yet unassigned sub-species was found at Butterwick Marsh-Freiston Shore in Nov 2006. It wandered around adjacent areas of the Wash and inland fields and was seen at Benington, Holbeach Marsh, Welland Mouth and Frampton Marsh. It spent the winter along the north Norfolk coast and was then seen at Gibraltar Point Feb 2017 before flying back to Norfolk and was last seen mid-Apr. It was thought to be of the eastern race of Black-eared Kite *M. m. lineatus* but the taxonomy of the eastern Black Kites is still in flux and this record awaits further adjudication.

Site Name	First Date	Last Date	Count	Notes
Freiston Shore	02/11/2006	21/11/2006	1	Juvenile
Gibraltar Point	17/02/2007		1	Same
Nene Mouth	12/04/2007		1	Same

Pye's Hall, Donna Nook

White-tailed Eagle *Haliaeetus albicilla*

Very scarce visitor. Breeds in Scotland following reintroduction in 1975.

Lorand and Atkin (1989) list 17 confirmed records of White-tailed Eagles. The first was a bird 'obtained' at Nocton in Jan 1732 and there were another 15 from 1819-1933. These historical records were as follows: Ancholme Carrs 1819; Great Coates autumn 1862; immature male, shot, Edenham, Nov 2nd 1883; Beesby, Dec 8th-17th 1896; immature, shot, North Somercotes Oct 10th, 1902; immature, Grainsby Park, Feb 27th-28th 1904; one at Manton Warren, Brigg Feb 9th 1916; one shot, Norton Place, Gainsborough mid-Nov 1916; immature, Grainsby Park, Feb 28th-Mar 2nd 1904, and another there Dec 25th 1920; one shot, Kirton Marsh, Dec 1921; one at Grainsby, late Feb 1923; Skegness, Jan 1925; one in-off the sea, Grainthorpe, Jan 5th 1927, later seen in the vicinity late Feb and early Mar 1927; immature male, shot, Aswarby, Mar 16th, 1933.

An immature bird, the first record of the modern era and 17th overall, flew south along the shoreline at Grainthorpe on Oct 27th, 1985 and was seen later that day at Saltfleetby-Theddlethorpe and Gibraltar Point.

A 2CY+ bird in 1989 was present at Revesby Reservoir from May 12th-17th at least, perhaps longer, and was presumed to be the bird seen over Manton and Scawby Woods on 21st and later that day at Brigg sugar beet factory, and Scawby Park on May 22nd.

A probable 1CY bird which had flown south at Spurn on Oct 23rd, 1990 was seen to make landfall in the Donna Nook area and was presumably the bird seen over Wyberton and Frampton Marsh on Oct 27th, and which later wintered in the Horsey area of Norfolk.

In 2005, an immature bird was first seen in the Mablethorpe area Feb 11th, later at Oxcombe Feb 21st-27th, and then flew to Kirkby-on-Bain before heading west into Derbyshire. It returned to the county on Mar 8th and was seen at Laughton village before the last sighting at Humberston Yacht club pools on Mar 23rd.

One in Mar 2011, which had wintered in Hampshire 2010-11 was seen over Burwell Wood, Louth and in the Farforth-Ruckland area on Apr 3rd before drifting across to Brancaster, Norfolk later the same day. It returned via Gibraltar Point on Apr 6th and was back at Ruckland Apr 10-15th. It flew over to Yorkshire on 18th, was back on the coast at Grainthorpe on 22nd-23rd before returning to Ruckland on 24th where it remained until Aug 7th, where it may have met its end.

An adult soared high over Gibraltar Point on Apr 25th 2015 before disappearing to the west. The last and 23rd record was found on the coast on Mar 23rd 2018 by two brave souls who ventured out during the severe weather popularly christened the "Beast from the East".

There have also been records of released, satellite-tracked birds in the county in 2010 (one) and 2020 (two), at least. These are not included in the county totals.

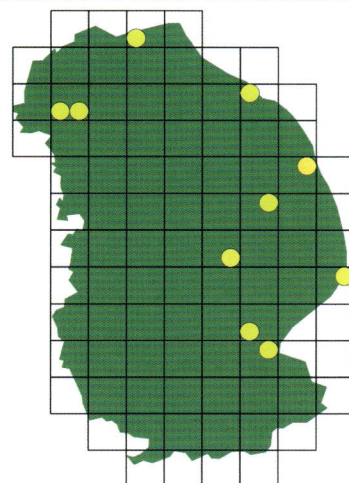

Rough-legged Buzzard *Buteo lagopus*

Very scarce/scarce coastal passage migrant and winter visitor, chiefly Oct-Nov and occasionally Dec-May.

In some years Rough-legged Buzzards may be absent, but influxes are noted at irregular and sometimes long intervals. Historically, influxes were recorded in 1839, 1875, 1880, 1891, 1903, 1915, 1962, 1966, 1973 and 1974 (Lorand and Atkin, 1989). In more contemporary times, larger numbers than usual were seen in 1994 (17), 1998 (14), 2010 (16), 2011 (16), 2014 (16) and 2015 (10).

The pattern of Rough-legged Buzzard seasonal occurrence is one of autumn arrivals, mainly in Oct and early Nov, with some new arrivals or birds remaining in diminishing numbers through to Mar, followed by a small pulse of spring passage in Apr and into May. One of the earliest autumn records was a bird found dead at Ulceby on 15th Sep 1977; it was just over four years of age and had been ringed as a nestling in Sweden on Jul 6th 1973. An unusual spring record was of three birds together near Caistor between mid-Mar and mid-Apr 1975 which were observed engaging in twig-carrying and tumbling displays. The latest recorded in spring was a 2CY bird at Laughton Forest on May 17th 1996.

The annual totals vary greatly, and there were several years in the 1980s and 1990s with none. Historically, larger arrivals in Britain were associated with low Arctic Lemming *Dicrostonyx torquatus* numbers in the Arctic but it is not clear whether that association persists or is very strong. There are also two factors that affect assessment of the numbers occurring. First, it can be difficult to know whether individuals seen at different sites or intermittently are the same or not. Second, especially with increasing numbers of pale Common Buzzards *B. buteo*, there is concern that misidentifications may be made by less experienced observers. Records during 2016-19 since the last influx in 2015 have remained low with five, one, four and one record, respectively. Another good year is well overdue.

Buzzard *Buteo buteo*

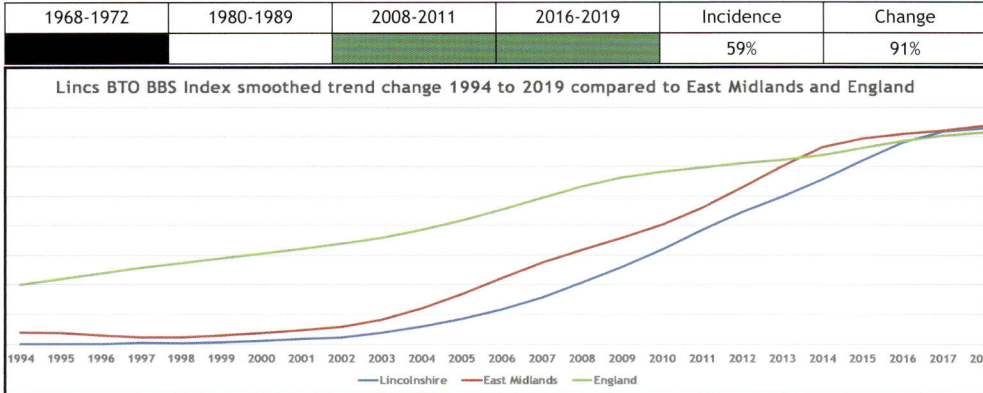

Bred to 19th century before local extirpation. Now common breeder, passage migrant and winter visitor.

Common Buzzard was still a scarce migrant and occasional winter visitor in the 1970s but birds spread into the county from the late 1980s. Breeding was proven in at least two areas in 1996 since when the increase and spread has been nothing short of phenomenal and it is now the commonest breeding raptor in the county. Numbers are hard to estimate due to the presence of a large stock of immature non-breeders but as an example in the Laughton Forest none bred in 2000 but by 2012 there were 21 occupied territories with ten pairs fledging 14 young. A study tracking the increase in 10km square TF02 (Fleming, LBR 2017) showed the scale of the increase in the south-west with a rise from 19 annual occupied territories in 2003 to 59 in 2017. Given these sorts of numbers it is entirely possible that the county population now exceeds 1,500 occupied territories, an incredible increase from less than 10 in 1996. Pairs now breed throughout the county even in the Fens and open arable areas with few trees and are moving into more urban habitats. Persecution remains at low levels. With such a high breeding population and surplus of non-breeding birds it is difficult to detect movements of birds, but some coastal passage is still noted. Pale plumaged birds have become more numerous in the breeding population as numbers have increased. The chart below shows the Lincolnshire and East Midlands BBS index rebased against the BBS England trend. By 2018 population growth in Lincolnshire and East Midlands had been such that the incidence of Buzzard and the number of birds counted was virtually identical to the mean of England as a whole.

1968-1972	1980-1989	2008-2011	2016-2019	Incidence	Change
				59%	91%

Lincs BTO BBS Index smoothed trend change 1994 to 2019 compared to East Midlands and England

— Lincolnshire — East Midlands — England

Barn Owl *Tyto alba*

Common and widespread resident with good breeding years coinciding with peaks in the vole population. Dark-breasted Barn Owl *T. a. guttata* is a vagrant.

The Barn Owl is a characteristic bird of the Lincolnshire farming landscape. The *BTO Atlas 2007-11* shows it present in virtually every 10km square in the county in both breeding season and winter. A glance at the abundance map in this volume makes it clear that Lincolnshire is one of the most important parts of Britain for this species. The *LBC Atlas* estimated that there were 400-450 pairs in the county in the late 1980s, remarkably consistent with a survey in 1932 that estimated around 410. Its nocturnal habits make it difficult to monitor them with BBS type surveys but its ready adoption of nest boxes for breeding and roosting make it an ideal species for ringers to study. In Lincolnshire there are several ringing groups who have erected around two thousand Barn Owl boxes across the county. This helps ensure Barn Owls are not limited by lack of breeding sites as old farm buildings and old trees with holes are lost. The *LBC Atlas* estimated that around one third of pairs were nesting in tree holes during the 1980s; it is now thought 85% of nests are in boxes. Nest box studies show that the numbers of pairs nesting is cyclic following the breeding success of vole numbers. Poor spring weather also plays an important role. LBR reports that between 2011-19 the number of broods ringed by one ringing team who monitor 700 boxes, ranged from 77 in 2015 to 394 in 2014, with an average of 254 per year. The current population extrapolated from this is 1,200 pairs or 10% of the British population, (Bob Sheppard pers. com.). Clutch size varies enormously too, from one to 12. With thousands of chicks ringed many are recovered, sadly, usually through road casualties. Around 97% are recovered within 100 km of their nest site. Interestingly, a bird ringed in Belgium and one in Netherlands of the nominate *T. a. alba* race which breeds in Britain and Western Europe have been found in the county in 2009 and 2016 respectively.

Dark-breasted Barn Owl *T. a. guttata* which breeds in central and eastern Europe is a vagrant. Lorand and Atkin (1989) reported seven records between 1962 and 1979 but since then there has been just one, a bird picked up dead at Goxhill on Nov 20th 2009.

1968-1972	1980-1989	2008-2011	2016-2019	Incidence	Change
				10%	-38%

Scops Owl *Otus scops*

Vagrant. Southern Europe.

There have been no further records, unsurprisingly, since Lorand and Atkin (1989) published details of what remain the only two records for the county. The bird 'obtained' at Dunston in 1895 became a specimen in Lincoln Museum at the time. The other record was of a dead bird found in a very emaciated condition at Saltfleetby-Theddlethorpe in Apr 1977, its fate thereafter is unknown.

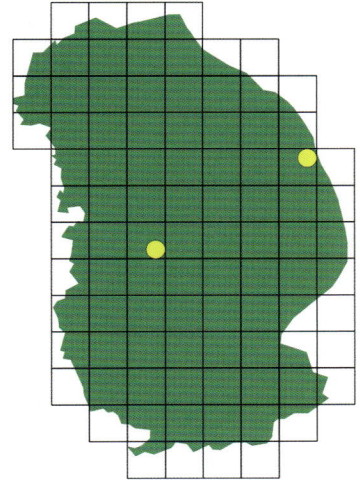

There have been 44 British records, 41 of them 1950-2018, 13 of them since 2000; it is still a very sought-after bird. The most recent British record was a long-stayer in County Durham, Sep 27th-6th Oct 2017, and in fact only the second for the north-eastern sector of England along with the 1977 Saltfleetby record.

Autumn birds are especially rare, and the numbers and distribution of British records suggest it is overlooked in that season. They are a long-distance migrant with a European population estimate of 227,000-381,000 pairs (Birdlife International 2020), so surely, we can be optimistic for one in the coastal scrub in the near future?

Site Name	First Date	Last Date	Count	Notes
Dunston	01/01/1895		1	Obtained in 1895
Saltfleetby-Theddlethorpe	07/04/1977		1	

Snowy Owl *Bubo scandiacus*

Vagrant. Arctic.

A 1CY male seen on farmland at Thornton Curtis on Dec 13th 1990 was the first confirmed county record, staying for two hours before the attention of the local Great Black-backed Gulls *Larus marinus* became too much and it flew off. It was not relocated in the vicinity despite several days of searching but a few days after on Dec 17th, it was reported being mobbed by gulls on allotments on the outskirts of Cleethorpes before flying off to the south-west. It was then reported again, by a farmer on his fields at Wainfleet on Dec 24th. It stayed for nearly three months, dining on hares and was last seen on Mar 18th 1991. It was subsequently seen in north Norfolk Mar 23rd-25th and then briefly at Spurn on Mar 30th before flying off north.

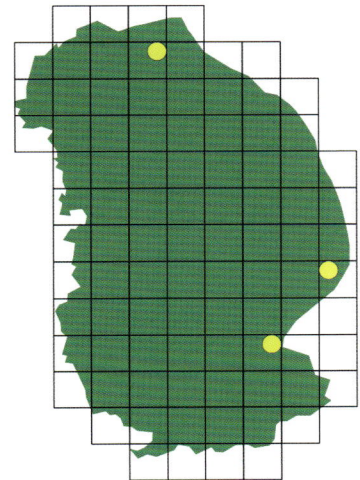

Occurrences of Snowy Owls south of Shetland and mainland Scotland are infrequent and the last English record at that time was in Nov 1981, an adult female in county Durham.

Perhaps an unlikely species to turn up in the county, but lightning struck twice when a 2CY female was found in Mar 2018, and again at Wainfleet, after being seen in north Norfolk earlier in the month. The bird was present Mar 21st-22nd and then promptly disappeared, presumably into the wide-open spaces of The Wash saltmarshes, as what was presumed to be the same bird was seen at Frampton Marsh and Freiston Shore May 1st-10th 2018. The year 2018 was an unusually good one for Snowy Owls with 11 records. During 1950-2018, the British total stood at 235 records. Snowy Owls bred on Fetlar 1967-75; events not yet repeated.

The origins of British Snowy Owls remain unknown as they are rare in Europe with few breeding in Scandinavia. The westerly bias to records suggests a Nearctic origin and there are several instances of these birds making use of ships.

Site Name	First Date	Last Date	Count	Notes
Thornton Curtis	13/12/1990		1	Immature male
Wainfleet	24/12/1990	18/03/1991	1	Same immature male
Wainfleet	21/03/2018	22/03/2018	1	Second calendar year female
Freiston Shore	01/05/2018	10/05/2018	1	Same second calendar year female

Tawny Owl *Strix aluco*

Common resident, widespread but predominantly in wooded areas.

Always thought of as having a very stable breeding population, Tawny Owls are reckoned to have declined over the last couple of decades, hence BoCC Amber conservation status from 2015.

The results of an extensive BTO survey from 2018-19 are awaited and as yet the causes of the decline are uncertain. A ringing scheme based in South Kesteven monitors a constant number of nest boxes. Reports in LBR indicate that between 2011 and 2019 the numbers of nests in the boxes ranged from six in 2015 to 60 in 2014 with an average of 27. The total number of chicks ringed each year ranged from only five in 2018 to 112 in 2014.

The extremely poor productivity in 2018 was attributed to the "Beast from the East" weather event in late Feb early Mar of that year. In some years many and variable numbers of pairs of Tawny Owls choose not to nest at all suggesting that food availability may be part of the problem in the recently observed decline. It is thought that global warming may be dampening down vole population fluctuations and the lack of food dissuades an increasingly ageing population from breeding, leading to long term gradual decline.

1968-1972	1980-1989	2008-2011	2016-2019	Incidence	Change
				4%	176%

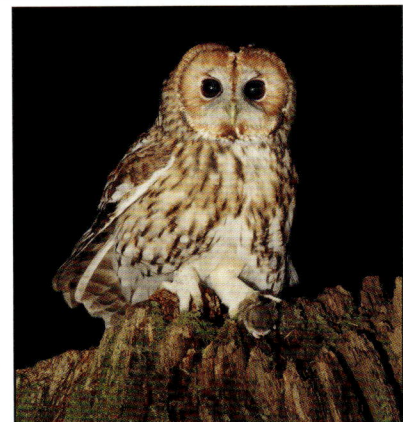

Little Owl *Athene noctua*

Common resident, very sedentary.

Little Owls were introduced to Britain in the 1870s and have only bred in Lincolnshire for just over 100 years. During the *LBC Atlas* period of the 1980s the population was estimated at 300-400 pairs. It has remained widely, but more thinly spread than Barn Owl *Tyto alba*. It is an engaging little bird beloved of birders. With a catholic diet embracing small mammals, birds, insects and worms the Little Owl population is much less vulnerable to food availability cycles than small mammal specialist owls. A ringing nest-box scheme operated in Lincolnshire by Alan Ball and Bob Sheppard is thought to be the largest in Western Europe. LBR reports show that between 2011-19 the number of broods ringed ranged from 59 in 2012 to 91 in 2018 with an average of 75 per year. Most recoveries are from within the county. Despite detailed specialist knowledge held it is not thought the current Little Owl population in the county can be accurately estimated on present information. Interestingly, no Little Owl ringed in Britain has ever been recovered outside the country and none ringed in Western Europe have been recovered in Britain. This suggests that left to nature Little Owls would be missing from our landscape and unable to exploit the obvious niche available to them.

1968-1972	1980-1989	2008-2011	2016-2019	Incidence	Change
				4%	49%

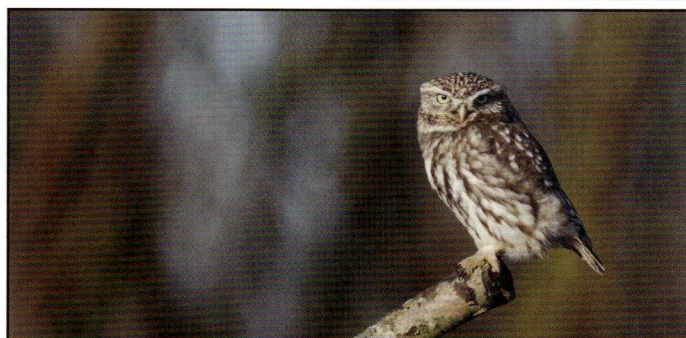

Tengmalm's Owl *Aegolius funereus*

BBRC

Vagrant. Northern Europe.

An adult was shot on the dunes north of Saltfleet Haven on Oct 22nd 1880. One can only imagine the mentality of the person whose first reaction on spotting this magnificent small Owl was to shoot it. The bird was described thus by Cordeaux (1880); It is a mature bird, the plumage somewhat injured in shooting; sex undetermined. This bird must have arrived with the same winds (N and NW) which brought the great flight of Woodcock *Scolopax rusticola* on our coast from Flamborough to Cromer on the night of Oct 18th and morning of the 19th and again on 22nd, along with large numbers of Short-eared Owl *Asio flammeus* and other immigrants. One wonders how many Woodcock have arrived on the Lincolnshire coast since then with apparently no vagrant owls.

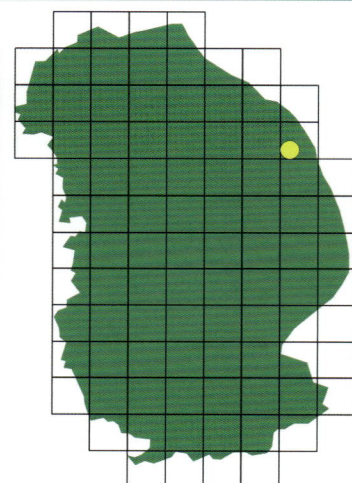

To label this as a county "vagrant" seems a little understated and given its great rarity countrywide it's anyone's guess as to whether we will see another one in the county. Pre-1950 there have been 51 records, most of them shot or 'obtained', the first being in 1812 in Northumberland. There have only been nine records since 1950 though, with six of them in the Orkney Islands. The most recent one on Shetland in Feb 2019 understandably attracted a lot of attention after the short-lived stay of the bird on Copinsay, Orkney in Nov 2018 which was not twitchable. If only the unannounced Spurn Point bird in Mar 1983, whose occurrence was extensively discussed by Roadhouse (2016), had managed to slip across the Humber.

Site Name	First Date	Last Date	Count	Notes
Saltfleet	22/10/1880		1	Adult

Long-eared Owl *Asio otus*

Very scarce and declining breeding bird in county and scarce winter visitor.

It is quite likely that Long-eared Owl is currently the most under-recorded bird in Lincolnshire. RBBP records show an average of two pairs per year bred during the period 2013-17. The *LBC Atlas* estimated the population at 50-100 pairs in the 1980s. It is possible that they have significantly declined, but it is equally possible they are being missed and/or under-reported. The woodland area of Lincolnshire has changed little over the last 40 years, though "tidying up" has resulted in fewer big old scrubby hawthorns being available as nesting and roosting sites and it may be that less early-stage conifer plantation is available. With green conservation status, if there has been any decline in Lincolnshire it is not being repeated across the country.

The numbers wintering in the county have declined too. The *LBC Atlas* reported a summed winter peak in the 1990s of 106 birds. LBR reports for the five years to 2018 show a range from three in 2014 and 2018 to 11 in 2016, with an average of five to six. Many former well-known winter roost sites are no longer occupied. Birds ringed in The Netherlands (one) and Germany (one) have been recovered in the county, while birds ringed in the county have been recovered in Scandinavia (two) and Germany (one). Spring and autumn migration averaged out at two and six birds per annum respectively. It is possible that wintering and passage has been affected by climate change but there is no firm evidence to support that conjecture. All in all, a grim picture.

1968-1972	1980-1989	2008-2011	2016-2019	Incidence	Change
				0%	

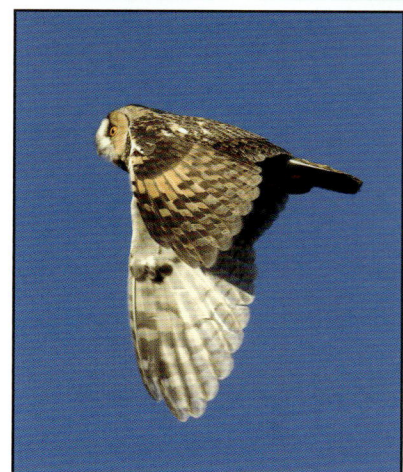

Short-eared Owl *Asio flammeus*

Scarce winter visitor and occasional breeder.

Recent advances in radio-tracking have shown that Short-eared Owls can roam large distances in different directions across Europe looking for suitable feeding habitat where rodents are abundant. These owls are perhaps the original European migrant moving between countries to breed depending on prey availability. Sadly, over the last five years it would appear that Lincolnshire has become less attractive for Short-eared Owls. RBBP records show an average of just one pair per year bred during the period 2013-17. It has never been a big breeder in Lincolnshire and back in the 1980s the *LBC Atlas* indicated there were only up to 10 pairs per year, declining to five a year in the 1990s. Even as far back as the 1860s, Cordeaux (1872) considered them a winter visitor. LBR reports for the seven years from 2012 to 2018 show that total peak counts across the county ranged from 24 in Nov 2014 to 69 in Apr 2012, with an average of 44. Interestingly, the peak count can arise in any month from Oct-May indicating the flexibility of Short-eared Owl movements. The largest aggregations were 13 observed at Worlaby Carrs in Jan 2013 and 12 at the same site in Nov-Dec 2012. The next largest was 10 at Hawthorpe in Jan 2017.

1968-1972	1980-1989	2008-2011	2016-2019	Incidence	Change
				0.6%	

Hoopoe *Upupa epops*

Very scarce migrant, mainly Apr-Oct, rarely to Dec.

Lorand and Atkin (1989) noted that there had been around 125 records of Hoopoe in the county up to that date. Since then, birds have continued to occur in regular but small numbers, mainly on the coast but also at inland locations, often in private gardens where they are usually noticed probing for invertebrates on areas of lawn. One or two Hoopoes per year was the average number found from 2000-19. They are more common in spring than in autumn. Analysis of all records since 1950 shows that most occur Apr 21st-May 10th with the spring peak being the first 10 days of May. There is a long tail-off through Jun and into the first half of Jul with no records between Jul 13th-Aug 15th. The extreme arrival dates since 1950 are Apr 9th 1958 (Denton) and Nov 11th 2008 (North Somercotes). The autumn occurrences are spread over a much wider date range with fewer birds. Four birds in 2007 and 2014 and five in 2008 being the most recorded since 2000. Most records have been of single birds but two have been seen together in 1957, 1977 and 1980, all on the coast in spring; there has been no suggestion of any breeding attempts.

Roller *Coracias garrulus*

Vagrant. Southern and eastern Europe.

An immature shot at Keddington near Louth in Oct 1863 was the first county record and those birds found in May 1871, 1900 (no specific date) and Jun 1962 were all shot; the one near Louth in Aug-Sep 1901 apparently escaped the gun. The immature bird in Oct 1983 obligingly stayed for 18 days and was the last chance county birders had to catch up with this increasingly rare visitor to Britain. The BBRC report for 1983 noted that 17 of the 49 Rollers recorded in the 15 years to 1983 had been in Scotland; of the 12 in Sep-Oct in that period, seven had been in Scotland and the other five were on the east coast, suggesting that autumn Rollers in Britain come from the east rather than from the south.

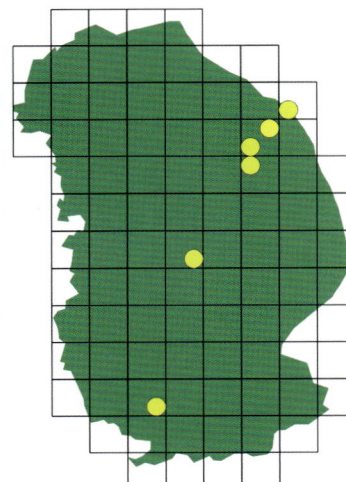

Rollers have declined across their European breeding range over the last 60 years, with countries such as Germany, Switzerland, Denmark, Finland and the Czech Republic, losing their breeding populations. This decline is reflected in the numbers recorded in Britain, and from 2000-18 there have been just 22 British records. A third of all British records have occurred in Jun, with a spread from Mar to Oct.

Site Name	First Date	Last Date	Count	Notes
Louth	01/10/1863		1	Immature
Elsthorpe	10/05/1871		1	Male
Marshchapel	01/01/1900		1	
Brackenborough	29/08/1901	26/09/1901	1	
Grainthorpe	13/06/1962		1	
Woodhall Spa	02/10/1983	19/10/1983	1	First winter
Donna Nook	28/05/2008		1	

Kingfisher *Alcedo atthis*

Fairly common resident and partial migrant. Gradually increasing but vulnerable to hard winters.

The *LBC Atlas* put the Lincolnshire population at around 60-80 pairs in the late 1980s with a wintering population of 60 birds. It is a thinly-spread bird and although there are many miles of apparently suitable habitat along drains the 60-degree grading of most banks is unsuitable for breeding Kingfishers which prefer a vertical cliff to nest in. Such sites offer better protection from predators. Thus, natural streams and rivers and gravel pits are favoured. The BBS does not pick up Kingfishers in sufficient numbers to produce an index for the county. The long-term UK trend from 1994-2019 is down 12%. The *BTO Atlas 2007-11* showed a slight net increase in the number of 10km squares occupied. The estimated Lincolnshire population was a minimum of 180 pairs in 2016. These numbers look unrealistically high, even though with no hard winters since 2010-11 the population will have had a chance to grow. Information that suggests the population may be lower comes from WeBS Online, which reports a rolling five-year mean of only 40 birds per winter over the period to 2018-19. Data from LBR 2017 indicates a peak monthly count of 60 in Aug 2017 was a high one so it is difficult to be certain about the overall situation currently.

1968-1972	1980-1989	2008-2011	2016-2019	Incidence	Change
				1%	51%

Blue-cheeked Bee-eater *Merops persicus*

BBRC

Vagrant. Central Asia.

One county record in Jul 1989. The bird flew south along the west side of the Wash stopping briefly on wires by the seawall. Seen by two observers, it was presumed to be the bird present for three days at Cowden, East Yorkshire two days earlier, where it was seen by multitudes and was at the time reckoned to be the ornithological event of the year. In the same summer two others graced these shores, one in Jun in Cornwall and another later in Jul in Kent. These three records were thought to have been the product of a small influx rather than the successful tracking of just one bird through mid-summer Britain.

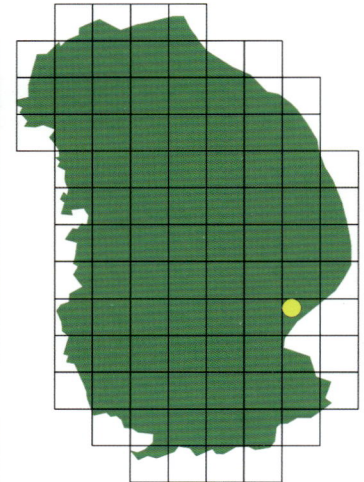

Blue-cheeked Bee-eater was added to the British List in 2003 on the basis of a record from 1921. *M. persicus* is polytypic, the race accounting for British records being the nominate form *persicus*. Since 1950 there have been just nine British records of this exotic Asian visitor with the last occasion being 2009 when there were two.

Site Name	First Date	Last Date	Count	Notes
Leverton	12/07/1989		1	

Greenshank Creek, Gibraltar Point

Bee-eater *Merops apiaster*

Very scarce migrant. Mediterranean Europe.

Predictably the early records of this species involved birds shot in 1879 and 1880. There is a gap in the historical archive with no further records until one at Gibraltar Point in Sept 1957. Since then there have been another 39 records involving around 89 birds. Although there were 26 records of single birds there were also flocks of five in 2002, six in 2005 and 10 in 2015. There was a breeding attempt in nearby Nottinghamshire in 2017 but there have been no signs of breeding in Lincolnshire.

Bee-eater was devolved from BBRC in 1991, a reflection of the large number of records involved with some 369 individuals logged between 1950-90. The species has since been reported in the Reports on Scarce Migrant Birds in Britain series, published in *British Birds*. In the most recent one, White and Kehoe (2019) the annual mean numbers occurring in Britain has been shown to have risen over successive time periods from 20 in 1980-89 to 75 in 2010-17.

Site Name	First Date	Last Date	Count	Notes
Grantham	01/07/1872		1	
Tetney	15/08/1880	16/08/1880	1	
Gibraltar Point	02/09/1957		1	
Gibraltar Point	02/09/1958		3	
Barrow upon Humber	29/04/1968		1	
Gibraltar Point	10/06/1973		2	
Gibraltar Point	26/08/1985	29/08/1985	2	
Gibraltar Point	28/08/1985		1	Juvenile
Gibraltar Point	05/05/1987		3	
Holbeach St Matthew	13/06/1987		3	
Saltfleet	19/08/1987		4	
Tetney	26/05/1988		1	
Friskney	31/07/1988		1	
Kexby	07/08/1988		1	
Gibraltar Point	27/06/1991		1	
Gibraltar Point	29/05/1997		1	
Gibraltar Point	21/05/1998		1	Adult
Gibraltar Point	07/05/2000		3	
Saltfleetby-Theddlethorpe	25/06/2002		3	
Ewerby	03/07/2002		5	
Gibraltar Point	03/08/2003		1	
Gibraltar Point	22/05/2005		1	
Chapel St Leonards	21/05/2006		1	
Gibraltar Point	31/05/2009		1	
Saltfleetby-Theddlethorpe	23/05/2010		3	
Gibraltar Point	29/05/2010		1	
Scremby	14/04/2011	15/04/2011	2	
Goxhill	20/05/2012		1	
Scartho	24/05/2012		1	Same
Ancaster	12/05/2015		2	
Alkborough	15/05/2015		2	
Chambers Farm Wood	05/06/2015		6	
Saltfleetby-Theddlethorpe	05/06/2015		1	
Tetney	16/06/2015		10	
Dowsby	19/06/2015		1	
Gibraltar Point	01/08/2015		2	
Alkborough	05/08/2015		1	
Gibraltar Point	07/08/2015		1	
Gibraltar Point	13/05/2017		1	
Gibraltar Point	22/05/2017	23/05/2017	1	Same
Anderby	23/05/2017		1	Same
Gibraltar Point	27/05/2018		1	
Gibraltar Point	04/07/2018		4	
Gibraltar Point	07/06/2019		2	

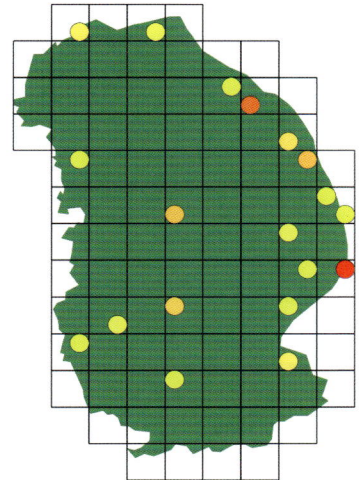

Sea Lavender

121

Wryneck *Jynx torquilla*

Very scarce/scarce declining passage migrant Apr-May and Aug-Oct, mainly coastal. Bred to early 20th century.

Wryneck was always a local breeder in Lincolnshire. In the 19th century it was absent from the north-east and neither was it recorded on passage by Cordeaux (1872). Lorand and Atkin (1989) quote Blathwayt (1915) suggesting it was a scarce breeder in woodlands in the south west. The last definite breeding was reported in 1914 but it may have bred in the Lincoln area up to 1918. Smith and Cornwallis (1955) reported only four records in the 20 years previous. It became a much commoner passage migrant in the 20 years to 1989 but has subsequently become much scarcer. Since 2000, LBR has 16 spring records (Apr 10th-May 29th) and 61 autumn records (Aug 17th-Oct 29th). The most in one autumn was in 2015 when there were 16, all in the 15-day period Aug 23rd-Sep 6th, with seven at Gibraltar Point. The majority have been coastal but there have been sporadic inland records often in observers' back gardens.

1968-1972	1980-1989	2008-2011	2016-2019	Incidence	Change

Lesser Spotted Woodpecker *Dryobates minor*

Very scare and declining resident.

Now one of our rarest breeding birds, Lesser Spotted Woodpecker doesn't even register for BBS on the UK scale. The catastrophic decline in the population led to it being considered by RBBP from 2010. RBBP records show an average of four pairs per year bred in Lincolnshire during the period 2013-17. It wasn't always like this.

The *LBC Atlas* estimated the population at 60-80 pairs in the late 1980s and considered it had increased through the 1990s by some 25%, suggesting a population around 1997 of 75 to 100 pairs. By 2010 only 15 possible pairs were reported and in 2018 that had fallen to three. What went wrong between 1997 and 2010 is unknown. Theories include competition and predation from Great Spotted Woodpecker *Dendrocopos major* and starvation of young through declining food quality. Sims (LBR 2015) reported detailed observations of a breeding pair in Lincoln across two years in 2015 and 2016. It is thought the sex ratio of the species is currently skewed with more males than females. This may lead females to abandon their first brood to their mate to raise the chicks so she can raise a second brood with a different unpaired male. This is what happened to the Lincoln pair in 2016. The female disappeared in late Apr and the male continued to incubate the eggs, feed the young after hatching with three eventually fledging. It will be a great shame to lose this species from Lincolnshire without fully understanding why these changes are occurring.

1968-1972	1980-1989	2008-2011	2016-2019	Incidence	Change
				23%	-22%

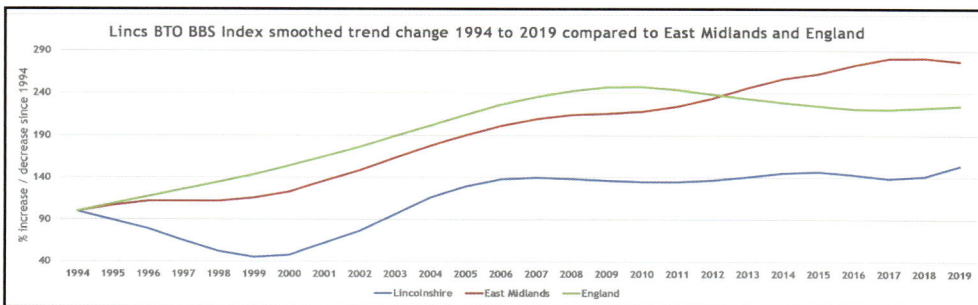

Great Spotted Woodpecker *Dendrocopos major*

Common resident and scarce partial migrant.

The *LBC Atlas* fieldwork in the 1980s took place against the growth in population and spread of "Great Spots". At that time a breeding population of 500 pairs was estimated. The *BTO Atlas 2007-11* showed the spread continued to cover most of the east side of Lincolnshire that had previously been unoccupied. The Lincolnshire BBS index shows that there was a fall in the 1990s before, as in the rest of the country, the population recommenced increasing, growing by 50% over the period 1994-2019. The estimated population for Lincolnshire in 2016 was 1,500 pairs.

In some years there is pronounced autumn passage along the coast. In the 10 years to 2018 the best passage site was Gibraltar Point which had a peak of 11 south on Oct 2nd 2012 and 15 birds on Sep 30th 2017.

1968-1972	1980-1989	2008-2011	2016-2019	Incidence	Change
				23%	-22%

Lincs BTO BBS Index smoothed trend change 1994 to 2019 compared to East Midlands and England

Green Woodpecker *Picus viridis*

Fairly common resident and partial migrant.

Green Woodpecker went through a population boom in Lincolnshire from 1994-2010. The Lincolnshire BBS index shows that the boom flattened out and the population has fallen back a little recently. The *LBC Atlas* estimated a breeding population of 100-200 pairs. The estimated population for 2016 is around 500 pairs. There are still parts of the north and east of the county that Green Woodpecker has yet to reach. What caused the spread and limits their current distribution is not understood but ants, their main food source, are likely to be a factor.

1968-1972	1980-1989	2008-2011	2016-2019	Incidence	Change
				19%	-12%

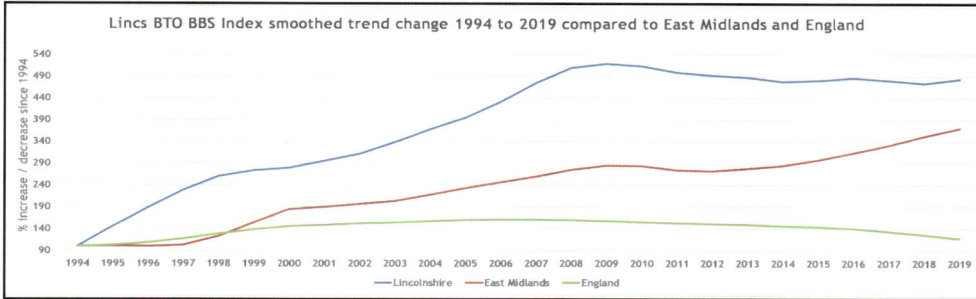

Lincs BTO BBS Index smoothed trend change 1994 to 2019 compared to East Midlands and England

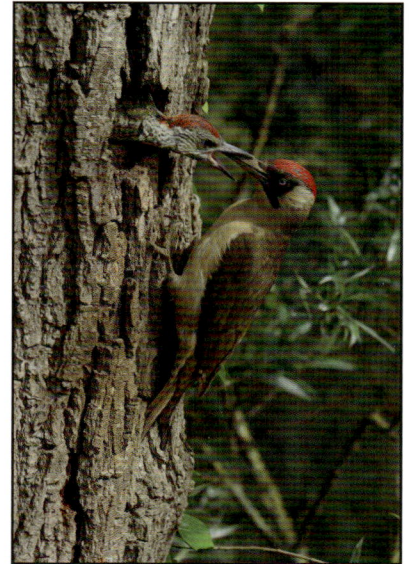

Kestrel *Falco tinnunculus*

Common resident and passage migrant.

The *LBC Atlas* suggested a breeding population of 1,000-1,500 pairs in the late 1980s. BTO survey data covering the most recent 25 years show a decline of 21% in England over this period and this is clearly reflected in the intensively farmed county of Lincolnshire. Intensification of arable farming and particularly the losses of rough grass field edges and earlier cultivation of fields combined with early autumn mowing of rough grass impacts food availability that may have a more serious effect upon newly-fledged young. A long running Lincolnshire study based on ringing pulli, by Alan Ball and Bob Sheppard, confirms the variation in productivity based on the four-year cycle of the Field Vole *Microtus agrestis* and also confirms the increasing importance of nest box provision to this species in the county. The highest concentration of birds, mostly juveniles, occurs on the Humber coast and the Wash strip in Aug and Sep when birds concentrate to feed on insects attracted to flowering Sea Aster *Tripolium pannonicum*. Spring and autumn passage occurs along the coast but is more difficult to detect inland.

1968-1972	1980-1989	2008-2011	2016-2019	Incidence	Change
				40%	8%

Kirkby Moor Farm LWT

Red-footed Falcon *Falco vespertinus*

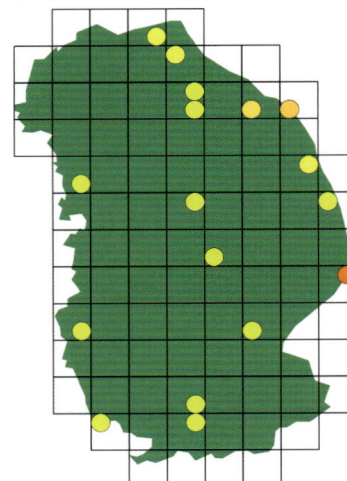

Rare. Eastern Europe.

The first authenticated county records of this attractive little falcon were of birds that were shot from a vessel at the mouth of the Humber in Nov 1864 and at Panton, near Wragby in May 1902. There have been 39 records in all, involving 37 birds averaging about six per decade 1960-2019, and of those which were sexed, 16 were males and 10 were females. Peak spring arrival is in the second half of May with fewer in the autumn: nine records in Aug-Sep, three in Oct and two in Nov. Extreme dates are Apr 30th 1996, an adult male at South Witham NR, to Nov 2nd 1979, an immature at Donna Nook which is also one of the latest ever UK sightings. Most of the sightings have been brief, but some birds have made protracted stays such as the long-staying 2CY male at Willow Tree Fen in Aug-Sep 2015. This was presumed to be the same as that seen in Northumberland Sep 9th-10th, and later found shot dead near Whittlesey, Cambridgeshire on Sep 19th.

The species ceased to be considered by BBRC from 2006. White and Kehoe (2020) note that there have been 919 British records, 1958-2018. Occurrences are highly variable and show no particular trend.

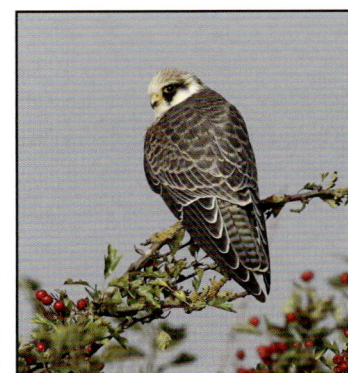

Site Name	First Date	Last Date	Count	Notes
Lincolnshire	01/11/1864		1	Shot on vessel in Humber
Panton	15/05/1902		1	
Marston	08/09/1963		1	Male
Cuxwold	22/05/1967		1	Female
Gibraltar Point	19/06/1967		1	Male
Gibraltar Point	08/10/1967	14/10/1967	1	Female
Donna Nook	30/06/1968		1	Female
Gibraltar Point	25/07/1969	05/08/1969	1	Immature male
Huttoft	04/10/1969		1	Immature
Boston	15/09/1971		1	Male
Keelby	26/05/1973		1	Female
Tetney	10/05/1976		1	Male
Donna Nook	12/06/1977		1	Female
Donna Nook	02/08/1977		1	Male
Killingholme	09/05/1978		1	Male
Saltfleetby-Theddlethorpe	24/06/1979		1	Male
Donna Nook	02/11/1979		1	Immature
Killingholme	16/05/1982		1	Female
Goxhill	19/05/1986		1	First summer
Gibraltar Point	30/05/1989	04/06/1989	1	First summer male
Gibraltar Point	04/06/1990	15/06/1990	1	Female
Kirkby Moor	14/05/1994	19/05/1994	1	
Donna Nook	22/05/1994	25/05/1994	1	Female
North Cotes	26/05/1994		1	Same female
South Witham	30/04/1996		1	Male
Tetney	17/10/1998		1	Same juvenile
Gibraltar Point	11/05/1999		1	Adult male
Deeping Fen	12/06/2003	02/09/2003	1	
Barrow upon Humber	25/07/2003	26/07/2003	1	First summer male
Willingham by Stow	06/04/2011		1	Female
Saltfleetby-Theddlethorpe	18/08/2013		1	Second calendar year female
Willow Tree Fen	12/08/2015	02/09/2015	1	First summer male
Donna Nook	25/08/2015		1	
Gibraltar Point	23/09/2015	30/09/2015	1	Juvenile
Tetney	21/05/2017		1	Adult male
Gibraltar Point	25/05/2017		1	Adult male
Gibraltar Point	28/05/2017		1	Second calendar year male
Gibraltar Point	10/09/2018	14/09/2018	1	Juvenile
Gibraltar Point	30/09/2019		1	Juvenile

Merlin *Falco columbarius*

British form *aesalon* is a Red listed scarce winter visitor and passage migrant, very scarce in summer. Bred sporadically in the 19th century. Icelandic form *subaesalon* is amber listed and may also occur.

The bird-hunting Merlin still occurs along the coastal and Wash saltmarshes in autumn and winter but numbers on the Humber have declined markedly in the last 20 years. Fewer birds now seem to winter inland perhaps as a result of the fall in the numbers of their prey species on intensive arable farmland where winter stubbles, a former favoured feeding area for finches, buntings, larks and pipits have been replaced with winter cereals, oilseed rape and brown fields. Inland, extensive areas of open farmland with a general lack of tall hedgerows and clear views over wide vistas are preferred. Adult males remain a rarity but occasionally, presumably first-summer birds sometimes occur in Jun and Jul with the first fledged juveniles appearing from mid-Aug. Most wintering birds are solitary but up to four may gather in winter roosts. Winter hunting in association with Hen Harriers *Circus cyaneus* is frequently observed.

1968-1972	1980-1989	2008-2011	2016-2019	Incidence	Change
■			■		

Hobby *Falco subbuteo*

Scarce summer visitor and passage migrant.

Bred to early 1900s and sporadically in the 1970s, regularly from the mid 1980s. The lack of proven breeding records of Hobby in the county belies the true status of this elusive but widespread breeding species. From unpublished fieldwork around 2000, Graham Catley, extrapolating from an observed density of 1.3 pairs per 10km square in the north of the county, estimated that there may be 80-100 pairs in Lincolnshire. This figure probably still holds good though there does appear to be considerable annual variation in the number of birds breeding in some areas and RBBP records show an average of 69 pairs per year bred during the period 2013-17. Established breeding pairs typically return in the second half of Apr with young fledging from the first week to the third week of Aug. Young remain within 1km of the nest for up to five weeks and males have been seen still feeding young in early Oct. In breeding pairs females typically depart first and males last but odd juveniles may linger to mid-Oct. Favoured breeding habitats range from the edge of mature woodlands to heathlands but are in the main scattered across the arable landscapes of the county with nests typically in trees within hedgerows.

1968-1972	1980-1989	2008-2011	2016-2019	Incidence	Change
				2%	-21%

Gyr Falcon *Falco rusticolus*

Vagrant. Arctic.

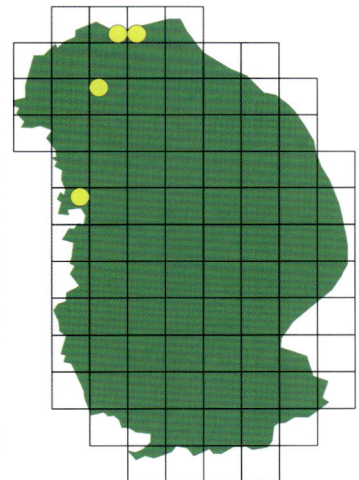

Two historic records, one caught in a trap at Twigmoor (date not specified) and an immature female shot at Saxilby in Dec 1900. The third is the only modern record and was found on the Humber at Read's Island Mar 9th and, it was later discovered, had been photographed at Far Ings NR Feb 16th 2014. Careful perusal of internet images by the finder (mindful of falconer's hybrids) and contact with experts in the field identified it as a grey-type Gyr Falcon. It was assumed to have been the same as the accepted immature bird seen on the north bank of the estuary in Nov 2013. The status of non-white Gyr Falcons has been clouded by a lack of acceptance of such birds in view of the variable forms originating from falconers, but diligent observation and photographs of this individual confirmed its identity. There were 23 British records 2010-18, only one of which was a grey type Gyr. Grey-type birds are commonest in Oct-Jan, white birds in Feb-Apr. Interestingly although the two forms recorded in Britain are clearly distinct in the field, IOC taxonomy and Cornell Birds of the World report that the species is monotypic, but highly polymorphic, with no subspecies currently recognised. BBRC statistics show there have been 184 records 1950-2018, averaging about three records per year and stable.

Site Name	First Date	Last Date	Count	Notes
Twigmoor	01/01/1826		1	Trapped during 1826
Saxilby	01/12/1900		1	Immature female
Barton Pits	16/02/2014		1	
Reads Island	09/03/2014	10/03/2014	1	Same

Rimac at high tide

Peregrine *Falco peregrinus*

Scarce but widespread resident, passage migrant and winter visitor; has bred since 2003.

Still described as a scarce migrant and winter visitor at the time of Lorand and Atkin (1989) the change in status of this charismatic falcon has been totally unpredictable with regular breeding in the last 15 years being the result of birds adopting not only quarries but a variety of man-made structures from Cathedrals to power stations, industrial buildings and even electricity pylons. Use of the latter nest sites have allowed an amazing increase in the last 10 years in the north of the county with no less than seven pairs present on pylons in 2020. RBBP records show an average of 22 pairs per year bred during the period 2013-17. Productivity is highly variable with frequent losses of fledglings from buildings and pylons but purpose-built nest boxes in several sites have a better fledging record. Breeding pairs are often resident in their territories, but young and non-breeding immatures disperse around the county and there is evidence of arrivals of wintering and passage birds from other areas of Britain. In recent years there have been a number of records of birds showing characteristics of Tundra Peregrine *F. p. calidus*. A female caught in a plover trap at Humberstone Fitties on Sep 28th 1910 and formerly accepted as of the North American race *F. p. anatum*, was found on review by the BOURC in 2004 to be an example of the nominate race and as such the record was removed from the British List.

1968-1972	1980-1989	2008-2011	2016-2019	Incidence	Change
				2%	352%

Ring-necked Parakeet *Psittacula krameri*

Very scarce. Occasional visits from feral populations and local escapes.

The rise and rise of the Ring-necked Parakeet has been well covered in the literature though its occurrence in the county is still somewhat sporadic. The species is not mentioned in Smith and Cornwallis (1955) but Lorand and Atkin (1989) noted that there had been several records since 1966, mainly on the coast and in spring with single birds being seen at Gibraltar Point in Jun 1976 and Apr 1982, for example. Since then sporadic reports have appeared in the LBR in the 1980s and 1990s, mostly in the 'Escapes' section, and during 2000-19 only 2001 was a blank year. There has been a steady trickle of records from the Skegness area and the west side of The Wash including a flock of 20 at Gibraltar Point in Oct 2014. However, the picture is muddied by the presence of around 30 free-flying birds at the Lincolnshire Wildlife Park near Friskney. There are no established breeding populations as far as is known, although in 2013 a bird paired with another of unusual plumage colour was seen at a nest hole in Weelsby Wood. However, a small colony may be getting established on the west side of Peterborough, Cambridgeshire so their presence in Lincolnshire seems unlikely to diminish (I. Gordon, pers. comm.)

The species status on the British List is Category C1, covering birds which, although introduced, now derive from the resulting self-sustaining populations.

Red-backed Shrike *Lanius collurio*

Very scarce passage migrant. Former rare breeder, last confirmed in 1978.

During the late 19th century Haigh recorded only two birds in almost 50 years of regular observation on the north-east coast. Migrants began to appear fairly regularly from the 1950s onwards, often being trapped at Gibraltar Point. A steady increase in records thereafter culminated in a sharp peak in 1977 when more than 40 were recorded in autumn alone. Up to four were at Gibraltar Point in late Aug and a staggering nine at Donna Nook in the same period. Since 1977 the pattern has been one of decline. From 1979-95 there was an average number of seven to eight per year falling to three to four per year from 1996-2019. The earliest record was of a female at Saltfleetby-Theddlethorpe on May 5th 1985 and the latest a juvenile at Anderby Creek on Oct 31st 2019. Adults in spring are scarcer than juveniles in autumn.

Lorand and Atkin (1989) reported that there were few confirmed instances of breeding in the past and it was probably always a rare and sporadic breeder. Two pairs bred near Alford in the peak year of 1977 and four young fledged. Another pair bred at Bradley Woods near Grimsby in 1978. There have been no further records of proven breeding since.

1968-1972	1980-1989	2008-2011	2016-2019	Incidence	Change

Daurian/Turkestan Shrike *Lanius isabellinus/phoenicuroides*

BBRC

Vagrant. Central Asia.

The taxonomy of the 'red-tailed' shrikes is far from settled and much debate continues. In Jan 2018, BOURC adopted IOC taxonomy and the Isabelline Shrike group was split into Daurian Shrike *L. isabellinus* (polytypic) and Turkestan Shrike, *L. phoenicuroides* (monotypic at present).

The BBRC advice in their 2018 report (*British Birds* 112: 556-626) is that adults of the two taxa, especially males, can be rather distinctive but 1CY birds, which form the bulk of British records, are, on current knowledge, not safely assigned to one species or another in a vagrant context (Note that five of the six Lincolnshire birds were 1CY individuals). BBRC now report these two taxa as "Daurian/Turkestan Shrike", DNA analysis may not always confirm identification and although most British examples are more likely to be *isabellinus* the BBRC approach is followed here with notes on whether particular individuals showed characteristics of either species. The six county records have been as follows:

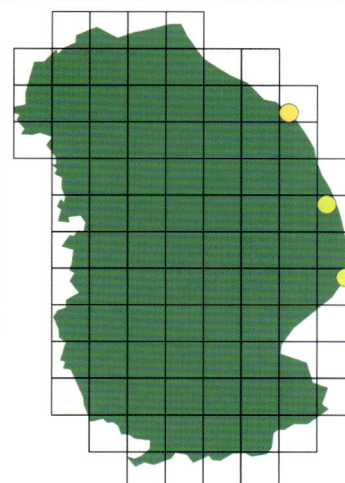

Oct 28th 1978: 1CY trapped at Donna Nook in Oct 1978 showing characteristics of *phoenicuroides*.

Nov 7th-Nov 15th 1982: Adult male seen at Anderby Creek (Nov 7th-8th) and then at Gibraltar Point (Nov 15th). It most resembled *phoenicuroides*, although the race *speculigerus* was also mooted at the time but this has since become synonymous with *isabellinus* and is no longer used.

Oct 14th-15th 1990: 1CY at Donna Nook was quite pale, and most reminiscent of *isabellinus*.

Oct 13th 2003: 1CY at Donna Nook, showing most similarity to *phoenicuroides*.

Oct 10th 2010: 1CY at Gibraltar Point was a pale bird resembling *isabellinus*.

Oct 20th 2013: 1CY at Donna Nook was a pale bird resembling *isabellinus*.

The natural desire to specifically identify a bird is frustrated by this group of birds which exhibit a particularly fluid taxonomy.

Lesser Grey Shrike *Lanius minor*

BBRC

Vagrant. Southern Europe

The first county record occurred at Gibraltar Point where a 1CY female bird was seen by just a single observer in Oct 1960. This species remains a vagrant in the County with just a further three records (a poor showing compared to neighbouring counties of Yorkshire and Norfolk). The remaining records are of one seen by several observers at Saltfleetby in Oct 1969, another at Donna Nook in May 1970 and a superb adult male that fed along the sea bank at Kirton Marsh in May 1990 for one day only.

In Britain, Lesser Grey Shrikes occur mainly in late spring and summer with smaller numbers in autumn. The species has declined over much of its European range, which has been occurring since the mid-19th century linked with lower summer temperatures and heavy rainfall. In Germany they last bred in 1987 having declined from 1,000 pairs in 1950. More recently, Lesser Grey Shrikes failed to breed in France in 2019 for the first time in recorded history. The drastic decline in French numbers, which have been closely monitored for more than 25 years, goes beyond France's borders and is linked to a combination of factors including: current agricultural practices, certain climatic factors, poaching on its long Mediterranean migration routes (6,200 miles) and the degradation of wintering habitats. A colour-ringed bird turned up at Long Nanny, Northumberland in Sep 2016 and was discovered to have come from a Spanish recovery project; this bird was captive-bred and released in Aug 2016 turning up on the Northumberland coast in mid-Sep 2016. Despite the declines in its breeding range, it remains an almost annual visitor to Britain although it is much less frequent averaging about two to three birds per year.

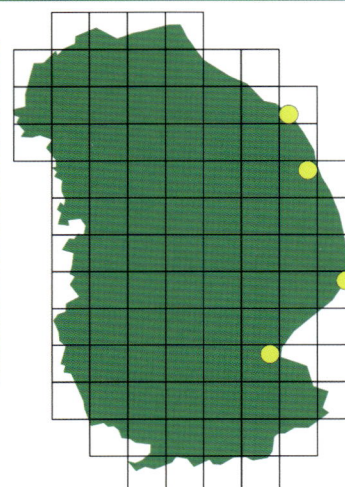

Site Name	First Date	Last Date	Count	Notes
Gibraltar Point	11/10/1960		1	First winter female
Saltfleetby-Theddlethorpe	05/10/1969		1	
Donna Nook	25/05/1970	26/05/1970	1	
Kirton Marsh	12/05/1990		1	Male

Great Grey Shrike *Lanius excubitor*

Very scarce/scarce passage migrant and winter visitor Oct-Apr. Exceptional in summer.

Great Grey Shrike numbers have fluctuated considerably over the last 50 years. The *LBC Atlas* reports an average of 12 per year in the 1970s, six in the 1980s and four in the 1990s. LBR reports show that numbers dropped to only two to three per year in the 2000s, but then from 2010 numbers increased to a peak of 27 in 2013 falling to 19 in 2014 and down to two in 2018. The average in the last decade being around 11 per year. Most birds are autumn migrants and turn up on the coast between Oct 9th-14th onwards. Of 110 birds in the last 10 years only 10 are known to have overwintered in Lincolnshire. There were no reports after Apr 21st in the last five years.

Steppe Grey Shrike *Lanius excubitor pallidirostris*

Vagrant.

This taxon has had a chequered history over the last 20 years. Originally treated as a Central Asian race of *L. excubitor* it was split by BOU as a race of the "Southern Grey Shrike" complex *L. meridionalis pallidirostris* and it was thought likely it would be elevated to full species status. When BOU decided to follow IOC taxonomy it was lumped with *L. excubitor* again from Jan 2019. If and when IOC and other authorities carry out a detailed review of the complex it may well be split again. There have been two Lincolnshire records of this vagrant Shrike. The first occurred at Nene Mouth in Nov 2005 and the second at Grainthorpe Marsh Nov 7th-26th 2008 which was one of the most twitched British birds of all times, as it followed the BOU split that made it "tickable" for the first time.

Woodchat Shrike *Lanius senator*

BBRC

Rare. Southern Europe.

The first Woodchat Shrike in Lincolnshire was seen at Gibraltar Point on Jun 7th 1960. Up to 2019, there has been a total of 14 individuals in the county.

Seven out of the 14 have been found in spring, and six in autumn, plus one bird at Saltfleet Haven on Jun 30th 2008. The latter was thought possibly to belong to the Balearic race *badius* but there was not enough evidence for a submission as such to BBRC. The earliest occurrence was of one at Chowder Ness, near Barton upon Humber, on May 1st 1994. However, there have also been birds at Donna Nook on May 2nd-3rd 1995 and May 4th 2003. The latest in autumn was one at Skidbrooke North End on Sep 21st 1974. Most have occurred along the coast, but with one on the Humber and two on The Wash. Most have been present for a single day but two have made prolonged stays, including the one at Frampton Marsh-Witham Mouth which was present for 46 days.

Woodchat Shrike ceased to be considered by BBRC in 1991 with 505 British records 1950-90. The trend has been for an increase in records and in the last four the annual mean number has risen steadily from 16, 1980-89 to 28, 2010-18 (White and Kehoe 2019).

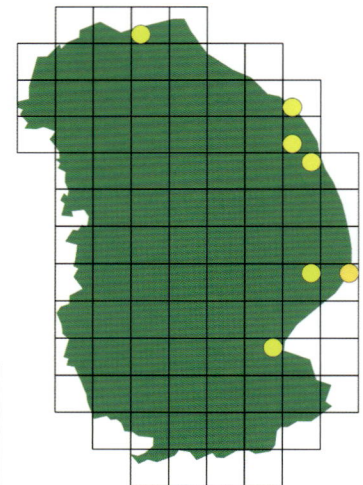

Site Name	First Date	Last Date	Count	Notes
Gibraltar Point	07/06/1960		1	Adult
Gibraltar Point	26/05/1968		1	Adult male
Skidbrooke	21/09/1974		1	Immature
Saltfleetby-Theddlethorpe	21/08/1977	22/08/1977	1	Immature
Donna Nook	06/08/1978		1	Immature
Gibraltar Point	16/05/1992		1	Female
Barton Pits	01/05/1994		1	Second calendar year male
Witham Mouth	30/07/1994	13/09/1994	1	Female
Donna Nook	02/05/1995	03/05/1995	1	Male
Donna Nook	04/05/2003		1	Female
Saltfleetby-Theddlethorpe	09/06/2005		1	Female
Friskney	29/07/2006	23/08/2006	1	Adult female
Saltfleet	30/06/2008		1	
Gibraltar Point	17/08/2013		1	Juvenile

Golden Oriole *Oriolus oriolus*

Very scarce migrant, mainly spring/early summer, but recorded Apr-Aug.

Lorand and Atkin (1989) considered that the earliest reliable record of Golden Oriole in the county was one killed after flying into telegraph wires near Gainsborough in Aug 1908 and since then they note that there had been around 40 records. In the ensuing decades the number of records (birds) has increased slightly: 1990-99, 20 (21); 2000-09, 23 (23); 2010-19, 33 (36). This brings the total records to around 116 up to 2019. Whether this is better detection or an increased number coming through is unknown; certainly, the latter reason looks less likely given the annual toll taken on them by hunters in the Mediterranean. The majority of birds found are males, typically just heard in song. Many of those actually seen prove to be immature birds, yet to attain full adult black-and-yellow plumage. Spring records predominate, peaking in the last 10 days of May. The earliest spring record was Apr 27th 1996 with a long tail-off into early Jul. Records in Aug-Sep are rare but in the successive years 2001-02 there were three records in each year, all juveniles, between Aug 17th-Sep 28th, the latter date being the latest ever. There was a record on Aug 28th in 2003, but since then there have been no Aug-Sep records with the latest being one on Jul 15th 2013.

Jay *Garrulus glandarius*

Common resident, irregular migrant and winter visitor.

An essentially woodland bird, Jay does less well in Lincolnshire with its low woodland cover. The *LBC Atlas* estimated 1,500 pairs in the late 1980s and surprisingly enough the estimated population was 1,500 pairs in 2016. That tallies with the England BBS index which shows an overall 3% change since 1994. The species is not frequent enough in Lincolnshire or the East Midlands for a BBS index chart to be produced.

Jays exhibit interesting but irregular movements along the coast, where they do not breed. Lorand and Atkin (1989) mentioned a particularly significant one in autumn 1983 with flocks of up to 50 birds on the coast. More recently in Oct 2012 flocks of up to 80 south were reported at Gibraltar Point with 70 south at Saltfleetby-Theddlethorpe. A bird trapped and ringed at Gibraltar Point at this time was found dead in West Yorkshire the following month. The mother of all movements came on Apr 30th 2013 when Gibraltar Point had 164 south. A smaller event was reported in May 2015 when Gibraltar Point had 26. The causes of Jay movements are thought to be related to acorn crop failure.

1968-1972	1980-1989	2008-2011	2016-2019	Incidence	Change
				9%	19%

Magpie *Pica pica*

Common resident.

The *LBC Atlas* put the Lincolnshire Magpie population at 18,500-21,000 pairs in the late 1980s and despite the Lincolnshire BBS index suggesting an increase of 55% over the period 1994 to 2019, the estimated population was 20,000 pairs in 2016, illustrating the difficulties with such estimates. Although Magpie was found in every 10km square in the county during the *BTO Atlas 2007-11*, it remains scarcer in the fens, possibly limited by the availability of large hedgerow bushes and trees in which it nests. It was difficult to confirm breeding in TF37 in the Wolds, almost certainly because Magpie still suffers from persecution by shooting interests. Larsen traps can still be encountered in quiet areas of the countryside. Ironically, Pheasant *Phasianus colchicus* carrion helps boost the Magpie population through periods of winter food shortages which would otherwise reduce their levels.

1968-1972	1980-1989	2008-2011	2016-2019	Incidence	Change
				75%	13%

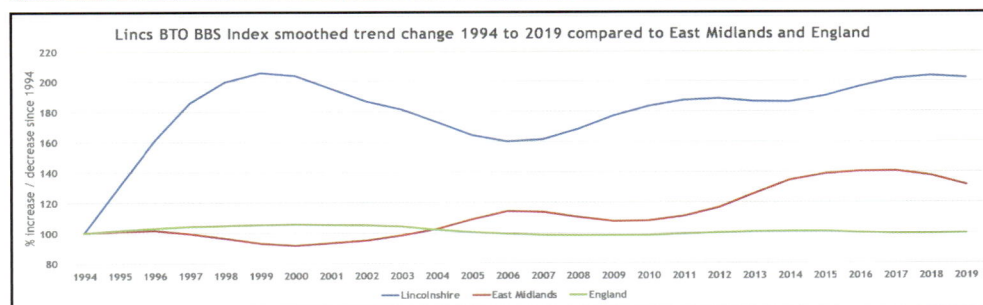

Lincs BTO BBS Index smoothed trend change 1994 to 2019 compared to East Midlands and England

Donna Nook

Nutcracker *Nucifraga caryocatactes*

Vagrant. Western Europe

The first county record in Mar 1883 (no precise date), was a male killed near Sleaford and the specimen was retained in Lincoln Museum for many years but is now held at the Natural History Museum at Tring, Hertfordshire. After that another was shot in Nov 1888, but the next records were not until 80 years later. These were seen during the well-documented irruption which occurred in 1968-69 when 339 were recorded in Britain, four of which were in Lincolnshire (Hollyer 1970). The lucky observers of the bird at Gibraltar Point noted that it landed in Elders *Sambucus nigra*, in the ringing hollow, giving excellent views before heading away north-west along the east dunes. The seventh and most recent was Sep 12th 1976 at Donna Nook. The last confirmed British record up to 2019 was in Kent in Sep 1998.

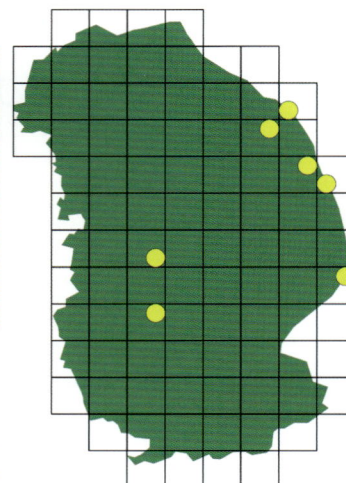

IOC taxonomy currently recognises eight races. The nominate *N. c. caryocatactes* breeds in northern and eastern Europe but it is the slender-billed race *macrorhynchos* of north-east Russia and Siberia which has periodically irrupted westwards. This race inhabits the taiga zone and every so often suffers a shortage of pine seeds forcing large numbers to disperse widely in search of food. Irruptions have occurred in Europe on about 25 occasions in the past 250 years, since the first reference in 1733 (Madge *et al* 2017).

Site Name	First Date	Last Date	Count	Notes
Sleaford	01/03/1833		1	Male
Marshchapel	06/11/1888		1	
Saltfleetby-Theddlethorpe	12/09/1968		1	
Gibraltar Point	17/09/1968		1	
Sutton on Sea	17/09/1968		1	
Metheringham	09/04/1969		1	
Donna Nook	12/09/1976		1	

Chapel Six Marshes

Jackdaw *Coloeus monedula*

Western European form *spermologus* very common resident, also passage migrant and winter visitor. Nordic Jackdaw *C. m. monedula*, rare or overlooked winter visitor.

This hole-nesting crow is a versatile breeder nesting in trees as well as buildings, especially chimney pots. The *LBC Atlas* estimated a population in the range of 7,500-11,900 pairs in the late 1980s and since then the Lincolnshire BBS chart shows an increase of 102% over the period 1994-2019, in line with the trend across England. The estimated 2016 Lincolnshire population was 28,000 pairs which seems high but could be correct. Post breeding birds congregate and the largest single flock reported in LBR in the five years to 2018 was 1,800 at Gibraltar Point in Oct 2017. Flocks of more than 1,000 are exceptional. Nordic Jackdaw reports in LBR are less than annual. Records in the 10 years to 2018 were 2009 (four), 2010 (one), 2011-12 (one), 2016 (nine). Most records were in Nov-Mar, with one on Apr 17th 2017 at Marston STW.

1968-1972	1980-1989	2008-2011	2016-2019	Incidence	Change
				95%	7%

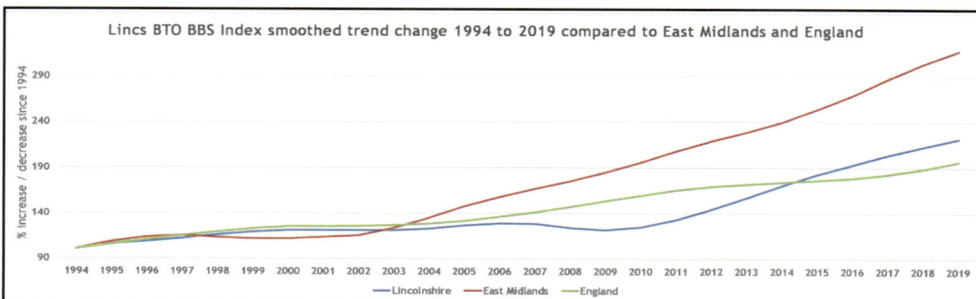

Lincs BTO BBS Index smoothed trend change 1994 to 2019 compared to East Midlands and England

Rook *Corvus frugilegus*

Very common resident also passage migrant and winter visitor.

Against the increasing agricultural intensification over the last 50 years which has seen many farmland species decrease, some alarmingly, Rook has managed to maintain a steady increase through the whole period. The *LBC Atlas* put the population at around 26,000 pairs in the late 1980s and by 2016 the estimated population was put at 38,000 pairs. It is the third most common bird in the county based on total bird count in the 2019 BBS. It is, however, to some extent patchily distributed being absent or much less frequent in some of the more intensively farmed areas that lack sufficient tree cover to provide nesting opportunities, especially near the coast.

1968-1972	1980-1989	2008-2011	2016-2019	Incidence	Change
				69%	19%

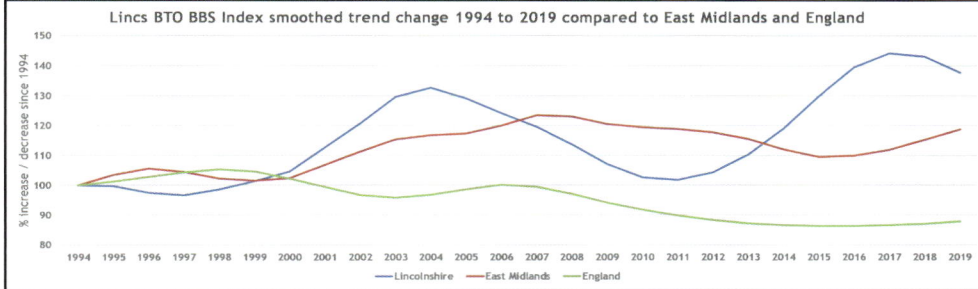

Lincs BTO BBS Index smoothed trend change 1994 to 2019 compared to East Midlands and England

Lincolnshire Wolds

Carrion Crow *Corvus corone*

Very common resident, also passage migrant and winter visitor.

Our most widespread crow species confirmed to breed in every 10km square in Lincolnshire during the last *BTO Atlas 2007-11*. The *LBC Atlas* estimated a population of 17,000 pairs in the late 1980s but thought that was too high. The Lincolnshire BBS index chart shows the population has been through two growth cycles since then in the period 1994-2019 and is now 65% higher. The population estimate for 2016 is 20,000 pairs which seems a likely figure. Why is it so common? Carrion Crows are extremely adaptable and will exploit any feeding opportunity from the seashore to the Pheasant *Phasianus colchicus* carrion the shooting community provides as a by-product of the current industrialisation of what used to be a sport. Should this industry ever be regulated, the Carrion Crow population will likely fall, benefiting other birds whose eggs and nestlings form another part of their catholic diet during Apr-Jul. The only factor limiting their spread is availability of tree nest sites and for this reason they were scarcer in the Fens during the 1980s. Tree planting in open environments is a boon for this species but a negative for ground breeding farmland birds.

1968-1972	1980-1989	2008-2011	2016-2019	Incidence	Change
				95%	7%

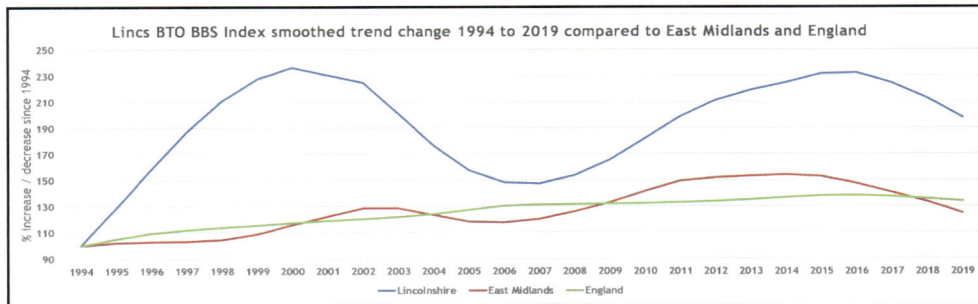

Lincs BTO BBS Index smoothed trend change 1994 to 2019 compared to East Midlands and England

Hooded Crow *Corvus cornix*

Very scarce migrant and winter visitor, formerly much more common in winter.

The status of Hooded Crow in Lincolnshire was given detailed treatment by Henderson in LBR 2018. It was a fairly common passage migrant and winter visitor up until the 1950s and it declined steadily from the 1960s onwards. It is thought that as the British and Irish population is sedentary, Lincolnshire visitors must be arriving from the more migratory Scandinavian population. Climate change has reduced that necessity and there are, in fact, three historic ringing recoveries from Sweden (two) and Norway (one) between Oct-Jan in 1938, 1947 and 1951. Over the last 20 years annual numbers have fluctuated between none in 2008-09 to 19 in 2018-19 with an average of three to four per year. Birds generally arrive in Oct-Jan with a small return spring migration in Apr. The peak site numbers reported were three at Gibraltar Point during Dec 2018.

Raven *Corvus corax*

Resident to 18th century but was extirpated, now a scarce but increasing resident.

The return of the Raven as a breeding bird in Lincolnshire has been welcomed by all birders. It was listed as a vagrant in the *LBC Atlas* up to the late 1990s. The only 20th century record was a bird at Donna Nook in Feb 1980. That changed when a bird was seen flying over Louth in Jul 2003 and then a pair turned up at Belton Park in Nov 2003. It was long anticipated, as breeding had become more widespread in adjacent Nottinghamshire and Leicestershire. The species had gradually been spreading eastwards from Wales and the border counties since the 1990s and the relatively rapid spread is well documented by the breeding change chart for Raven in the *BTO Atlas 2007-11*. The first confirmed breeding at Belton Park took place in 2005 when two young were successfully fledged, and a second pair nested near Temple Wood in 2010. The last *BTO Atlas 2007-11* showed breeding was confirmed in three 10 km squares, with probable breeding in two squares all in the extreme south-west. By 2014, eight territories were known. The pace of colonisation has been much slower than for Red Kite *Milvus milvus*, or Buzzard *Buteo buteo*, as emphasised by Fleming (LBR 2017). In the following five years breeding reports in LBR came from a wider area, including Whisby, the north and the Wolds. In 2018 Raven was reported from 48 sites across the county and 57 in 2019. There are probably still fewer than 20 pairs but the chances of seeing a Raven in Lincolnshire are now better than they have ever been.

1968-1972	1980-1989	2008-2011	2016-2019	Incidence	Change
				5%	

Waxwing *Bombycilla garrulus*

Passage migrant and winter visitor, most in Oct-Apr. Scarce most winters but fairly common in irruption years.

Waxwing invasions are memorable events but, as the birds can be extremely mobile and flock sizes can change from hour to hour and turn up anywhere with suitable berries, they can be very difficult to count. A particularly large invasion in 1965-66 was reported in Lorand and Atkin (1989) when Waxwings were reported from all over the county and there were thought to be 500-1,000 in the Louth area alone by Nov 1965. During the 1980s-90s the *LBC Atlas* reported that the largest single invasion involved some 1,300 birds in Jan-Apr 1996. In the last 10 years to 2018 the maximum flock size reported in LBR each year ranged from one in 2014 to 400 in 2012.

The year 2012 was probably the biggest invasion of the county in the 21st century so far. Burton Stather had 400 birds on Dec 18th-19th and the highest day count across sites was 691 on Dec 9th. It is thought that upwards of 3,800 birds passed through.

The second biggest year in that 10-year period was 2011 in which a total of 1,740 birds were reported in Jan with 150 in Bourne on Jan 1st-3rd. Discounting the extremes, in an average year the biggest flock size reported is around 50 birds.

Coal Tit *Periparus ater*

British form *brittanicus* is a common resident and passage migrant. Nominate continental form a very scarce irruptive visitor in autumn but no accepted records as yet.

Coal Tits use broadleaf and coniferous woodland but prefer the latter, and gardens with large conifers are usually essential for this characterful little bird. It is fairly widespread in Lincolnshire but missing from much of the Fens due to habitat limitations. The *LBC Atlas* estimated around 3,500 pairs in Lincolnshire in the late 1980s. This Tit is too scarce in Lincolnshire to produce a reliable BBS index, but the figures for East Midlands suggest a long-term increase since 1994 of only 5%. Surprisingly, the estimated population came in at 1,800 pairs in 2016 but this apparent fall may be a random result arising from the small number of BBS squares in which this bird appears in Lincolnshire. The LBR reports very few large flocks and the largest over the five years to 2018 was 25 at Osgodby Moor in Jan 2015. During this period there were no reports of significant coastal movements.

1968-1972	1980-1989	2008-2011	2016-2019	Incidence	Change
				12%	-5%

Marsh Tit *Poecile palustris*

Fairly common but local resident, mainly in the south-west.

At first glance the status of Marsh Tit has remained fairly local throughout the last 50 years, but this masks a definite retreat of the species from the north and east of the county to its woodland strongholds in the Kesteven forests of the south-west. This was observed as a result of the detailed survey work done for the *LBC Atlas* from which some 200 pairs were estimated to be in the county in the late 1980s. The population in Lincolnshire and the East Midlands is too small to show up on the BBS index, but the England index shows a long-term decline of 37% since 1994 for this BoCC4 red data species. The *BTO Atlas 1988-91* showed a net loss from 25 10 km squares, with a further net loss of two squares by the time of the *BTO Atlas 2007-11*. That decline has probably continued and the estimated population for Lincolnshire was 140 pairs in 2016. Over the six years to 2018 the maximum counts reported have been eight at Anwick STW in Aug 2014 and eight at Callan's Lane Wood in Dec 2013. Marsh Tit is not entirely sedentary and over the period reliable reports of singles on the coast came from Alkborough Flats where one was seen on and off from Nov 2013 to Apr 2015, Gibraltar Point in Aug 2014, and Sep 2018, Freiston Shore Oct 2016 and Frampton Marsh Sep 2017.

1968-1972	1980-1989	2008-2011	2016-2019	Incidence	Change
				1.5%	-37%

Willow Tit *Poecile montanus*

Once common and widespread, now seriously declining and scarce.

The *LBC Atlas* through the 1980s found Willow Tit confirmed or probably breeding in 58 10km squares in the county and the population was estimated at 400 pairs. The *BTO Atlas 2007-11* showed it had been lost from 26 squares mainly in the south and east and was found in only 22 10km squares. RBBP records show an average of 22 pairs per year bred during the period 2013-17 and LBR reports that records now come from only eight to 12 sites per year. This 95% crash, which is ongoing, is the severest decline experienced by any formerly common resident bird in the county in the last 30 years. The reasons are not fully understood but may include lack of scrub habitat with rotting willow, birch or hawthorn branches and stumps which meet critical nest hole excavation requirements, predation of nestlings by Great Spotted Woodpecker *Dendrocopos major*, and competition from the commoner tit species driving them from their excavated nest holes. Willow Tits continue to do well in the river valleys of South and West Yorkshire only 25 miles to the west of north-west Lincolnshire but there is little evidence so far of dispersion in this direction that would augment our rapidly depleting population. The Willow Tit race in Britain, *P. m. Kleinschmidti*, is endemic and saving it should be a top priority. National surveys were organised by RSPB in 2019-20 and have been extended into 2021 to try and understand what can be done to stem and, hopefully, reverse the decline.

1968-1972	1980-1989	2008-2011	2016-2019	Incidence	Change
				0.3%	-83%

Blue Tit *Cyanistes caeruleus*

British form *obscurus* a very common resident.

The *LBC Atlas* considered that the population was around 100,000 pairs in the late 1980s and the indications are that the population has increased since then, particularly over the last 10 years. The BBS Index suggests a long-term trend increase over 1994-2019 of 21%, but the estimated population in 2016 was still 100,000 pairs. The resilience of the Blue Tit compared to some of our other small birds is probably due to its preference for feeding in gardens. The last report of the Garden Bird Feeding Scheme (Goodall LBR 2011) showed it to be one of our most consistent year-round garden feeders. Broadleaf woodland remains its natural habitat. In 1957 the famous "tit irruption" was recorded across the east and south coast of England (Cramp *et al*, 1960). On Oct 6th that year 200 were reported at Gibraltar Point. There has been nothing similar since.

1968-1972	1980-1989	2008-2011	2016-2019	Incidence	Change
				90%	13%

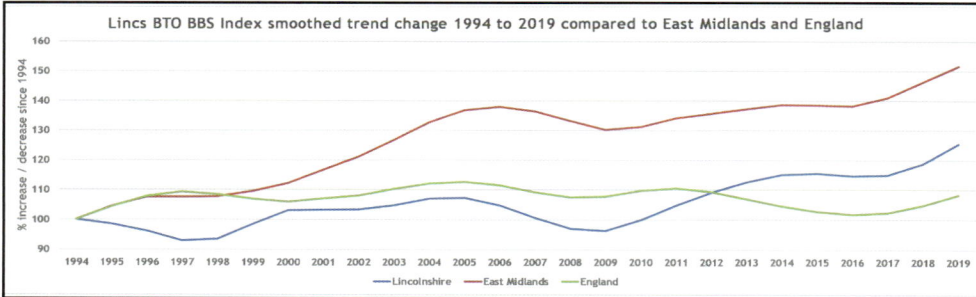

Lincs BTO BBS Index smoothed trend change 1994 to 2019 compared to East Midlands and England

Great Tit *Parus major*

British form *newtoni* a very common resident.

The *LBC Atlas* estimated that there were 46,000 to 52,000 pairs in Lincolnshire in the late 1980s. It also showed up interesting gaps in Great Tit distribution in the more intensively farmed parts of the Fens, reclaimed coastal areas and the Marsh. The Lincolnshire BBS index chart shows that the population has increased by 115% during the period 1994-2019 in line with a general increase across England. However, the estimated population for Lincolnshire in 2016 was 50,000 pairs. One might think that given the BBS increase it would be of the order of 100,000; it is difficult to square these two different numbers. Interestingly, for what is a very sedentary bird as demonstrated by ringing recoveries, the largest numbers in the county each year tend to be recorded at Gibraltar Point. With 85 reported there in Jan 2017, the highest count reported in LBR in the five years to 2018.

1968-1972	1980-1989	2008-2011	2016-2019	Incidence	Change
				83%	9%

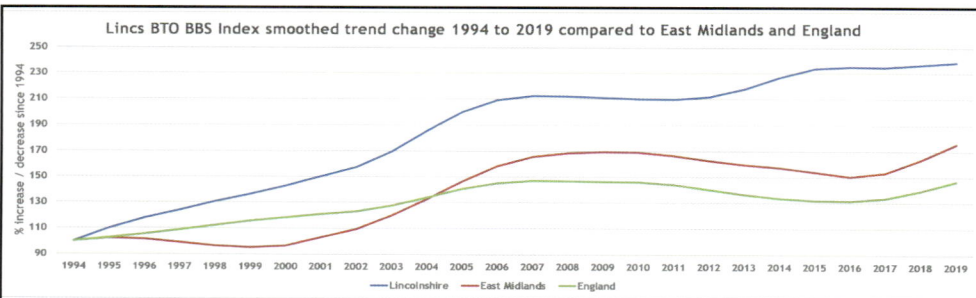

Lincs BTO BBS Index smoothed trend change 1994 to 2019 compared to East Midlands and England

Brinkhill

Penduline Tit *Remiz pendulinus*

BBRC

Vagrant. Southern and Eastern Europe

An adult male was found in the late afternoon of Oct 14th 1991 in the reed bed at Wolla Bank Pit; it was seen only briefly and was presumed to have gone to roost. The next day other observers visited the same site to search for the male and to their surprise saw both this and a female. They were observed there for only 35 minutes and once flew up very high before dropping back into the reed bed but flew off shortly afterwards. These were the first records for the county. The second record involved a brief visit of a male at Shoveler's Pool, Gibraltar Point in Apr 1994. The third, which was also the second for Gibraltar Point, was again seen only briefly (but excellently photographed) in Jun 2000. The final record involved yet another male less than a year later, this time at Barton Pits in Apr 2001. There have been none since despite regular occurrences elsewhere in Britain with about 10 records per year up to 2014, which was the last year it was considered by BBRC.

From the 1970s the species has expanded across central Europe and birds ringed in France (two), Sweden (one) and The Netherlands (one) have been found in Britain. A male was seen at a nest in Kent in Apr-May 1990 and another adult with a brood patch was trapped in northern England in a reed bed in Jul 1992, but there has been no other evidence of breeding since and they have not featured in the RBBP reports up to 2018.

Site Name	First Date	Last Date	Count	Notes
Chapel St Leonards	14/10/1991	15/10/1991	1	Male
Chapel St Leonards	15/10/1991		1	Female
Gibraltar Point	27/04/1994		1	Male
Gibraltar Point	13/06/2000		1	Adult male
Barton Pits	10/04/2001	11/04/2001	1	

Bearded Tit *Panurus biarmicus*

Scarce and localised, but increasing, resident, passage migrant and winter visitor.

Formerly bred in the Fens but had gone by around 1840. Recolonised from around 1968. Lorand and Atkin (1989) suggested that around 30 pairs a year bred in the 1980s. The *LBC Atlas* indicated that breeding was confined to reedbeds at the Humber Clay Pits and the Sea Bank pits from Huttoft south.

In the early 90s bad winters reduced the population to around three pairs. The population stayed low until 2012, with none at all 2002, 2003, 2007 and 2008. It began to pick up as the Alkborough Flats EA-managed retreat reedbeds developed. RBBP records show that an average of 34 pairs per year bred during the period 2013-17.

The average masks a steady increase to 65 pairs in 2018. Most records are from the coast and inland records, which tend to be in winter, are very scarce.

1968-1972	1980-1989	2008-2011	2016-2019	Incidence	Change

Woodlark *Lullula arborea*

Scarce summer visitor and passage migrant, rare in winter. Bred to 1959 and again from 1984 with numbers peaking around the year 2000, although now steadily declining.

The most recent National Woodlark survey produced a total of 81 males and 47 confirmed pairs in seven broad areas of Lincolnshire in 2003; of this total 60 males and 37 pairs were on the north-west Coversands. In Laughton Forest-Scotton Common, the former main area of the county for this highly localised species, the peak numbers actually occurred earlier in 2000 when 34 territories were occupied. Woodlarks occupy two distinctly different habitats in the county, natural short-grazed heathland and managed forestry sites where they rely upon clear felling to produce suitable breeding habitat. As the area of clear felling was reduced at Laughton, numbers fell to a peak of 17 territories in 2012 since when changes in forest management have removed clear felling as an option and replanting of fast-growing species like *leylandii* mean that open areas are only suitable for Woodlarks for one or two years. This has led to a massive decline in the Laughton-Scotton area where recently there have been no more than four pairs and six territorial males. The natural heathland sites in the north-west appear to have stable populations but numbers have declined in the Market Rasen forests and on the Kirkby on Bain area heathlands of late. The total breeding population of the county is considered to now be in the region of 25-30 pairs. Breeding sites are occupied from early Feb through to Aug or Sep. Small parties have occasionally wintered around the periphery of Laughton Forest but it is presumed that most birds winter further south. Coastal passage migrants occur from Feb through to May and less regularly in Oct and Nov.

1968-1972	1980-1989	2008-2011	2016-2019	Incidence	Change
				0%	

Skylark *Alauda arvensis*

Very common resident, passage migrant and winter visitor.

The quintessential farmland bird in Lincolnshire remains very widespread throughout the county. The *LBC Atlas* estimated 76,000 pairs in the late 1980s but there was a suspicion that numbers had fallen by 25-50% by the late 1990s. The Lincolnshire BBS index chart reflects this but there have been subsequent ups and downs which have left the population down some 14% over the long term since 1994. That compares reasonably well to the overall position in England where Skylark is down 24%. Looking back a bit further the population had already fallen by more than half from the late 1970s to the early 1980s onwards. The estimated population in 2016 was 70,000 pairs which ties in well with the *LBC Atlas* estimate and the BBS index. Four figure counts in the *LBC Atlas* in the 1980s were in the range 1,000-4,250. Such wintering flocks are not reported these days. In the five years to 2018 the largest wintering flock reported in LBR was at Frampton Marsh which had 700 in Feb 2017. The average peak flock was 440. Migration can still throw up big flocks occasionally. In the five years to Dec 2018 there was one such big movement in Oct 2014 at Gibraltar Point when on two separate days numbers heading south totalled 1,765 on 11th and 1,600 on 25th.

1968-1972	1980-1989	2008-2011	2016-2019	Incidence	Change
				90%	0-5%

Shore Lark *Eremophila alpestris*

Scarce, formerly fairly common winter visitor. Currently low numbers compared to the 1960s and 1970s. Exceptional inland.

Lorand and Atkin (1989) reported that Shore Larks were rare and irregular visitors to Lincolnshire from the first in 1837, which was the second for Britain at the time, until the late 1940s when they began wintering regularly in the county. The *LBC Atlas* reports that numbers were high from the mid-1960s to the mid-1970s with flocks of over 100 in several years. There was then a steady decline to zero by the end of the 1980s, after when numbers began to steadily increase again. In the 10 years to 2018 LBR shows that the total of peak annual site counts ranged from a low of six in 2015 to a high of 51 in 2016 with an average of around 27 birds per year. Three sites held 90% of the birds. These were Brickyard Lane, Theddlethorpe with summed peaks across 10 years of 103, Donna Nook with 77 and Gibraltar Point with 69. Nine other sites held 25, all on the North Sea coast between Cleethorpes and Gibraltar Point apart from singles at Frampton Marsh and Covenham Reservoir. The only flocks of more than 20 were at Brickyard Lane with 28, Oct 24th 2016 and 24 on Nov 11th 2018, Gibraltar Point with 26 on Apr 16th 2011 and 21 on Dec 19th 2010; and Donna Nook with 22 on Feb 4th 2017.

Short-toed Lark *Calandrella brachydactyla*

Vagrant. Europe.

The first record was at Gibraltar Point in Sep 1971 and the greyish plumage tones suggested it belonged to one of the (numerous) eastern races. There were probably five other autumn records in Britain that year one of which, on Fair Isle Sep 27th-28th, was also a greyish, eastern bird. There have been only two records since, both autumn birds, in Oct 1991 and Aug-Sep 2013.

The species ceased to be considered by BBRC in 1994, by which time there had been 395 British records. White and Kehoe (2020) in the latest *Report on Scarce Migrants in Britain* reported the annual average number of records as between 17-27 over the three decades 1990-99 to 2010-18. This inconspicuous small lark can easily be overlooked in the expanses of ploughed fields and expansive saltmarshes in the county and is likely to be under-recorded. Its status in Lincolnshire contrasts sharply with its frequency in neighbouring Norfolk where there have been 12 records 2011-18 alone.

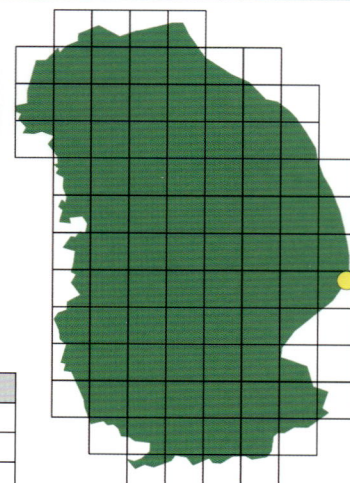

Site Name	First Date	Last Date	Count	Notes
Gibraltar Point	18/09/1971	26/09/1971	1	
Gibraltar Point	11/10/1991	20/10/1991	1	
Gibraltar Point	25/08/2013	01/09/2013	1	

Calandra Lark *Melanocorypha calandra*

Vagrant. N. Africa, southern and south-eastern Europe, to western Asia.

One county record at Gibraltar Point May 11th 2011. The finders flushed this impressively chunky lark from the saltings in the north of the reserve. Good views were obtained despite the bird's elusive nature and frequent flights. It was eventually mobbed by a pair of Skylarks *Alauda arvensis* and flew low north not to be seen again. The observers were struck by the black neck patches and underwing and broad white trailing edge to the secondaries and inner primaries of the upperwing. Its stay was brief and it could not be relocated after late morning that day, a tragically short stay.

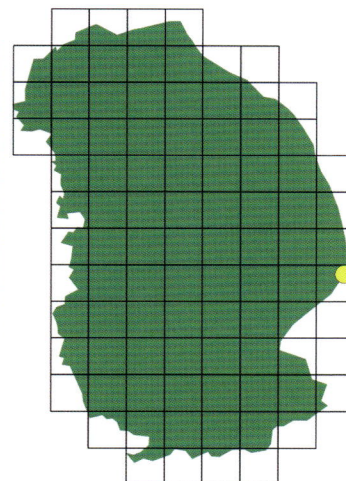

This was the 16th British record with 19 in all to 2018. All but two of these records have been in spring and it characteristically turns up on islands (Fair Isle has had six) and stays only briefly. Until the mid-1990s, Calandra Lark was a major rarity in Britain, but has featured in no fewer than 13 BBRC annual reports since 1994. This recent upturn may be associated with various factors including more successful breeding seasons and changing weather patterns as well as improved observer coverage. Mediterranean populations are largely resident but wander outside the breeding season. Eastern populations, on the other hand, are migratory or partially migratory and winter from southern Russia to North Africa and the Middle East; passage in Russia mainly Oct and Mar.

Site Name	First Date	Last Date	Count	Notes
Gibraltar Point	11/05/2011		1	

Gibraltar Point and Wainfleet Marsh

Gibraltar Point Foreshore

Sand Martin *Riparia riparia*

Common but declining summer visitor and passage migrant.

The *LBC Atlas* estimated around 3,000-6,000 pairs of this tunnelling martin in the late 1980s and provides a good account of the woes this species had suffered in the previous 20 years. Its distribution is limited by the availability of suitable nest sites, near vertical sandy cliffs. Many breeding sites in sand and gravel quarries have been lost due to concerns about the health and safety of the remaining excavations. Natural sites are scarce these days and limited to riverbanks and occasional soft sandstone outcrops. Providing artificial banks has met with mixed success. Frampton Marsh had a good initial take up, but then predatory *mustelids* caused mass desertion.

Serial *BTO Atlases* have reflected a steady loss in distribution in the county from the first in 1968-72 to the second, with net losses in 1988-91 from ten 10km squares and a further net loss from seven squares in *BTO Atlas 2007-11*. The estimated population for Lincolnshire in 2016 was 3,000 pairs. Very little information on breeding colonies gets reported in LBR. Sand Martin is generally one of our earliest migrants arriving in early Mar. It usually leaves earlier too, most birds having passed through by early Sep. The most spectacular passage of recent years was 10,000 estimated at Frampton Marsh on Aug 28th 2017 and there were 8,000 at Gibraltar Point on Aug 3rd 2014.

1968-1972	1980-1989	2008-2011	2016-2019	Incidence	Change
				4%	50%

Swallow *Hirundo rustica*

Very common summer visitor and passage migrant. Exceptional in winter but recorded in most months.

In the late 1980s the *LBC Atlas* put the Lincolnshire Swallow population at some 25,000 breeding pairs. It was thought that the population had declined by up to 50% by the late 1990s. It remains a very widespread bird. The long-term trend in the Lincolnshire BBS index since 1994 shows a small increase; that may have been coming off a sharp decline in the 1990s.

The estimated population for the county in 2016 was 19,000 pairs. Substantial movements of Swallows take place in spring and late summer. The LBR reports of peak daily counts at various sites in spring and autumn during 2014-18 featured Gibraltar Point, Covenham Reservoir and Witham Mouth. The spring peak in that period was 1,005 at Gibraltar Point in May 2017 and the autumn peak 7,000 at Gibraltar Point on Aug 4th 2014 and Sep 29th 2016. This pales into insignificance compared to the peak of 30,000 there on Sep 9th 1995.

1968-1972	1980-1989	2008-2011	2016-2019	Incidence	Change
				81%	0%

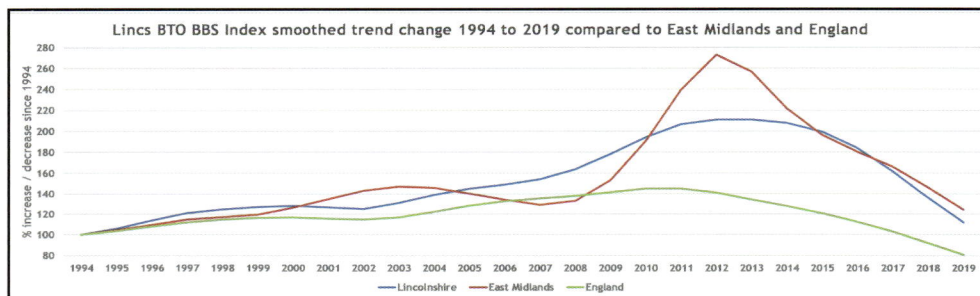

Lincs BTO BBS Index smoothed trend change 1994 to 2019 compared to East Midlands and England

House Martin *Delichon urbicum*

Common summer visitor and passage migrant.

Who are these people who don't like House Martins breeding under their eaves? The cheerful chattering of these colonial nesters is one of the most enchanting sounds of a summer evening. They are heavily dependent on the right kind of overhanging eaves for nesting locations and in the late 1980s the *LBC Atlas* estimated we had somewhere between 7,500-20,000 pairs. A wide range of uncertainty. The Lincolnshire BBS index has an insufficient sample size to produce a long-term trend for Lincolnshire but in England the population is down 40% over the period 1994-2019.

The estimated population for 2016 was 11,000 pairs indicating what many will have observed; though still very widespread the House Martin has declined. One can't help but feel that the mean-minded house holders who poke down nests and dangle deterrent ribbons from their eaves are a part of the problem. The peak autumn migration of the last 70 years at Gibraltar Point was 7,840 on Sep 7th 1970 and in recent years the peak has been 5,078 on Sep 13th 2018.

1968-1972	1980-1989	2008-2011	2016-2019	Incidence	Change
				37%	27%

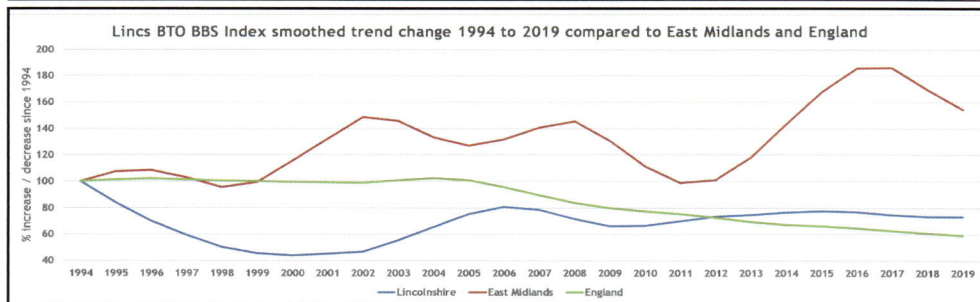

Lincs BTO BBS Index smoothed trend change 1994 to 2019 compared to East Midlands and England

Red-rumped Swallow *Cecropis daurica*

BBRC
British Birds Rarities Committee

Rare migrant, almost annual since 2003. Southern Europe

There had been more than 50 records of Red-rumped Swallow in Britain before the first county bird was seen at Gibraltar Point in Oct 1977. It was another decade before any more were seen when one in May 1987 and four in Oct 1987 were part of an influx of 61 that year, which included seven together at St Mary's rubbish dump on the Isles of Scilly on Oct 27th. Following this record number, the numbers continued to rise with double figures in most years thereafter and more than 20 in the years 1990 (29), 2002 (24), 2003 (47), 2004 (53) and 2005 (24) after which it ceased to be considered by BBRC.

The total number of Lincolnshire records to 2019 is 44 involving around 59 birds although the latter total may involve some duplication. During the period 2003-19 there have only been two blank years, 2008 and 2009. Two birds have appeared together on around 10 occasions and four were recorded at Gibraltar Point in 1987 and 2003.

The species was formerly a hard one to see in Britain, but it has spread as a breeding species north and east across Europe. It extended north from the 1920s from Iberia (where it first bred in central Spain 1951-53 and northern Spain 1960) followed by a large expansion into Portugal 1970-90. Northwards expansion occurred at the same time into mainland France and east to Romania, Slovenia and Hungary. It has become an annual vagrant in The Netherlands and Sweden, is regular in Austria and also reaches Ireland, Iceland, Norway, Denmark, Finland and Poland and has turned up in Madeira and the Azores. White and Kehoe (2020) reported that there have now been more than 1,000 British records during 1958-2018.

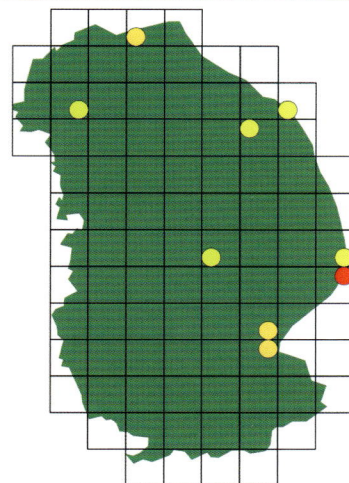

Site Name	First Date	Last Date	Count	Notes
Gibraltar Point	29/10/1977		1	
Gibraltar Point	02/05/1987	06/05/1987	1	
Gibraltar Point	24/10/1987	28/10/1987	2	
Gibraltar Point	27/10/1987		4	
Gibraltar Point	14/11/1988		2	
Messingham	23/04/1994		1	
Gibraltar Point	12/05/1999		2	
Freiston Shore	26/04/2003		1	
Fishtoft	27/04/2003		1	
Gibraltar Point	29/04/2003	30/04/2003	4	
Gibraltar Point	02/05/2003		1	
Gibraltar Point	17/04/2004		2	
Barton Pits	04/05/2004	09/05/2004	1	
Barton Pits	05/05/2004		1	
Gibraltar Point	01/05/2005		1	
Donna Nook	05/11/2006		1	
Gibraltar Point	22/04/2007		1	
Gibraltar Point	06/05/2007	08/05/2007	1	
Gibraltar Point	26/04/2010		1	
Gibraltar Point	30/04/2010		1	
Donna Nook	05/10/2010		1	
Gibraltar Point	06/05/2011		1	
Barton Pits	28/04/2012	29/04/2012	1	
Gibraltar Point	30/04/2012		1	
Frampton Marsh	11/05/2012		2	
Gibraltar Point	11/05/2012		1	
Freiston Shore	18/05/2012		1	
Gibraltar Point	09/05/2013		1	
Gibraltar Point	07/05/2014		2	
Covenham Reservoir	07/05/2015		1	
Covenham Reservoir	29/05/2015		1	
Freiston Shore	15/04/2016		1	
Gibraltar Point	13/05/2017		2	
Gibraltar Point	23/04/2018	26/04/2018	2	
Skegness	23/04/2018		2	
Frampton Marsh	27/04/2018	30/04/2018	1	
Gibraltar Point	27/04/2018	29/04/2018	1	
Kirkby on Bain	29/04/2018		1	Adult
Gibraltar Point	02/05/2018	03/05/2018	1	
Frampton Marsh	03/05/2018		1	
Barton Pits	17/04/2019		1	Adult
Gibraltar Point	24/05/2019		1	Adult
Gibraltar Point	27/05/2019		1	Adult
Gibraltar Point	29/05/2019		1	

Cetti's Warbler *Cettia cetti*

Scarce and increasing breeding resident and passage migrant/winter visitor; first proven breeding in 2008. RBBP records show that an average of 40 pairs per year bred during the period 2013-17

The colonisation of the county by this species has been phenomenal from first territory holding in 1995 to first proven breeding in 2008 and at least 130 singing males by 2020. The distribution of breeding birds is still highly concentrated with a minimum of 93 males along the south bank of the Humber and a further 12 males on the adjacent River Freshney in 2020. The sea-bank clay pits are another stronghold, but many inland wetlands and river corridors remain unoccupied and there is clearly still room for expansion. Outside of the breeding season many males remain in their territories, but post-breeding dispersal sees birds appear in other habitats particularly along the coastal strip during Sep-Nov in particular. Non-vocal birds are particularly difficult to locate and recorded winter numbers are clearly lower than the number of birds present.

1968-1972	1980-1989	2008-2011	2016-2019	Incidence	Change
				2%	532%

Long-tailed Tit *Aegithalos caudatus*

British form *rosaceous* a common resident and partial migrant.

The Long-tailed Tit is known for its vulnerability to hard winters, but the population can spring back relatively quickly. The *LBC Atlas* estimated that there were between 3,000-6,000 pairs in Lincolnshire in the late 1980s and the species was thought to be increasing. It continued to spread and in the *BTO Atlas 2007-11* it occupied the nine coastal 10km squares from which it had been hitherto absent. The spread may have been caused by a run of mild winters that allowed the population to grow and an increasing tendency to feed in gardens during the winter reflected in the results of the Garden Bird Feeding Scheme up to 2012 as reported by Goodall in the LBR 2011.

Unfortunately, Long-tailed Tit was not sufficiently common throughout the period to produce a reliable long term Lincolnshire BBS index. The estimated population was 8,000 pairs in 2016. However, following the "Beast from the East" which struck in Mar 2018, Lincolnshire BBS results showed Long-tailed Tit decreased by 27% between 2013-18. It recovered by 7% in 2019.

1968-1972	1980-1989	2008-2011	2016-2019	Incidence	Change
				33%	20%

Wood Warbler *Phylloscopus sibilatrix*

LBRC

Very scarce and declining passage migrant. Formerly bred up until 1951.

Cordeaux (1872) reported that "this exquisite and delicate bird" was a rare and declining summer visitor in the north east "found only in old woodlands and parks". Smith and Cornwallis (1955) report it was formerly local across the county in suitable beech and oak woods. By the 1920s it was confined to the north-west and five pairs bred in Manby Woods near Scunthorpe in 1920. Singing males were reported there till 1951 and since then occasional spring males stop and sing for a few days. Burton Pits was good for them in the 1970s.

The species has gradually grown scarcer with the occasional good year. From 1979-95 there were 10-11 birds, but later from 1996-2019 it was only four to five. Through the same period the species has tended to occur less in spring and more in autumn; 59% of birds were found before mid-Jul during 1979-95, but during 1996-2019 this fell to 39%.

1968-1972	1980-1989	2008-2011	2016-2019	Incidence	Change

Western Bonelli's Warbler *Phylloscopus bonelli*

BBRC

Vagrant. Southern Europe.

One record in May 2016 at Gibraltar Point. First heard in song at the north end of the reserve where it favoured varied scrub with Hawthorn *Crataegus monogyna*, and the occasional Pine *Pinus sp*, not too dissimilar to its habitat in the Mediterranean. The call note (see below) excluded Eastern Bonelli's Warbler *P. orientalis*. The bird showed on and off to about 60 birders before disappearing into the mass of vegetation to the north and adjacent to the golf course at 09.45. It returned early afternoon for about an hour but was last seen at 14.00 that day.

Western and Eastern Bonelli's warbler were split in 1997 with DNA analysis and significant vocal differences supporting the split with both species being monotypic (BOU 1997). Following the split BBRC reviewed 121 British records and published 51 accepted as *bonelli*. The remainder were assigned as 'indeterminate' (BBRC 1997).

The first accepted British record of *orientalis* was on the Isles of Scilly Sep-Oct 1997 (Rogers *et al* 1998). To 2019 there have been 150 records of *bonelli*, averaging three to four per year, and just eight of *orientalis*, the last in May 2016 on the Calf of Man. The world population of *orientalis* has been estimated at 1% of *bonelli* so it is likely it will remain much the rarer of the two.

Plumage details overlap and even trapped birds may not be assigned; recording the characteristic calls remains crucial in confirming identity – a sparrow-like 'chup' in *orientalis* and a rather Willow Warbler-like 'hooeet' in *bonelli*.

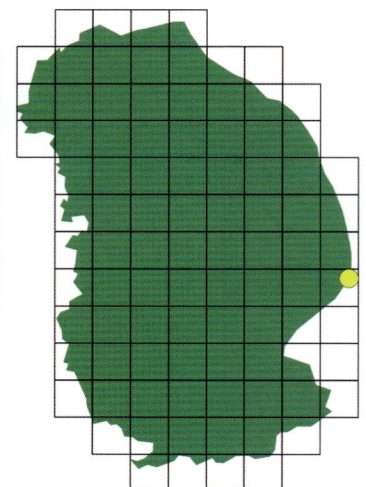

Site Name	First Date	Last Date	Count	Notes
Gibraltar Point	08/05/2016		1	

Hume's Warbler *Phylloscopus humei*

BBRC
BritishBirds Rarities Committee

Vagrant. Central Asia, winters mainly Indian Subcontinent.

There have been just three county records. The first of the two in 2003 was seen by just one lucky observer in the Sycamores *Acer pseudoplatanus*, at Anderby Creek on Oct 18th. The second was found at Gibraltar Point on Oct 23rd 2003 and often showed very well as it flitted about some of the Sycamores and Hawthorns *Crataegus monogyna* remaining until Oct 29th. The third record was also at Gibraltar Point for eight days in 2013 and similar to the second in 2003, spent most time feeding in Sycamores.

As is typical for a far eastern warbler, this species has a late arrival date in Britain with most from the middle of Oct into Nov; several overwintering individuals have also now been noted.

It was as long ago as 1955 that attention was drawn to the fact that the (then) Hume's Yellow-browed Warbler *P. inornatus humei* was a serious contender for addition to the British List. It was granted full species status in 1997 and formally added to the British List in 1997 (BOURC 1997). The first accepted British record was a bird at Beachy Head, Sussex, Nov 13th-17th 1966, and since then the total number of records 1950-2018 stands at 154. Along with some other eastern warbler species, there was an influx in the 'big year' of 2003 (28) but more usually the species averages about five records per year.

Site Name	First Date	Last Date	Count	Notes
Anderby	18/10/2003		1	
Gibraltar Point	23/10/2003	29/10/2003	1	
Gibraltar Point	27/11/2013	04/12/2013	1	

Yellow-browed Warbler *Phylloscopus inornatus*

Scarce but increasing passage migrant Sep-Nov, very scarce inland.

Over the last 10 years, this small *Phylloscopus* warbler has become almost too frequent to accurately record their annual arrival pattern in autumn. Always treated as a scarce migrant in Britain, White and Kehoe (2019) in the *British Birds* Scarce Migrants report for 2017 announced that henceforth they would no longer be considering the species. In 2017 there were nearly 2,000 British records, half the 2016 record arrival of 4,000 or so. Writers in LBR shared the difficulty faced by White and Kehoe and gave up recording autumn arrivals in detail after the 2016 autumn season, that saw a suspected 148 individuals in Lincolnshire. This followed what had seemed big years in 2014 (45) and 2015 (94). This species has been expanding its range west of the Urals and increasing numbers appear to be wintering in south-western Europe. Formerly it was thought the entire population migrated south-east to winter in South-East Asia.

Their autumn arrival in Britain corresponds with easterly airflows and the phenomenon of reverse migration was invoked to explain their occurrence here. It now appears we have been watching evolution in action over the last 50 years as part of the breeding population has switched migration strategy through optimising their chances of winter survival by shorter migration routes. To emphasise this point a bird ringed at Gibraltar Point on Oct 3rd 2015 was retrapped at Nanjizal, Cornwall seven days later. Did it end up wintering in Iberia or drowning in the Atlantic Ocean?

There is a good chance given the large numbers involved that the birds have adapted their migration route possibly very successfully, but birders have always been concerned that the latter option of mass death by drowning is more likely. In terms of big counts Gibraltar Point has seen most of the action with 24 on Oct 2nd 2016 and 18 on Oct 7th 2018. There were spring records at Gibraltar Point on Apr 20th 2004 and May 7th 2017 strongly suggesting the autumn migrants don't all end up in the drink.

Pallas's Warbler *Phylloscopus proregulus*

LBRC
Lincolnshire Bird Records Committee

Very scarce but increasing passage migrant Sep-Nov, rare inland. Siberia.

First recorded in 1968 when there were singles at Anderby Creek and Gibraltar Point. This species has become much more familiar to birders but still one of the great pleasures of an autumn easterly is to encounter one hovering on the edge of a coastal sycamore, hawking insects. Occurrences since 1968 have been patchy but generally show an upward trend with six or more records in several years and peaks of 17 in 2003, 12 in 2016 and seven in 2019. Of particular interest was the autumn of 2003 in Britain when there was a record arrival of both Pallas's (303) and Yellow-browed Warblers *P. inornatus* (853).

Surprisingly in view of their increased status in Britain, there were blank years in Lincolnshire in both 2017 and 2018. The peak arrival period in Lincolnshire for these amazing little warblers tends to be Oct 11th-31st with a significant minority going into Nov. The extreme dates for the county in autumn are Oct 7th 2016 and Nov 24th 2019. There has been a single record of a spring bird: a male found singing at Freiston Shore May 7th 2007. This was one of two spring records that year, the other being in Kent, the third and fourth spring records in Britain at the time.

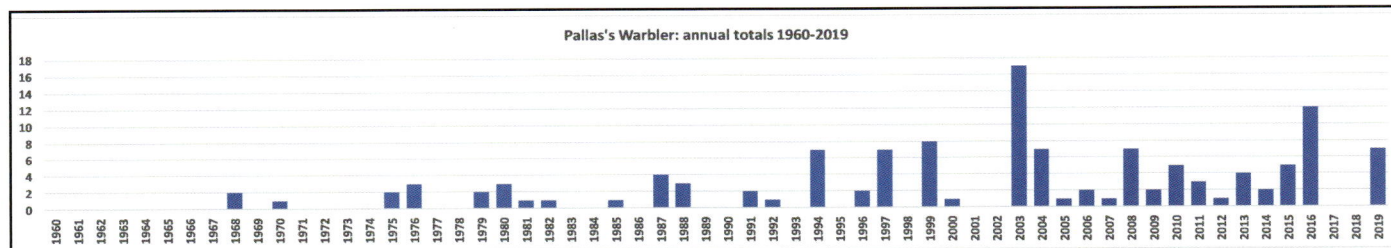

Pallas's Warbler: annual totals 1960-2019

Radde's Warbler *Phylloscopus schwarzi*

LBRC

Rare. Siberia.

Gustav Radde, a Prussian apothecary and keen naturalist found and collected a leaf-warbler he had never seen before near Lake Tarei Nor in eastern Transbaikalia, Oct 5th 1856. He named it after himself and later gave it the Latin name *Sylvia (Phyllopneuste) schwarzi* in honour of his friend the astronomer *Schwarz* (Neufeldt 1961). Later of course it became Radde's Warbler *P. schwarzi*. Only 42 years later Caton Haigh shot the first county and British record in Oct 1898, a female, at North Cotes. There have now been 14 county records to 2019, all at coastal sites. Every one of the records has turned up in the period Oct 1st-24th. Six of the 14 were trapped including two of the three birds which occurred in 2016.

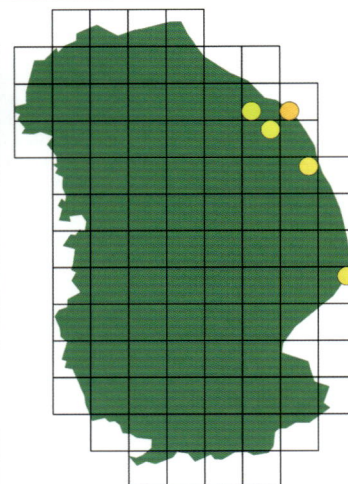

Radde's Warbler ceased to be a BBRC description species from 2006. White and Kehoe (2020) noted that there have been 444 records, 1958-2018 with an average of 12 records per year, 1990-2018. Along with the six birds ringed in the county, 112 others have been ringed in Britain, 1909-2019, but not unexpectedly perhaps, there have been no recoveries or controls.

Site Name	First Date	Last Date	Count	Notes
North Cotes	01/10/1898		1	
Saltfleetby-Theddlethorpe	16/10/1988		1	Trapped
Donna Nook	21/10/1990	22/10/1990	1	First winter
Saltfleetby-Theddlethorpe	24/10/1990		1	Trapped
Donna Nook	01/10/2000	03/10/2000	1	
Grainthorpe	01/10/2000		1	
Gibraltar Point	13/10/2003		1	
Marshchapel	12/10/2006		1	
Donna Nook	04/10/2007		1	Trapped
Donna Nook	11/10/2014	14/10/2014	1	
Gibraltar Point	14/10/2014		1	
Saltfleetby-Theddlethorpe	15/10/2014		1	Trapped
Donna Nook	15/10/2016		2	Trapped
Gibraltar Point	18/10/2016		1	

Dusky Warbler *Phylloscopus fuscatus*

LBRC

Rare. Siberia.

One ringed at Huttoft in Nov 1964 was the first county record. There have been 15 more records involving 12 birds, six of which were trapped and ringed. This is a similar picture to its close congener, Radde's Warbler *P. schwarzi*, also a skulking species and just as likely to be trapped as seen.

The earliest bird in the autumn arrived on Sep 25th 2001, the only Sep record, and the latest Nov 16th 2019. The longest stay concerned a wintering bird found at Huttoft Bank on Jan 1st 2019, remaining in the area until Feb 3rd at least. It is fair to assume that it had arrived there or somewhere nearby on the Lincolnshire coast in autumn 2018 remaining undiscovered until New Year's Day listing got underway in Jan 2019.

As with Radde's Warbler, Dusky Warbler ceased to be a species considered by BBRC in 2006. In the *Report on scarce migrants in Britain in 2018* (White and Kehoe 2020) the total number of British records 1958-2018 was reported to be 590. The annual average number of records rose from 13 during 1990-99 to 27 in 2010-18, although the report adds that the annual variability during 2000-18 was high.

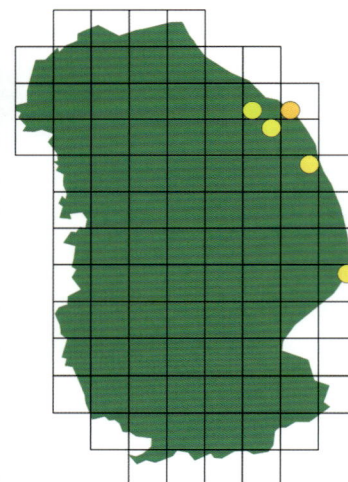

Site Name	First Date	Last Date	Count	Notes
Huttoft	01/11/1964		1	Trapped
Donna Nook	03/11/1980	09/11/1980	1	Trapped
Donna Nook	12/10/1988	13/10/1988	1	
North Cotes	19/10/1990		1	
Saltfleet	19/10/1990	20/10/1990	1	
Donna Nook	25/09/2001	26/09/2001	1	
Saltfleetby-Theddlethorpe	12/10/2013	14/10/2013	1	Same
Saltfleetby-Theddlethorpe	12/10/2013		1	Trapped
Gibraltar Point	15/10/2015	21/10/2015	1	Trapped
Saltfleetby-Theddlethorpe	15/10/2015		1	Trapped
Saltfleetby-Theddlethorpe	19/10/2015		1	Same
Gibraltar Point	12/10/2016	13/10/2016	1	
Gibraltar Point	15/10/2016	18/10/2016	1	Same
Huttoft	01/01/2019	03/02/2019	1	
Anderby	16/11/2019		1	

Willow Warbler *Phylloscopus trochilus*

Nominate form a very common but declining summer visitor and passage migrant. Northern form *acredula* scarce passage migrant to the UK.

It is a fact of life that the population of species changes through time for all sorts of reasons. In the *LBC Atlas* period in the 1980s, Willow Warbler was by far and away the commonest breeding warbler in Lincolnshire with an estimated population of 50,000-59,000 pairs. The estimate in 2016 was 14,000 pairs which sees Willow Warbler sliding down the scale to number five in the Lincolnshire warbler league table, from its former number one position. It retains number one spot for Britain as a whole. The BTO Lincolnshire BBS index shows that the decline of the species here is worse than other parts of the country. It remains widespread but has thinned out and become scarce in many areas. It is interesting that this decline is mirrored across south-east Britain but not in the north-west, nor in Scotland. The reason for the decline is thought to be related to decreasing productivity of young in south-east Britain but problems on migration routes have not been ruled out.

Why Chiffchaff *P. collybita* is increasing in Lincolnshire while Willow Warbler is suffering, points to the intriguingly delicate balances that exist between success and failure for seemingly similar species. It seems likely that as a trans-Saharan migrant the pressures on Willow Warbler are greater. Will it adapt its lifestyle and bounce back? Readers 50 years hence will hopefully know the answer. There is no accepted record of the northern form of Willow Warbler, *P. t. acredula* having occurred in Lincolnshire.

1968-1972	1980-1989	2008-2011	2016-2019	Incidence	Change
				28%	-24%

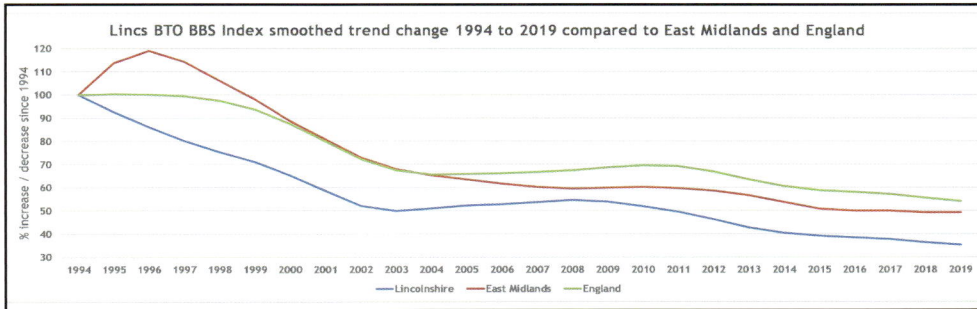

Lincs BTO BBS Index smoothed trend change 1994 to 2019 compared to East Midlands and England

Chiffchaff *Phylloscopus collybita*

Common summer visitor and passage migrant. Rare/scarce in winter. Siberian Chiffchaff *P. c. tristis*, very scarce migrant and winter visitor. Scandinavian/Baltic subspecies *P. c. abietinus* vagrant.

Chiffchaff has become much commoner in Lincolnshire over the last 30 years as reflected by the Lincolnshire BTO BBS index and its range increase between the *LBC Atlas* of 1980-99 to the *BTO Atlas 2007-11*. The *LBC Atlas* estimated the population at 6,700 pairs in the 1980s and the estimate for 2016 was 21,000 pairs. It is now our commonest *phylloscopus* having switched places with Willow Warbler *P. trochilus*. Interestingly, BTO Bird Trends reported that it suffered badly along with other migrants in the Sahel drought of the early 1970s. It then rebounded strongly. It may be that this increase has been driven by a northerly shift in its wintering area which has improved its winter survival success. As the earliest arriving warbler species in spring, LBR reflecting arrival from Mar onwards, it may also maximise its chick productivity through more reliably producing two broods of chicks.

1968-1972	1980-1989	2008-2011	2016-2019	Incidence	Change
				54%	40%

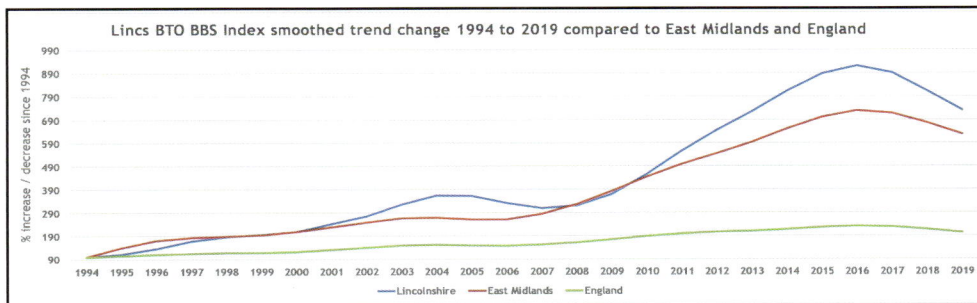

Lincs BTO BBS Index smoothed trend change 1994 to 2019 compared to East Midlands and England

Lorand and Atkin (1989) had no records of either *P. c. tristis* (Siberian Chiffchaff) or of the Scandinavian/Baltic subspecies *P. c. abietinus*. Shipilina *et al* (2017) showed that there is a *tristis/abietinus* zone of hybridisation in Russia. Although abietinus averages paler than *collybita* most cannot be identified on plumage alone. Such intergrades might explain those Siberian Chiffchaff candidates which show lots more green and yellow in the plumage than 'traditional' *tristis* and have mixed call types. Annual claims of *abietinus* in the 1980s and 1990s cannot be considered to be sound in the light of this and there have been no claims of *abietinus* since the late 1990s.

Valid records of *tristis* have been reported from 1980 onwards with between one and 10 records per year, mostly in Oct-Dec and at coastal sites. A few remain into Mar and early Apr and a notable record was of a singing male at Goxhill Marsh Mar 11th-23rd 1989. Catley (2000) stated he was attracted to the bird because of its unusual song which he described as "a mix of 'chiff' notes given with a delivery recalling the song of Willow Warbler which I noted as 'chueet to tweet tweet'".

Iberian Chiffchaff *Phylloscopus ibericus*

BBRC

Vagrant. South-western Europe. One county record in May 2019, a second pending record May 2020.

The first county record was of a bird heard singing in the east dunes at Gibraltar Point, May 7th-10th 2019. The bird remained in the same area of dune over the next two days and mostly sang while fairly well concealed. The weather was generally damp with cold northerly winds, but it sing regularly during the early morning periods, even when it was raining. Later, on the early morning of 10th May it delivered full song repeatedly, having moved over to a more open Sallow *Salix caprea*, and became bolder, moving almost out of cover, nearing the tops of the foliage. Shortly after this the bird disappeared. A second singing male was found in May 2020 at Croft, Skegness, well seen and photographed, pending BBRC acceptance.

The BOURC added Iberian Chiffchaff to the British List in their 27th Report (Oct 2000) and in 2002 the Taxonomic Sub-committee of the BOURC recommended that the correct Latin name of Iberian Chiffchaff should be *P. ibericus* (Knox *et al*, 2002). Following a review of historic records, the first British record was a bird seen and sound recorded at Brent Reservoir, Greater London on Jun 3rd 1972. Amazingly, the species bred for the first time in Britain in 2015 in the Nedd Valley, Gower (Hunter 2018).

After the first record the second wasn't until Apr 14th 1992 on the Isles of Scilly. To the end of 2019 there had been 80 accepted records, averaging about six per year.

Site Name	First Date	Last Date	Count	Notes
Gibraltar Point	07/05/2019	10/05/2019	1	Adult Male

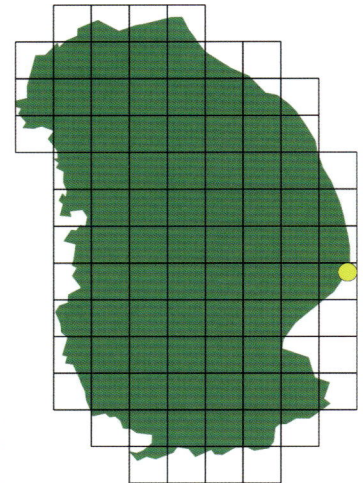

Greenish Warbler *Phylloscopus trochiloides*

LBRC

Rare. The Baltic eastwards to western Siberia, central-eastern Himalayas and central Asia.

Caton Haigh shot the first county record, a female, in Sep 1896. The next record wasn't until 1958 and the total up to 2019 now stands at 18 records in all, five of which were birds trapped and ringed on the coast at or near Gibraltar Point. Sixteen turned up in Aug-Sep, just two in May.

Explaining the relative dearth of records of this species and some other rare and scarce species in the county is something of an annual chore. Whether down to geography, weather patterns or coverage, Greenish Warbler records in neighbouring coastal counties are usually significantly higher. For example, in 2007 when there were 42 British records in all, an incredible 34 were in Norfolk; none were seen in Lincolnshire that year.

The taxonomy of this species, Two-barred (Greenish) *P. plumbeitarsus* and Green Warbler *P. nitidus* has been somewhat fluid with IOC splitting this complex in 2017. There have been no claims of the latter two species in the county. Greenish Warbler ceased to be considered by BBRC from 2006. The British total 1958-2017 stood at 741 records with a mean of 26 records per year 2010-17 (White and Kehoe 2019); the other two taxa remain the province of the BBRC.

Site Name	First Date	Last Date	Count	Notes
North Cotes	05/09/1896		1	Female
Gibraltar Point	03/09/1958		1	Trapped
Anderby	14/08/1976		1	Immature
Saltfleetby-Theddlethorpe	22/08/1976		1	
Donna Nook	23/08/1976		1	
Saltfleetby-Theddlethorpe	21/08/1977	22/08/1977	1	
Seacroft	23/08/1977		1	Trapped
Saltfleetby-Theddlethorpe	31/08/1981		1	
Gibraltar Point	02/09/1981	04/09/1981	1	
Gibraltar Point	22/09/1984		1	First winter trapped
Donna Nook	10/09/2002	13/09/2002	1	
Saltfleet	11/09/2005		1	
Gibraltar Point	13/09/2009	17/09/2009	1	
Saltfleetby-Theddlethorpe	18/09/2009		1	
Gibraltar Point	28/08/2012		1	
Gibraltar Point	29/05/2013		1	Male trapped
Gibraltar Point	31/05/2013		1	Same
Donna Nook	30/05/2014		1	Trapped
Gibraltar Point	16/08/2016		1	Trapped

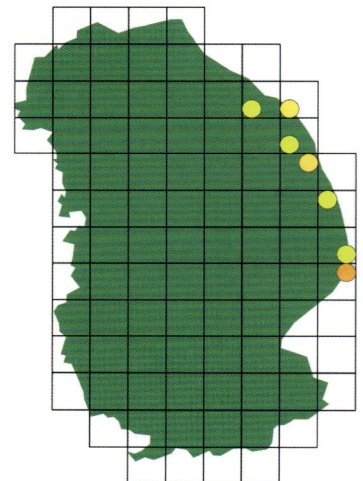

Arctic Warbler *Phylloscopus borealis*

Rare. Scandinavia, North-eastern Europe.

One shot at North Cotes on Oct 24th 1932 was the first county record. Fortunately, times have changed, and the succeeding 10 records were either well-documented sight records (seven) or were trapped and ringed (three). All were recorded between Sep 3rd and Oct 24th. A contentious bird trapped in May 1983 at Gibraltar Point and submitted as this species was assessed at length by BBRC but finally considered not proven and is not included here.

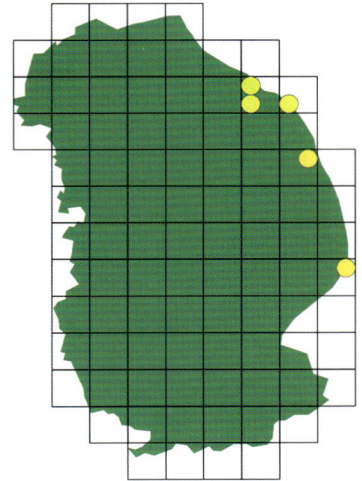

Between 1950 and 2018 there were 398 British records and the BBRC ceased to consider this species after 2018.

Site Name	First Date	Last Date	Count	Notes
North Cotes	24/10/1932		1	
Gibraltar Point	19/09/1976	27/09/1976	1	
Humberston	10/10/1978		1	
Saltfleetby-Theddlethorpe	10/09/1985		1	First winter trapped
Gibraltar Point	20/09/1986	21/09/1986	1	First winter trapped
Tetney	26/09/1986	29/09/1986	1	First winter
Donna Nook	03/09/1995		1	
Saltfleetby-Theddlethorpe	18/09/1995		1	
Donna Nook	11/10/2014	14/10/2014	1	
Gibraltar Point	11/10/2018		1	
Saltfleetby-Theddlethorpe	05/10/2019		1	Juvenile trapped

Great Reed Warbler *Acrocephalus arundinaceus*

Vagrant. Western Europe.

One was present in a reed bed near Tetney Lock for several weeks in Jul 1897 and several people including Cordeaux heard it singing. A long period elapsed before the second in May 1969, which was trapped at Huttoft Bank. A mobile singing male turned up in May 1976 at Chapel Pit, Chapel St. Leonard's and stayed until the end of Jun 1976. It then made a brief appearance at nearby Wolla Bank on Jul 11th before disappearing until late Jul when it was found at Huttoft Pit, Jul 26th-28th. Another, the fourth, was at Burton Pits in Jul 1979 after which there were singles, all in May, in 1980, 1990 and 2000. Four further records occurred, including one trapped at Gibraltar Point in May 2014, which was a first for the reserve, with the last in 2016 bringing the total to just 11 birds. Note that all of the records were in spring and summer, extreme dates being May 3rd-Jul 28th, and that 10 of the 11 were males.

Great Reed Warbler is a classic spring overshoot and over 80% of records fall between early May and mid-Jun, most in the southern half of England. They often appear at inland reedbeds as well as on the coast. In Britain there have been 288 records, 1950-2018, with an average of about six per year.

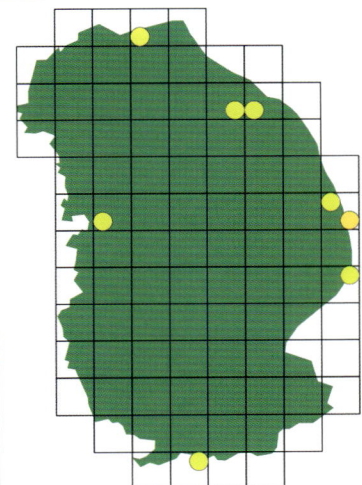

Site Name	First Date	Last Date	Count	Notes
Tetney Lock	01/07/1897	22/07/1897	1	Male,
Huttoft	03/05/1969		1	Trapped
Chapel St Leonards	23/05/1976	30/06/1976	1	Male
Chapel St Leonards	11/07/1976		1	Same male
Chapel St Leonards	26/07/1976	28/07/1976	1	Same male
Burton Pits	06/07/1979	22/07/1979	1	Male
North Cotes	20/05/1980		1	
Waithe	19/05/1990		1	
Chapel St Leonards	29/05/2000	01/07/2000	1	
Deeping St James	16/05/2001	22/05/2001	1	Adult male
Barton Pits	24/06/2006		1	
Gibraltar Point	23/05/2014		1	Trapped
Barton Pits	05/06/2016	06/06/2016	1	Second calendar year male

Aquatic Warbler *Acrocephalus paludicola*

Vagrant. Eastern Europe.

A 1CY bird was trapped at Theddlethorpe on Sep 8th 1971 and another was seen at Saltfleetby-Theddlethorpe on Oct 29th 1979.

Originally a BBRC species, it became numerous enough to devolve to county records committees from 1983-2014. Although the total number of British records 1950-2018 stands at 1,362 this masks the rapid decline of this species in eastern and central Europe and it returned to BBRC consideration in 2015.

There have been 45 records 2010-18, but only 15 since 2013 and it remains a Lincolnshire vagrant; another would be more than welcome.

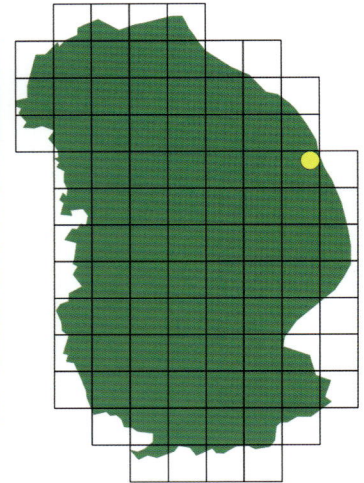

Site Name	First Date	Last Date	Count	Notes
Saltfleetby-Theddlethorpe	08/09/1971		1	First calendar year
Saltfleetby-Theddlethorpe	29/10/1979		1	

Sedge Warbler *Acrocephalus schoenobaenus*

Common summer visitor and passage migrant.

A look at the relative abundance map of breeding Sedge Warblers in the *BTO Atlas 2007-11* shows that Lincolnshire is at the heart of this species range in England with the Fens, Marsh, Ancholme and Trent Valleys and Isle of Axholme all relatively densely populated. This pattern is reflected in the *LBC Atlas* too and the distribution has stayed stable. Unfortunately, there is insufficient data to calculate a Lincolnshire BBS index for Sedge Warbler. The national BBS index shows that the population has declined by 26% in England since 1994. While the *LBC Atlas* put the population at around 7,500 pairs in the 1980s, the estimated population in Lincolnshire in 2016 was 18,000 pairs which seems to be on the high side. Either way the Lincolnshire population appears to be in a relatively healthy state currently compared to other areas of England.

BTO BirdTrends suggests that annual population changes are related to rainfall dependent adult winter survival in the sub-Saharan wintering grounds.

1968-1972	1980-1989	2008-2011	2016-2019	Incidence	Change
				26%	-2%

Blyth's Reed Warbler *Acrocephalus dumetorum*

Vagrant. North-east Europe.

A bird trapped and ringed in Sep 1991 was the first record, one of three in Britain that year. Another was trapped and ringed in 2006 and a third well seen in the east dunes at Gibraltar Point in Sep 2014. A very popular singing male which performed at Far Ings NR in Jun 2020 for more than two weeks will be the fourth but has not yet been formally accepted by BBRC.

The species ceased to be considered by BBRC from 2015. In the period 1950-2017 there had been 242 in Britain although all modern records have occurred since 1979; there are about six per year but with recent significant increases.

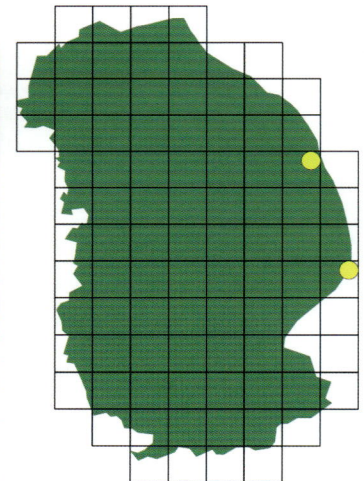

Site Name	First Date	Last Date	Count	Notes
Saltfleetby-Theddlethorpe	03/09/1991	04/09/1991	1	Trapped
Gibraltar Point	16/09/2006		1	Trapped
Gibraltar Point	05/09/2014	06/09/2014	1	

Reed Warbler *Acrocephalus scirpaceus*

Common summer visitor and passage migrant

What has Reed Warbler got that most other insectivorous trans-Saharan migrants lack? A good question to which no-one knows the answer. It migrates the longest distance into Africa and yet has shown the greatest long-term population increase on the BTO BBS index. Since 1994 there has been a 20% increase in England. Its status change throughout that period in Lincolnshire is difficult to assess as it appears in too few BBS squares to produce a local index. However, the *LBC Atlas* estimated a population of 6,000 pairs in the 1980s while the 2016 estimate is 10,000 pairs. It is very much a *Phragmites* specialist when it comes to breeding though it will venture out into surrounding habitat to feed. Lincolnshire has countless narrow reed-filled dykes and ditches and it exploits them to the full, down to the narrowest ditch. In such habitats it becomes easier to monitor. In large inaccessible reed beds it can present an impossible task. During passage it can turn up in all kinds of scrubby habitat and can present novices with an identification challenge.

1968-1972	1980-1989	2008-2011	2016-2019	Incidence	Change
				18%	4%

Marsh Warbler *Acrocephalus palustris*

RBBP LBRC ●

Very scarce migrant and potential breeder. First recorded in Lincolnshire in 1961.

First recorded on Oct 8th 1961 when one was trapped at Tetney, around 40 have now been recorded. Between none and two per year is typical but there were three in 1992, and a remarkable 12 in 2008. That year saw an arrival along the east coast between Shetland and Essex involving about 50 birds from May 27th-31st. The Lincolnshire complement included six in the Donna Nook area, two on Saltfleetby-Theddlethorpe, one at Skegness and three at Gibraltar Point. Most, if not all, were singing males, and that is a feature of the majority of records in the county. Indeed, they are probably very likely to be missed if they are not singing or trapped. Most Marsh Warblers have been found in a short period in late spring with 35 individuals being recorded between May 17th and Jun 25th.

Singing males have held territories in spring in four of the last 29 years. These were in 1997, 2012, 2019 and 2020. There was no evidence of a pair or nest building up to 2019. At one site in 2020 two birds were thought to be present. Marsh Warblers have declined dramatically as a breeding bird in Britain and 17 out of 18 singing birds reported by RBBP in 2018 were along the east coast from Kent to Shetland. There is a healthy breeding population along the Dutch, German and Danish coast which breeds in very similar habitat to the scrubby reedy dykes along the Lincolnshire coast. The European population takes a south-easterly migration route to winter in Africa. Birds returning to the North Sea coast to breed in spring can easily overshoot and end up in Lincolnshire.

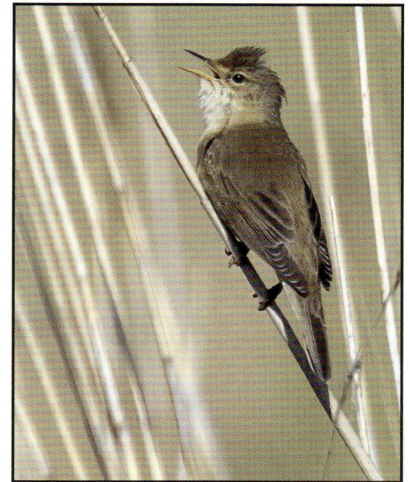

1968-1972	1980-1989	2008-2011	2016-2019	Incidence	Change
				0%	

Booted Warbler *Iduna caligata*

BBRC

Vagrant. Central Asia.

There are two county records: one trapped and ringed at Theddlethorpe in Oct 1980 and the second more than 20 years later at Donna Nook in Sep 2003. This bird frequented thick coastal scrub but eventually worked its way through to the seaward side of the dune where it showed well in a stand of Mayweed *Tripleurospermum sp.*

Formerly in the genus *Hippolais*, Booted Warbler was moved into the genus *Iduna* in 2009 and is monotypic. The first British record was shot on Fair Isle in 1936, the second only arriving 30 years later. Since then, the flood-gates have opened and the British total to 2018 is now 169 records, with around 50 records in each of the last three decades with most turning up in mid-Sep.

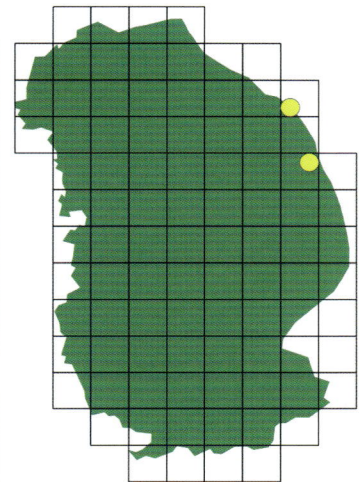

Site Name	First Date	Last Date	Count	Notes
Saltfleetby-Theddlethorpe	12/10/1980		1	
Donna Nook	06/09/2003		1	

Icterine Warbler *Hippolais icterina*

LBRC

Very scarce and declining coastal passage migrant mainly Aug-Oct.

Yet another county first that was recorded after being shot at North Cotes on Sep 4th 1922, with the second not until 1951. Spring birds arrive late May-Jun and autumn birds from the first week of Aug with most arriving in the last seven to 10 days of Aug and the first three weeks of Sep. Some of the influxes which occurred in the 1970s were remarkable compared to current times. For example, in autumn 1977 around 20 were reported (allowing for birds staying for more than one day) all at coastal sites, with up to six at Donna Nook, four at Gibraltar Point and one or two at Anderby, Saltfleet and Theddlethorpe. In total there have been around 120 records up to 1990. Since then, the species' fortunes have declined and there was an average of one to four per year during the 2000s.

There were none 2000-01 and the first birds of the new millennium were in 2002 (four) which included an over-wintering bird at Donna Nook, Sep 4th-Dec 31st. There were singles 2005-07, followed by eight in 2008 which included an influx on May 28th with five on the coast and another singing male inland on Jun 5th. Since then, 2009-19 there have been just one or two each year apart from 2017 when there were none. Most were seen in 2013 with a singing male in Jun at Gibraltar Point and six in the autumn. It is interesting to note the increased number of spring birds: just one during 1951-76, four during 1977-99 and 12 during 2002-19. Exceptional inland with one trapped at Hemswell Jun 20th 1970 and singing males at Crowle Moors Jun 5th 2008 and at Swanpool, Lincoln on Jun 11th 2019.

Lanceolated Warbler *Locustella lanceolata*

BBRC

Vagrant. Siberia.

The first record of this diminutive *Locustella* fell to the gun of Caton Haigh in Nov 1909 and at the time it was the second British record. The first was on Fair Isle in Sep 1908, and was also shot; as all birders know, the vast majority have turned up on the Shetland Isles which have seen 131 of the 161 British records to 2018, many of them trapped. The second Lincolnshire record was discovered lurking in the saltmarsh at the Rimac end of Saltfleetby-Theddlethorpe. As is often the case, very close views were obtained by a fortunate group of around 30 birders as, true to form, it crept mouse-like in the saltmarsh vegetation. It was nowhere to be seen the next day.

This Eastern Palaearctic species breeds across the taiga in damp valleys, on lake and marsh edges, open meadows with scattered bushes, damp woodland clearings and forest-edge scrub. It reaches as far west as eastern Finland. Will the next one turn up in a mist net or Heligoland trap on the Lincolnshire coast? Hope springs eternal.

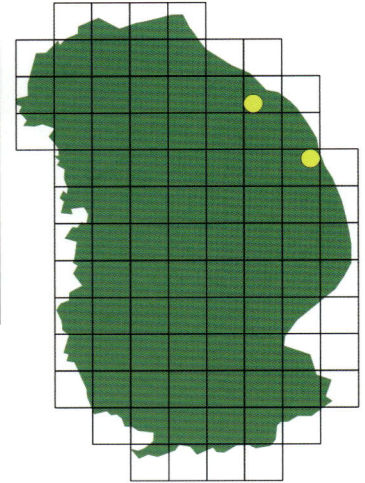

Site Name	First Date	Last Date	Count	Notes
North Cotes	18/11/1909		1	Male
Saltfleetby-Theddlethorpe	22/09/1996		1	

River Warbler *Locustella fluviatilis*

BBRC

Vagrant. Eastern Europe.

A singing male was discovered in Elder *Sambucus niger*, and Sea Buckthorn *Hippophae rhamnoides*, at the north end of Gibraltar Point on May 29th 1998. Conditions were conducive for an eastern vagrant with sea mist and a strong easterly breeze.

It proved frustratingly elusive, singing intermittently from within cover but provided sufficient views for the finder to note all relevant details. The song was a typical loud mechanical buzz, tze-tze-tze-tze-tze... producing a rhythm similar to that of a sewing machine. This was easily audible up to 100m away. The bird also gave a separate high-pitched ticking call heard at close quarters.

The first British record was not until Sep 24th 1961 and, typically on Fair Isle. There have been 47 records in all, the most recent being three in 2018 with two on Shetland and one in Caithness, thus averaging one to two records per year.

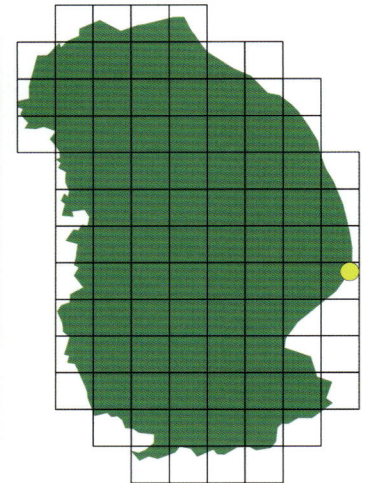

Site Name	First Date	Last Date	Count	Notes
Gibraltar Point	29/05/1998		1	Male

Savi's Warbler *Locustella luscinioides*

LBRC ●

Vagrant. Southern Britain, Europe

Savi's Warbler remains a vagrant in Lincolnshire with six records between 1967 and 1992. The first was trapped at Pyewipe Marsh near Lincoln and the BBRC report for that year revealed that the species had for several years been breeding in Kent and that "records from two other counties (Lincolnshire and Norfolk) encourage the hope that further suitable localities will be colonised in the course of time". The birds in 1986 and 1992 were singing males, the first present for eight days, the second more briefly. Optimism for colonisation has since dimmed and the species remains a very rare breeder in Britain with 10 pairs or fewer every year since 1993 (Holling 2019). In England they tend to prefer extensive managed reedbeds of a scale that is rare in Lincolnshire. Most of the records since 1986 have been in East Anglia and south coast counties like Kent and Hampshire.

Abroad, they breed in marshes, fens and lake edges, reedbeds over shallow water, especially with underlayer or clear areas of sedges and rushes, often with scattered bushes; also, in tall grass and bushes along riverbanks, and reedy canals within sparse forest. It is to be hoped that with the number of wetland regeneration schemes being undertaken in Britain that the species takes advantage of new habitat and increases its presence in both Lincolnshire and Britain. The European population was thought to be stable between 1980 and 2011.

The number of migrants recorded in Britain stands at 664 from 1950-2018 and contrasts sharply with the small breeding population. Savi's warbler is a long-distance migrant and the entire Western Palearctic population migrates to Africa.

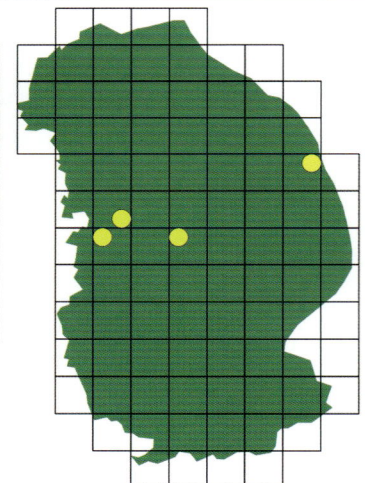

Site Name	First Date	Last Date	Count	Notes
Lincoln	09/05/1967	11/05/1967	1	Trapped
Saltfleetby-Theddlethorpe	03/08/1969		1	Adult
Bardney	22/08/1969		1	Adult
Saltfleetby-Theddlethorpe	05/09/1976	07/09/1976	1	
Boultham Mere	18/06/1986	25/06/1986	1	
Chapel St. Leonards	23/05/1992	24/05/1992	1	Male at Chapel Pit LWT

148

Grasshopper Warbler *Locustella naevia*

Scarce summer visitor and passage migrant. Decline in late 20th century.

The reeling of a "Gropper" is a delight to hear on a spring evening. Judging from Cordeaux (1872) and Smith and Cornwallis (1955), it seems it has always been scarce in Lincolnshire, but it may have gone through a temporary increase by exploiting young forestry plantations during the boom in coniferous afforestation from the 1920s through to the 1970s. It is also found along the coast in scrubby reed beds and buckthorn. The *LBC Atlas* put the population at 150-200 pairs in the 1980s and suggested it had declined since the 1960s. In the *BTO Atlas 2007-11* it was confirmed breeding in only six 10km squares, down from 11 in the 1980s. Overall the BBS index for England shows it has declined by some 44% since 1994. The Lincolnshire population is difficult to estimate but is probably less than 50 pairs. On arrival in spring it can turn up in any hedge reeling for a day before moving on which may cause it to be overstated in BBS. First birds arrive in early Apr and LBR shows that the best showing in the five years to 2018 was of eight singing at Alkborough Flats in May 2017. Other sites which report more than single birds regularly include Barton Pits, Donna Nook, the Rimac area of Saltfleetby-Theddlethorpe, Anderby and Gibraltar Point. Passage is usually over by late Sep, but Gibraltar Point had a late one on Oct 18th 2015.

1968-1972	1980-1989	2008-2011	2016-2019	Incidence	Change
				0.9%	-67%

Blackcap *Sylvia atricapilla*

Very common summer visitor and passage migrant. Scarce winter visitor Dec-Mar.

Something is going right for Blackcap. Since 1994 the Lincolnshire BBS index shows the species has increased by 205% and its soulful melodic song is heard increasingly throughout the county. Much of the increase has come in the last 10 years. The *LBC Atlas* estimated a population of 12,000 pairs in the 1980s and the estimate in 2016 was 28,000 pairs corresponding closely to the trend increase. Why has it been so successful? BTO BirdTrends states that the cause is unknown. Several factors might be involved. Amongst the warblers, Blackcap is very flexible in both the habitats it chooses to breed and the food it chooses to eat. In addition, it may have shortened its migration by adopting a more northerly wintering area within the Western Palearctic. At one time it appeared that Blackcaps might start to winter in Lincolnshire more frequently, but that behaviour remains relatively scarce. It is likely that rather than being local breeders, wintering birds may come from further east. A male ringed at Theddlethorpe on Oct 13th 2013 was re-trapped in the Czech Republic on May 19th 2018, presumably on its breeding grounds.

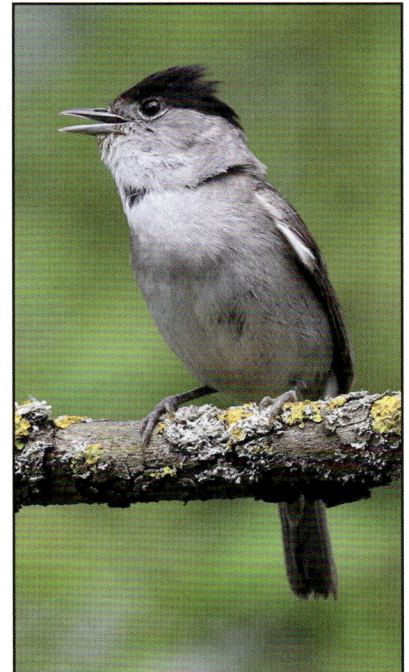

1968-1972	1980-1989	2008-2011	2016-2019	Incidence	Change
				71%	27%

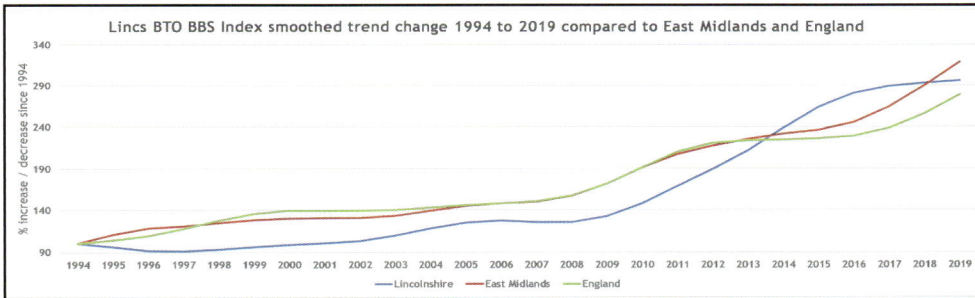

Lincs BTO BBS Index smoothed trend change 1994 to 2019 compared to East Midlands and England

Garden Warbler *Sylvia borin*

Fairly common summer visitor and passage migrant.

Garden Warbler is becoming scarcer in Lincolnshire. Its preference for deciduous woodland with good undergrowth makes it very scarce in much of the Fens, the Marsh and north-east edge of the Wolds. The *LBC Atlas* distribution map shows this pattern very well and it comes into sharp focus when viewed with the relative abundance map of this species in the *BTO Atlas 2007-11*. The *LBC Atlas* estimated the population at 7,200 pairs in the 1980s. The estimated population for 2016 was only 600 pairs. The BBS index for England shows a long-term decline of 27% since 1994. Clearly the Lincolnshire decline is much steeper and recalls that of Nightingale *Luscinia megarhynchos*, a species whose habitat it shares, and its decline may be for similar reasons. BTO BirdTrends suggests Garden Warbler decline is likely to be driven by reduced productivity or juvenile survival. Could another possibility may be excessive grazing of woodland undergrowth by deer?

1968-1972	1980-1989	2008-2011	2016-2019	Incidence	Change
				4%	-43%

Barred Warbler *Curruca nisoria*

LBRC

Very scarce/scarce coastal passage migrant mainly Aug-Oct, exceptionally to Dec.

One shot at North Cotes on Sep 5th 1898 was the first county record; eight more were reported by Lorand and Atkin (1989) up to 1927 although no specific dates were available. There were no more until regular reporting began in 1951. After that, Barred Warblers were regularly recorded and caught by ringers in small numbers on the coast each autumn. There have been no spring records.

In terms of numbers of records (not individuals) there have been 101 in Aug, 182 in Sep, 42 in Oct, eight in Nov and one in Dec. The heyday for Barred Warblers in the county was in the 1970s and 1980s when 10-15 birds were recorded annually and up to eight ringed per autumn at Gibraltar Point; exact numbers of individuals at coastal sites are difficult to determine with one to two birds present over a long date range.

The earliest autumn migrant was one at Tetney on Aug 4th 1966 and the latest an inland migrant at Covenham Reservoir on Nov 11th 2004 which stayed until Dec 9th. There have been at least six other Nov records between 4th-27th. It is debatable whether migrants are arriving later in the autumn with no Oct records until 1971 (Oct 26th, Gibraltar Point) with the first Nov record also in that year (Nov 2nd, Donna Nook). Total numbers have also fallen and in the 2000s averaged around four birds per year, although this average was inflated by a good year in 2010, when there may have been as many as 15. There have been blank years in 2016 and 2017 and the recent trend is one of decline.

Lesser Whitethroat *Curruca curruca*

Nominate *curruca* common summer visitor and passage migrant. *C. c. blythi* ('Siberian Lesser Whitethroat') vagrant (LBRC). *C. c. halimodendri* (Central Asian Lesser Whitethroat) vagrant (BBRC)

This fascinating little warbler is our only summer visitor that migrates on a north-west, south-east axis, passing through the Middle East to reach its wintering grounds south of the Sahara in north-east Africa. Lincolnshire is therefore on the extreme north-west of its range and the *BTO Atlas 2007-11* relative abundance map shows it does particularly well on the Marsh. Its song culminating in a little rattle coming from inside the top of a hedge is a regular feature of a May bike ride along the lanes of Lincolnshire, except at the top of the Wolds and the southern Fens from which it is largely absent. Its status appears to have been relatively unchanged in Lincolnshire over the last 50 years and its BBS long term trend in England shows the population has grown a modest 11% since 1994. The *LBC Atlas* estimated its Lincolnshire population at anywhere between 1,300-4,500 pairs and the estimated population for 2016 was 4,000 pairs.

1968-1972	1980-1989	2008-2011	2016-2019	Incidence	Change
				21%	-14%

Identifying the 'Asian Lesser Whitethroats' in Britain has been a long process, but the taxonomy has settled somewhat, and genetic analyses have revealed more about the birds reaching Britain. The Siberian subspecies *C. c. blythi*, a regular scarce migrant in mid to late autumn in Britain, is still a vagrant in Lincolnshire. Lorand and Atkin (1989) mention birds 'showing characteristics of *C. c. blythi*' but the first definitive mention of this race was in LBR 1981 when around five individuals were seen Oct 18th-Nov 14th that year, with good descriptive evidence. Two well-documented birds were found at Gibraltar Point in 1987 and at Donna Nook, Oct 2019. From published records in the LBR and the LBC database they average less than one per year, but the suspicion remains that they are under-recorded. There are two records of individuals showing characteristics of *C. c. halimodendri* although the recently accepted British records (eight in all) were all confirmed by DNA analysis.

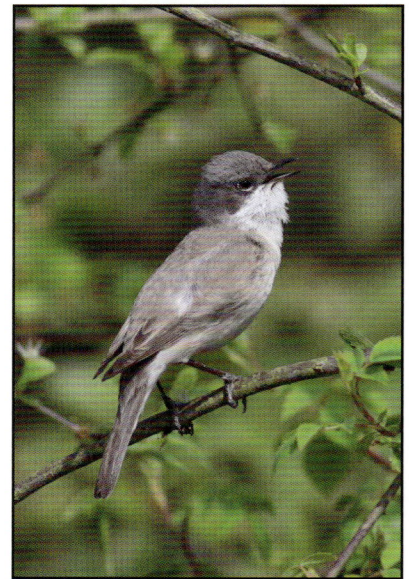

Sardinian Warbler *Curruca melanocephala*

BBRC

Vagrant. South-western Europe.

An adult male of this attractive Mediterranean species was trapped in Jun 1979 at Gibraltar Point and summered there, being retrapped on Jul 28th and Sep 6th and seen last on Sep 15th. The second record wasn't until Jul 1986 at the same site and again a long stayer, present for more than eight weeks. The third and fourth records sprang a bit of a surprise as first an adult male was discovered in Sea Buckthorn *Hippophae rhamnoides*, on the coastal strip at Skegness present Oct 2003 and was joined by a female in Nov 2003 nearby raising the possibility they may have summered. The female was last seen in Dec 2003, but the male remained to Jan 2004. Given their choice of habitat and skulking behaviour they may easily have gone undetected somewhere along the coast further north.

An adult male trapped on Lundy Island in May 1955 was the first of 81 British records. Sixty percent of records have been in the 1980s and 1990s, but since then, the trend has been downwards with just 25 British records 2000-18, still a genuine rarity.

Site Name	First Date	Last Date	Count	Notes
Gibraltar Point	30/06/1979	15/09/1979	1	Male trapped
Gibraltar Point	06/07/1986	31/08/1986	1	Adult male trapped
Skegness	02/10/2003	11/01/2004	1	Adult male
Skegness	10/11/2003	01/12/2003	1	Female

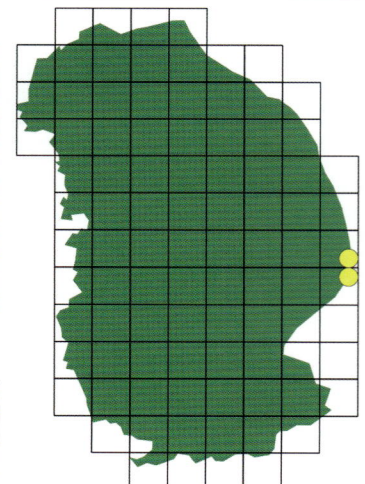

Subalpine Warbler *Curruca iberiae/cantillans*

The changes to the taxonomy of the 'Subalpine Warblers' are very recent and although they have been in the offing for some years now (Stoddart 2014) adoption by the BOURC is still awaited, and the older county records will be resubmitted to the BBRC for review. Not all of these records can be assigned to one of the 'new' species.

The split will see Western Subalpine Warbler *C. iberiae* (monotypic) which breeds in the Iberian Peninsula and southern France; Eastern Subalpine Warbler *C. cantillans* (*C.c. cantillans* and *C. c. albistriata*) which breeds in Sicily, southern Italy and the Balkans; and Moltoni's Warbler *C. subalpina* (monotypic) found in northern Italy, Corsica and Sardinia emerging as three separate full species. A review by the county records committee and other experienced county birders has excluded the possibility of Moltoni's Warbler from any of the records. Moltoni's Warbler is most reliably identified on the basis of its call often described as a dry "ttrrrrrr".

In their report of rare birds in Great Britain in 2019 BBRC summarised the position in terms of records of each new species since 1950 as follows: Eastern Subalpine Warbler 90 records, Western Subalpine Warbler 11 records and Moltoni's Warbler nine records. Records unassigned to species 723. The present Lincolnshire position is summarised below.

'Subalpine Warbler', unassigned, (*Curruca cantillans/iberiae*)

Six records of as yet unassigned 'Subalpine Warblers' with three in May, one in Jun and one in Sep. The adult female trapped at Gibraltar Point in May 1983 had a long wing length which may well fit with *cantillans*, but this awaits confirmation. Hindsight is a wonderful attribute but as four of the six birds which are unassigned were trapped, it's a great shame no material for DNA was obtained (although the availability of DNA analysis only became routinely available from the mid-1990s).

The details in the submission of the 1970 bird tend to suggest Western ("the underparts were pink – breast and flanks – with a white belly") as do those for the 1976 Tetney bird ("brick red underparts except for white belly"). Hopefully a full BBRC review will clarify at least some of these 'unassigned' records.

Site Name	First Date	Last Date	Count	Notes
Ingoldmells	11/05/1970		1	Male
Tetney	14/05/1976		1	Adult male trapped
Gibraltar Point	07/05/1983	14/05/1983	1	Adult female trapped
Grainthorpe	23/05/1985		1	Male
Saltfleetby-Theddlethorpe	28/06/1994		1	2CY male trapped
Saltfleetby-Theddlethorpe	15/09/1996		1	Juvenile female trapped

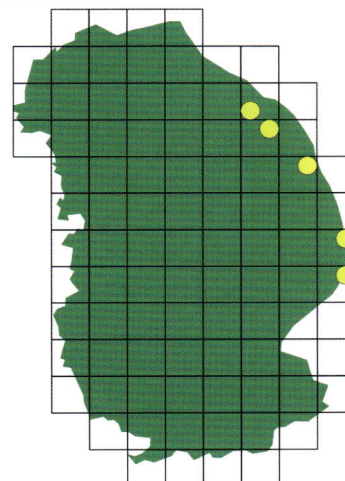

Eastern Subalpine Warbler *C. cantillans*.

Two records (at least), one in May 1981, the second in Apr 2019. That in 1981 was a male and well described by the observers: restricted dark orange/ brick red throat and upper breast sharply demarcated from the white to dull pinkish-buff belly and undertail coverts dirty white. Sub-specific identification not possible with no material collected for DNA analysis.

The second bird was at Gibraltar Point NNR Apr 29th-30th 2019, trapped Apr 29th, was confirmed as *C. c. albistriata* by DNA analysis. The photographs of this bird in the hand clearly show the main features of a spring male *albistriata* with a brick-red throat and upper breast forming a clear contrast with the rest of the underparts and the flanks which show a pale pinkish wash. (Image right)

The differences between *cantillans* and *subalpina* are subtle at best and in Britain DNA analysis seems to offer the best way to differentiate the two sub-species.

Site Name	First Date	Last Date	Count	Notes
Humberston	12/05/1981	13/05/1981	1	Male
Gibraltar Point	29/04/2019	30/04/2019	1	Adult male trapped

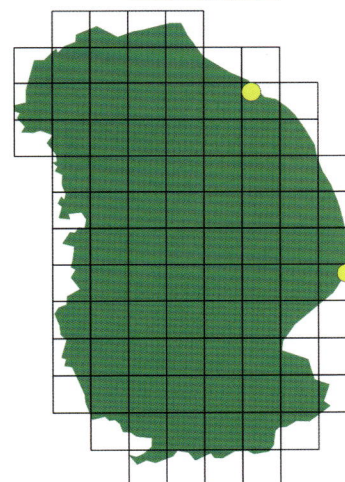

Western Subalpine Warbler *C. iberiae*.

Two records (at least) of male birds, one in Apr 2011 and another in Apr 2013. Both were well photographed and showed the typical reddish-pink underparts and paler bellies. (Image right)

In the opinions of the observers, and after scrutiny of the photographs afterwards the male at Humberstone Fitties in 2011 appeared to be a full adult. That in 2013 however was aged as a 2CY bird after forensic examination of the photographs. As expected in a 2CY, immature bird, it largely had old remiges and rectrices but showed some contrasting and newly grown inner greater coverts.

There is a complete post-nuptial moult in adults but 1CY birds have only a partial post-juvenile moult in late summer involving all head- and body-feathers, the majority of the wing coverts, and some tertials but the rectrices are not usually renewed until later in the 2CY (Shirihai and Svensson 2018).

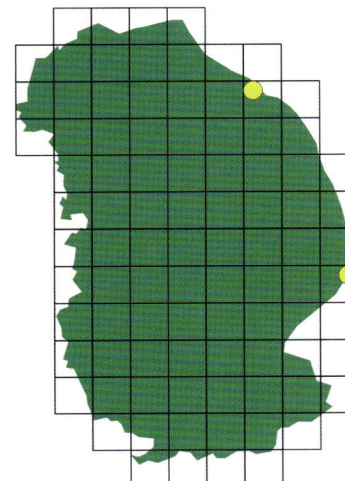

Site Name	First Date	Last Date	Count	Notes
Humberston	03/04/2011		1	Male
Gibraltar Point	20/04/2013	23/04/2013	1	Second calendar year male

Whitethroat *Curruca communis*

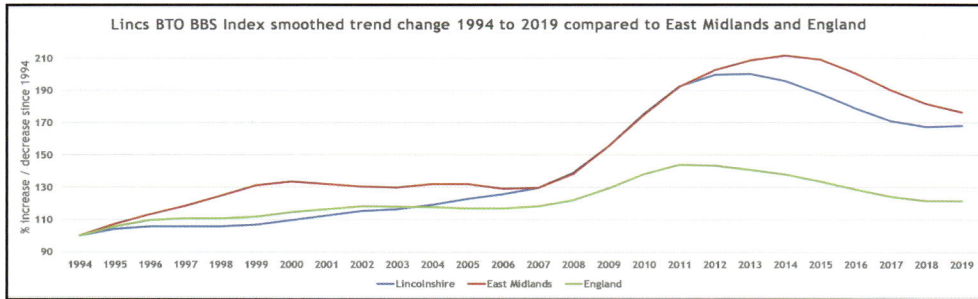

Nominate form a very common summer visitor and passage migrant.

Whitethroat is the warbler of the Lincolnshire countryside; both widespread and plentiful its song flight captures the joy of spring from mid-Apr onwards like no other. It has had its ups and downs, particularly during the infamous Whitethroat crash of the winter of 1968-69 caused by severe drought over several years in its Sahel wintering grounds, which brought it to an all-time low by 1974. Its population has still not fully recovered from that event, but the Lincolnshire BBS index shows that the breeding population has increased by 60% during the period 1994-2018. The *LBC Atlas* put the population at 25,000 pairs in the 1980s and the estimated population for 2016 was 53,000 pairs, making it Lincolnshire's most abundant warbler. The probable British population of 1.1 million ranks it the fourth commonest British warbler. In the five years to 2018 LBR reports some particularly large counts at Gibraltar Point, with breeding season peak numbers of 143 on Jun 15th 2015 and 110 on May 10th 2017. In autumn migration peak counts of 135 were recorded on Aug 23rd 2016 and 112 on Aug 20th 2018.

1968-1972	1980-1989	2008-2011	2016-2019	Incidence	Change
				86%	9%

Lincs BTO BBS Index smoothed trend change 1994 to 2019 compared to East Midlands and England

Dartford Warbler *Curruca undata*

Vagrant. Southern and eastern England and southern Europe.

Despite a relatively healthy breeding population of this delightful heathland warbler it remains a county vagrant with only four records, albeit the three most recent birds were all long stayers at Gibraltar Point. The first record was of a male bird trapped at Tetney in Jun 1984 and this showed characteristics of the Mediterranean race *S. u. undata*. The bird that over-wintered in 2004-05 was a 1CY male when it arrived in Oct 2004 and the female/immature in Nov 2007 frequented Rock Ridge, the same area as the bird in 2004-05. A female/immature was found in Nov 2009 at the very northern end of Gibraltar Point and stayed into Feb 2010. There have been no more records.

The British breeding population has had its highs and lows. In 2006 a national breeding survey reported that 3,214 territories had been located, an increase of 70% since the previous survey in 1994. This figure fell to a recent all-time low in 2011 when just 530 territories were found, a crash thought in large part to be due to severe winter weather in 2008-09 and 2009-10. Since then, the population has gradually recovered to 819 territories in 2014, doubled to 1,677 in 2015, but falling back to 1,148 in 2018 (Holling *in litt.*). Whether the next county record will be a vagrant from the Mediterranean or a dispersing British juvenile remains to be seen.

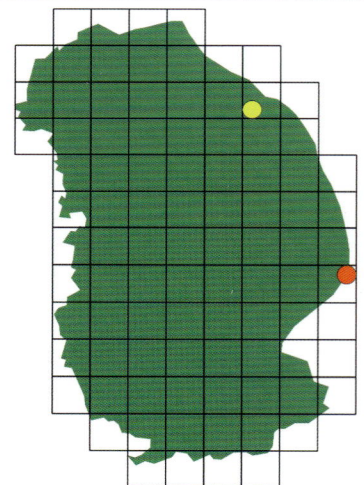

Site Name	First Date	Last Date	Count	Notes
Tetney	02/06/1984	03/06/1984	1	Male
Gibraltar Point	31/10/2004	01/11/2004	1	
Gibraltar Point	14/11/2004		1	Same
Gibraltar Point	23/11/2004		1	Same
Gibraltar Point	24/11/2004		1	Same
Gibraltar Point	07/12/2004		1	Same
Gibraltar Point	08/12/2004		1	Same
Gibraltar Point	09/12/2004		1	Same
Gibraltar Point	11/12/2004		1	Same
Gibraltar Point	18/12/2004		1	Same
Gibraltar Point	19/12/2004		1	Same
Gibraltar Point	30/01/2005		1	Same
Gibraltar Point	03/02/2005		1	Same
Gibraltar Point	29/11/2007	28/01/2008	1	
Gibraltar Point	28/11/2009	17/02/2010	1	Female

Firecrest *Regulus ignicapilla*

Scarce and increasing passage migrant and rare winter visitor.

The Firecrest has been spreading northwards from Hampshire since colonising the New Forest area from 1962 onwards. The *BTO Atlas 2007-11* showed it breeding in Breckland and the north coast of Norfolk and sparsely into South Yorkshire/Nottinghamshire to the west of Lincolnshire. By 2016 the estimated British breeding population was at least 2,000 pairs. Breeding evidence remains scant in the county, but males have been singing at Gibraltar Point for increasingly lengthy periods in Mar-May 2016-19 and at a couple of other inland locations. It is hard to know whether these are migrants or genuine prospecting territorial males. LBR reports spring migration in the six years to 2019 began between Mar 5th-19th and lasting into May with just one Jun record and none in Jul-Aug. The peak site counts across the county ranged from four to seven across this period. Autumn migration began from Sep 12th-Oct 10th extending into Nov with peak counts ranging from two to nine. There were just six winter records with two in each of the months Dec-Feb. The largest one-day counts were all at Gibraltar Point which gets well over 75% of all county records and had the all-time county peak count of 10 on Oct 23rd 1999. There were five here on Apr 5th 2016, and four on Apr 2nd, 6th and Oct 11th 2015, Mar 27th 2016 and Mar 30th and May 10th 2019. Elsewhere Saltfleet had three on Sep 17th 2014.

1968-1972	1980-1989	2008-2011	2016-2019	Incidence	Change
				0%	

Goldcrest *Regulus regulus*

Common resident, passage migrant and winter visitor.

Although Goldcrest is fairly widespread as a breeder across Lincolnshire it does not breed in large numbers. It is probably limited by the low proportion of coniferous tree cover throughout much of the county particularly in the Wolds, Fens and Marsh. The *LBC Atlas* put the population in the late 1980s towards the lower end of 3,000-12,000 pairs. Since then, the *BTO Atlas 2007-11* showed that it had spread a little more widely into the Fens and along the coast. The estimated population in 2016 was 4,000 pairs. The sample size is too small to produce a Lincolnshire BBS index but the long-term BBS index for England is up 33%. The population is boosted each autumn by a large-scale migration from northern Europe. Sometimes spectacular falls can occur. In the five years to 2018 LBR reports the largest such fall was on Oct 11th 2015 when 3,000 were estimated present at Saltfleetby-Theddlethorpe. A year earlier the same site had 2,000 on Oct 14th 2014. In 2016-18 there were no such large falls reported though Saltfleetby-Theddlethorpe had 500 on Oct 8th 2016. Ringing recoveries show that these birds are coming from and returning to, amongst other places, The Netherlands, Germany, Denmark and Sweden.

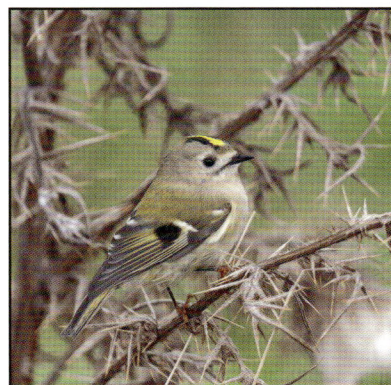

1968-1972	1980-1989	2008-2011	2016-2019	Incidence	Change
				14%	74%

Wren *Troglodytes troglodytes*

Abundant resident and partial migrant.

One of our top three commonest birds and perhaps now the most abundant. The *LBC Atlas* estimated a population of 200,000 pairs in the late 1980s. BBS data shows that the population has significantly increased by 39% during the period 1994-2018. Like other small passerines, Wrens are susceptible to hard winters and the last really bad ones were in 2009 and 2010, when the BBS chart for Lincolnshire shows a large population setback. More recently the "Beast from the East" in Mar 2018 had a negative impact too, but little Jenny Wren is nothing if not fecund and usually bounces back.

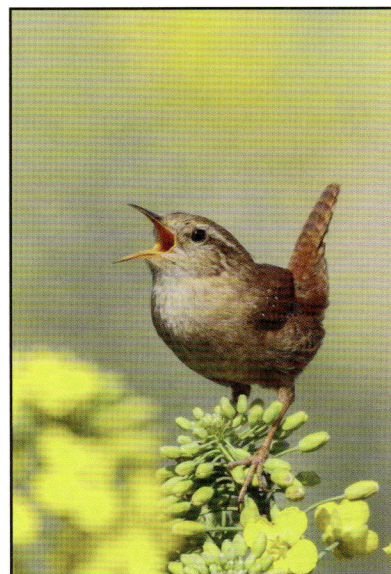

1968-1972	1980-1989	2008-2011	2016-2019	Incidence	Change
				92%	1%

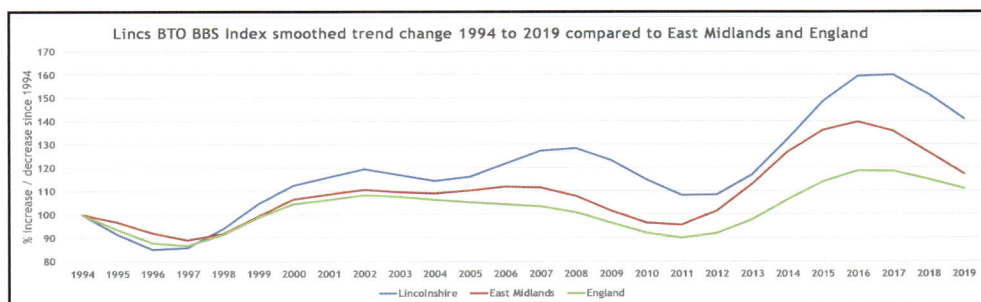

Lincs BTO BBS Index smoothed trend change 1994 to 2019 compared to East Midlands and England

Nuthatch *Sitta europaea*

Fairly common and increasing woodland resident.

The *LBC Atlas* showed that in the 1980s Nuthatch was largely confined to the south-west corner of the county but had been spreading from the extreme south west since the *BTO Atlas 1968-72*. The population in the 1980s was estimated at around 300 pairs. Nuthatch has continued to spread north and north-east through the Wolds to the eastern edge and up to the south bank of the Humber. This spread is well depicted in the *BTO Atlas 2007-11* maps. It is only fairly common on a local basis in areas of suitable deciduous woodland. Brocklesby Woods and the Hubbard's Hills area to the west of Louth are good locations colonised over the last 20 years. The estimated population for 2016 was 750 pairs. In the five years to 2018 the maximum counts received were of eight in Callan's Lane Wood in Jan 2015 and eight at Belton Park in Jan 2016. Both of these sites are in the original heartland of Nuthatch distribution. Counts of six birds came from Normanby Park north of Scunthorpe in Jan 2018 and Hubbard's Hills, Louth in Sep 2020.

1968-1972	1980-1989	2008-2011	2016-2019	Incidence	Change
				3%	-14%

Treecreeper *Certhia familiaris*

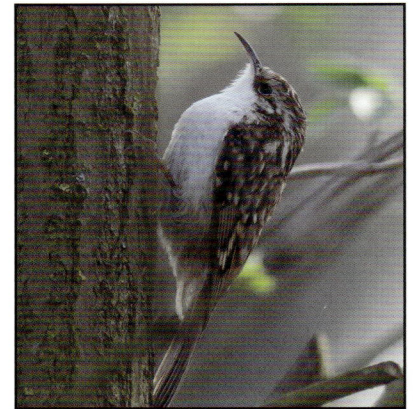

British form *britannica* a common resident and partial migrant. Nominate Scandinavian and eastern European form, vagrant.

Fairly widespread as a breeder but like many small woodland birds in Lincolnshire it is not as numerous as one might expect. The *LBC Atlas* estimated only 2,000 pairs in the 1980s. The *BTO Atlas 2007-11* showed it absent from the Fens and scarce in the Central Vale and Marsh. The estimated 2016 population was 2,500 pairs and the long-term BBS trend for England shows it up 8% from 1994-2019.

Records of the nominate race are exceptional. Lorand and Atkin (1989) report that one was shot at North Cotes on Mar 13th 1947 (the worst winter of the twentieth century). However, it is considered that there are no safe field criteria for separating the two races and birds thought to be nominate race are now described as "showing characters of *C. f. Familiaris*". Since precise identification criteria still need to be developed (Hudson *et al* 2016) submitted records will be "held until taxonomy or identification criteria are resolved". As such, the putative records at Donna Nook, Oct 14th-15th 2013, Crook Bank, Saltfleetby-Theddlethorpe on Oct 19th 2015 and Oct 15th 2016 await further elucidation.

1968-1972	1980-1989	2008-2011	2016-2019	Incidence	Change
				6%	115%

Rose-coloured Starling *Pastor roseus*

Rare. Eastern and south-eastern Europe, south-west Asia.

This gaudy visitor from the east first fell victim to someone's gun, reportedly prior to 1783, near Grantham (no specific date) and another was recorded in the county in Jul 1818 (no locality). An adult in Aug 1909 fed on cherries in a Cleethorpes garden and others were seen in the autumns of 1932 and 1947. A 1CY bird wintered in Skegness Dec 1955-Apr 1956, during which time it was trapped. A 1CY bird was trapped at Wainfleet Oct 1973 and a decade later a 2CY bird was present in Market Deeping, often near the fish and chip shop, Feb-Mar 1983. Two more turned up in the 1980s and just one in the 1990s. There then followed a run of 22 records 2000-18 bringing the total to 32 in all up to 2019. Given the long stays of several of these birds, records have accrued in every month of the year. The spring arrival dates covered May 6th-Jun 28th. The main autumn arrival period covers Aug 23rd-Nov 24th, although the Nov-Dec records may refer to birds having arrived earlier in the autumn.

There have been periodic invasions of Rosy Starlings into Britain, most recently in 2002 (182), 2018 (115) and 2020 (200+); more than 1,500 were recorded in France in 2020 and around 20 pairs bred in a colony in the Alpes-de-Haute-Provence department. Whether this will continue, and with a consquent increase of migrants into the county, remains to be seen.

Starling *Sturnus vulgaris*

Very common resident, passage migrant and winter visitor.

The English Starling population crashed around 90% between 1966 and 2012 which resulted in red status. A lack of breeding holes appears to be part of the problem. The Lincolnshire population has also been hard hit. BBS data shows that the Lincolnshire breeding population has fallen by 55% during the period 1994-2019. Whereas the *LBC Atlas* estimated a population of 60,000 pairs we probably now have half that number. The good news is that the population appears to have stabilised to some degree and there is little change in the incidence of breeding Starlings across Lincolnshire over the last 10 years. The problems here are not replicated across Europe so we still receive a large influx of wintering birds that provide the pre-roost murmuration spectacles that are so popular with the public and birders alike.

1968-1972	1980-1989	2008-2011	2016-2019	Incidence	Change
				63%	2%

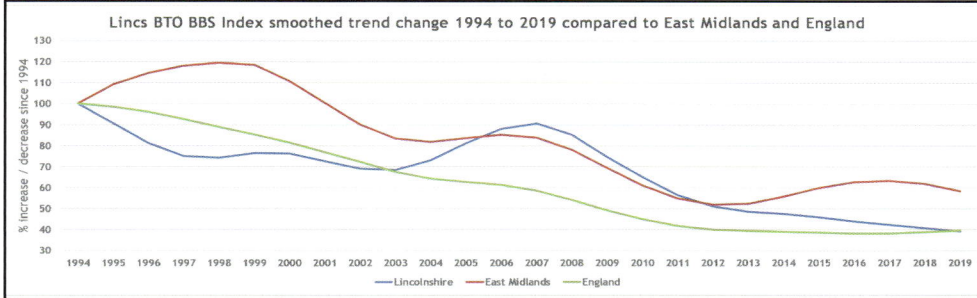

Lincs BTO BBS Index smoothed trend change 1994 to 2019 compared to East Midlands and England

Starling murmuration, Alkborough

Ring Ouzel *Turdus torquatus*

Scarce/fairly common passage migrant Mar-May and Sep-Nov; rare in winter.

Ring Ouzel is a bird that every inland patch birder has a chance of finding especially in spring when they can turn up anywhere on suitable pasture en route to their upland breeding grounds. The autumn migration tends to be more focused on coastal sites. During the five years to 2018 LBR reports showed that spring peak day counts at individual sites ranged from two in 2016 to 13 at Risby Warren on Apr 13th 2015, with an average of five to six per year. Autumn passage was much stronger ranging from five in 2017 to 54 at Gibraltar Point on Oct 13th 2014 with an average of 29 per year. In 2018 peak flock size was four in spring and 13 in autumn and a detailed analysis in LBR 2018 showed that there were over 215 birds in that year with 75% of them at Gibraltar Point. This suggests that in most years Ring Ouzel is probably a fairly common migrant. In Jan 2020 there was an exceptional record of a bird wintering in the Sea View/Rimac area at Saltfleetby-Theddlethorpe, probably of the alpine race *T. t. alpestris* found in the mountain ranges from northern Iberia eastwards towards the Balkans, Greece and Asia Minor.

Blackbird *Turdus merula*

Very common resident, passage migrant and winter visitor.

One of our top three commonest birds. Not quite found everywhere, but near enough. The *LBC Atlas* estimated the Lincolnshire breeding population at 200,000 pairs in the late 1980s and it likely remains around this level. BBS data suggests a 10% increase in the breeding population of this species from 1994-2018 but the increase is not statistically significant. Every winter the population is boosted by wintering individuals from the north European population and significant falls of migrating birds can occur along the coast in Oct and Nov.

1968-1972	1980-1989	2008-2011	2016-2019	Incidence	Change
				98%	4%

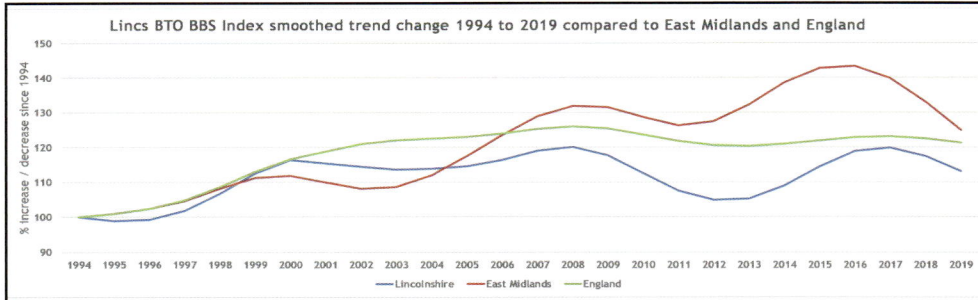

Lincs BTO BBS Index smoothed trend change 1994 to 2019 compared to East Midlands and England

Black-throated Thrush *Turdus atrogurlaris*

BBRC
British Birds Rarities Committee

Vagrant. Eastern European Russia to north-central Siberia, north-west Mongolia.

One record of this handsome thrush in Jan-Feb 2020 thought to be a 2CY male. An excellent find in suburban Grimsby frequenting a college green by a busy commuter road and finding plenty of earthworms to eat. The record is still under consideration by BBRC at the time of going to press. This species has been regarded as conspecific with Red-throated Thrush *T. ruficollis*, but has several very different phenotypic characters and considerable vocal differences; reports of mixed pairs are exceptional, and hybridization is, perhaps, reduced by difference in timing of breeding.

These thrushes breed in a variety of habitats: pure coniferous forests along rivers and streams, sparse dry woods, larch (*Larix*) clumps, semi-open willow (*Salix*) scrub, groves of poplar (*Populus*) and birch (*Betula*), buckthorn (*Rhamnus*) thickets and wooded bogs, to 2,200 m. They depart from mid-Aug and winter from Iraq and Arabia across southern Asia. There have been 84 British records, 1950-2018, with this and another under consideration in 2019. With seven previous records in Yorkshire, this was an overdue find, albeit not among the 1,000s of commoner thrushes arriving on the coast.

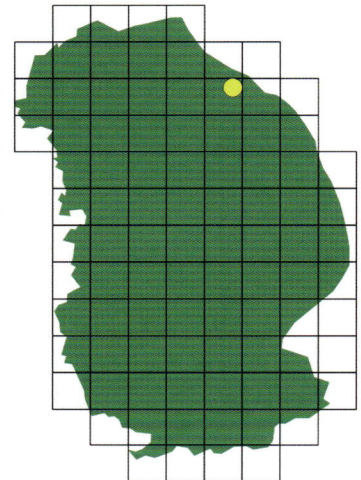

Site Name	First Date	Last Date	Count	Notes
Grimsby	30/01/2020	24/03/2020	1	

Fieldfare *Turdus pilaris*

Very common passage migrant and winter visitor, mainly Sep-Apr; rare in summer.

Thrush fortnight in early Oct sees the arrival of most of our wintering thrushes and the "chuk-uk-uk" call of a flyover Fieldfare is one of the sounds of autumn and winter in Lincolnshire. Many birds pass straight through heading west, but plenty stay for the winter and returning flocks boost numbers again in Apr. The *BTO Winter Atlas 1981-84* estimated the British wintering population at 720,000 birds and there has been no formal estimate since that time. Ringing recoveries reveal that a Russian Fieldfare wintering in Lincolnshire one year might well winter in Turkey the following year. Virtually every tetrad in the county will record a Fieldfare at some point between Oct-Apr; it is a very widespread wintering bird. This is one thrush that is not prone to falls on the coast. The three largest flocks reported in LBR over the five years to 2018 were 2,000 at Frampton Marsh on Dec 6th 2016, 1,500 there on Dec 5th 2017 and 1,672 at West Common, Lincoln on Oct 28th 2018. In the past there have been occasional reports of summering Fieldfares, but these are rare with none in the five years to 2018 during which period the last birds were reported from May 5th-24th. Late birds occasionally sing but none thought worthy of reporting to RBBP since 1990.

Redwing *Turdus iliacus*

Nominate Scandinavian and Russian form a very common passage migrant and winter visitor, mainly Sep-Apr, rare in summer. Icelandic form *coburni* rare but probably under-recorded.

A pasture full of several hundred worm-seeking Redwing and Fieldfare *T. pilaris* is a great sight of late winter. They tend not to move onto pasture until the last berries have gone from the hedgerows. Most pass through and winter further west. Over 99.5% of our birds come from Scandinavia and further east but Gregory in two notes in LBR 2018 showed that we get a regular trickle of Icelandic birds of the *coburni* race, (which can only be reliably identified on the measured wing length of trapped birds, or formerly shot ones) and peak flock size varies dramatically from year to year at Gibraltar Point. Using the long-term dataset collected by Gibraltar Point Bird Observatory from 1949 to 2018 Gregory showed that the mean highest annual flock size was 1,371 with a spike up to 9,595 in Oct 2012, the highest one-day count there in nearly 70 years. He found a mean annual increase in flock size of 50 birds per year over the period. LBR reports over the five years to 2018 showed that the largest annual flock size across the county ranged from 1,180 in 2016 to 8,000 at Crook Bank, Saltfleetby-Theddlethorpe on Oct 19th 2014. The largest numbers presumably depend on exactly where migrating birds touchdown on the coast on arrival. And someone being there to count them.

Song Thrush *Turdus philomelos*

British form 'clarkei' a very common resident. Nominate continental form a common passage migrant and winter visitor.

The Song Thrush is one of our commonest songsters and is found pretty much everywhere in Lincolnshire. The *LBC Atlas* estimated the population in the 1980s at around 38,000 pairs but noted that there was a steady decline through the late 1980s into the 1990s followed by a slight recovery in the late 1990s. The chart suggests that Song Thrush experiences ups and downs, but a long-term increase since 1994. Each autumn there is an influx of continental birds best observed at coastal sites and usually in the first half of Oct. It is difficult to estimate the numbers of birds passing through in the dense Sea Buckthorn *Hippophae rhamnoides* of coastal dunes, but peak day counts give a good indication and the largest usually come from Gibraltar Point. LBR reports indicate that in the five years to 2018 the peak day counts ranged from 117 on Oct 11th 2018 to 295 on Oct 14th 2014 with an average of around 170. The actual numbers of birds involved along the coast runs into many thousands. Some of these stay to winter and single birds with their cold grey tones can be found in field hedges throughout the county in the winter months.

1968-1972	1980-1989	2008-2011	2016-2019	Incidence	Change
				66%	11%

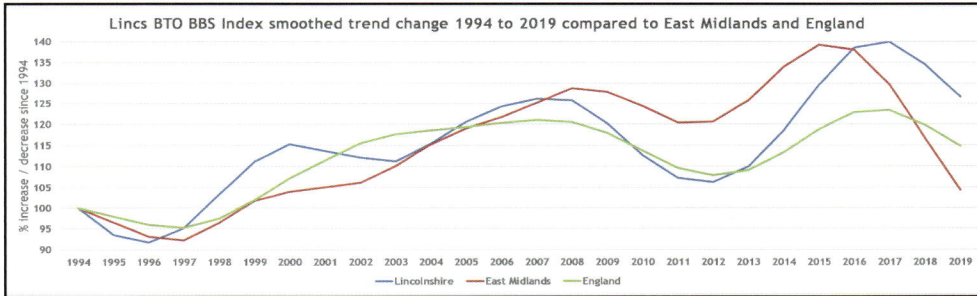

Lincs BTO BBS Index smoothed trend change 1994 to 2019 compared to East Midlands and England

Mistle Thrush *Turdus viscivorus*

Common resident and partial migrant.

Though common and widespread, the "Stormcock" has declined of late. The *LBC Atlas* put the population at between 3,200-6,800 pairs and the estimated population in 2016 was around 3,000 pairs. Given the recent decline since 2016 shown by the Lincolnshire BBS index for Mistle Thrush, there are probably fewer than that now. The causes for the decline are not understood but may relate to reduced survival of juveniles according to BTO BirdTrends. This decline is reflected to some extent in the largest flocks reported each year in LBR which were analysed in LBR 2015 for the period 1979-2015. This showed that maximum flock size grew to a peak of 70 in 2005 and has been falling steadily ever since. In 2016 it was 28, 2017, 34 and 2018, 24.

These figures are pretty much in line with numbers that prevailed for most of the last 40 years. Whether we are witnessing a cyclical change, or a definite decline remains to be seen.

1968-1972	1980-1989	2008-2011	2016-2019	Incidence	Change
				31%	-22%

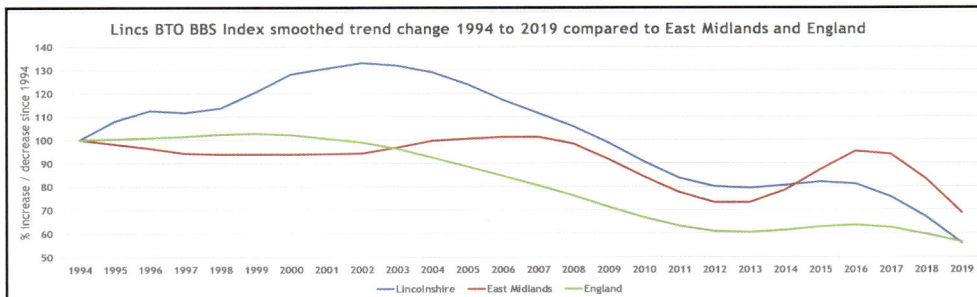

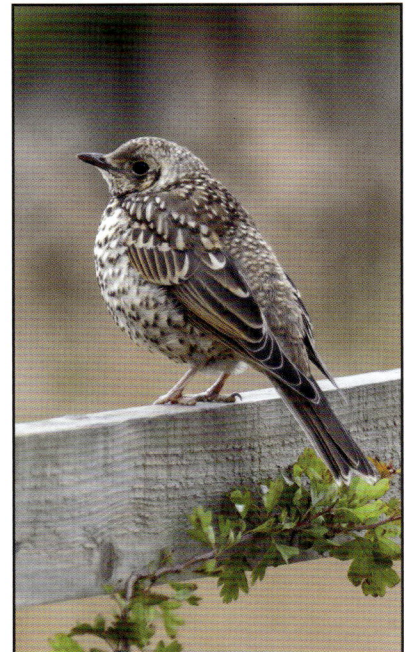

Lincs BTO BBS Index smoothed trend change 1994 to 2019 compared to East Midlands and England

American Robin *Turdus migratorius*

Vagrant. North America.

Local birders looking for Waxwings *Bombycilla garrulus* on the Pyewipe industrial estate on the outskirts of Grimsby were astounded to stumble across this county first on Jan 1st 2004. It became the focus of one of the biggest twitches in county history. It proved to be a 2CY female and frequented a small patch of scrub outside a busy industrial unit for nine weeks before succumbing to a Sparrowhawk *Accipiter nisus* on Mar 8th.

Two others were discovered in the late autumn and winter of 2003-04, one in Cornwall, one on Bardsey Island. Their appearance has been linked to a massive easterly displacement of this species and other late migrants from the North American Midwest to the eastern seaboard of the USA in the second week of Nov 2003. Although the arrival of American vagrants is a phenomenon usually associated with western and south-western Britain, once here they can clearly disperse within Britain and may turn up anywhere. These three individuals took the British total to 21 since 1950; in total there have been 28 records up to 2019.

Site	First Date	Last Date	Count	Notes
Pyewipe, Grimsby	01/01/2004	08//03/2004	1	2CY Female

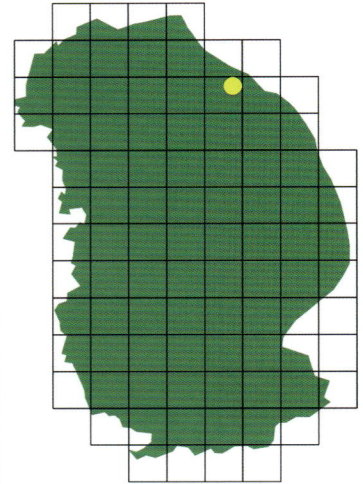

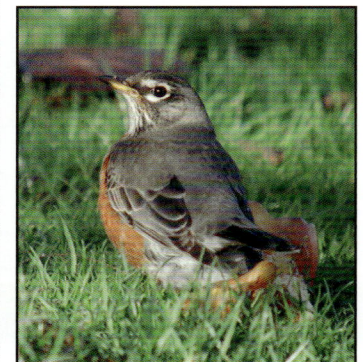

Rufous Bush Chat *Cercotrichas galactotes*

Vagrant. Southern Europe.

One, all too distant, record in Sep 1963 of one trapped near the sewage treatment works in the grounds of Butlin's Holiday Camp, Ingoldmells. Although it was thought to be the nominate western race, detailed examination of the photographs at a much later date showed that, based on plumage tone and tail pattern, it was one of the eastern races, most likely *C. g. syriaca* or *familiaris*. The fresh remiges and neat pale fringes and tips to the greater coverts show that it was a 1CY bird (Vinicombe and Cottridge 1996). A very interesting account by one of Britain's top birders in Slack (2009) tells of his receipt of the news of this bird by postcard and the subsequent tortuous journey from Norfolk to Skegness, only to be denied entry by the 'uniform' on the gate at Butlin's and ultimately failing to see the bird from a public footpath overlooking the sewage plant. In the event of a recurrence, news would travel slightly more quickly.

There have been three British records prior to 1949 and now six since (the last two in 1980 and 2020). This species was one of the 'Holy Grails' for British birders for 40 years until the Norfolk bird turned up at Stiffkey Marsh in Oct 2020. It is a polytypic species with five races, three in Western Europe and Asia and two in Africa. The first three are migratory and winter in the northern Afro-tropics including the Sahel.

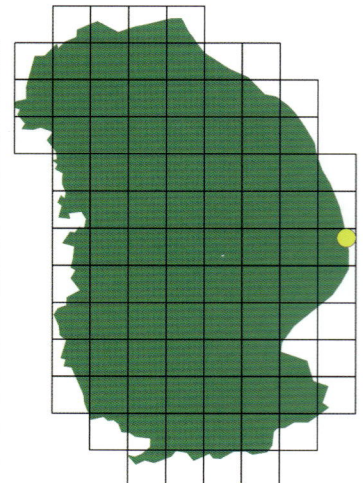

Site Name	First Date	Last Date	Count	Notes
Ingoldmells	02/09/1963	09/09/1963	1	Trapped

Spotted Flycatcher *Muscicapa striata*

Fairly common but much decreased and now very local summer visitor and passage migrant.

Spotted Flycatcher is one of our longest distance African migrants wintering right down into South Africa. It is also one of our latest arriving spring migrants. Both these factors need to be kept in mind when looking at the massive 93% decline this species has experienced in England over the 50 years since 1967 based on the CBC/BBS index. According to the *LBC Atlas* the population was estimated at 2,000-3,800 pairs in the 1980s but by the end of the 1990s it had already fallen to half that. That decline has continued with an occasional remission ever since. The estimated population was only 250 pairs in Lincolnshire in 2016 and it has now largely disappeared as a breeding bird from the north-eastern half of the county. In the southern Wolds in what appeared as a stronghold in the Relative Abundance map for the species in *BTO Atlas 2007-11*, it is now more difficult to find. BTO BirdTrends offers no explanation of the causes of the decline other than to note that it is likely the result of reduced juvenile survival. Another case of the young of insectivorous birds starving through reduced insect populations? Peak counts over the five years to 2018 ranged from six in 2017 to 22 on Aug 5th 2015 at Spridlington north of Lincoln. The autumn migration is on a broad front across the county and not confined to the coast.

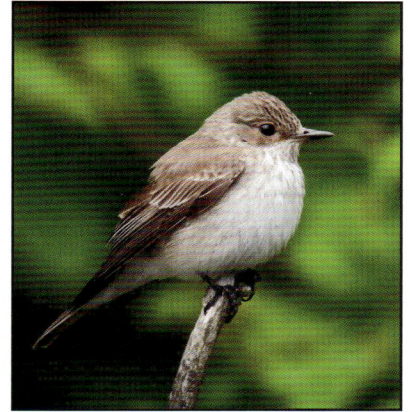

1968-1972	1980-1989	2008-2011	2016-2019	Incidence	Change
				2%	33%

Robin *Erithacus rubecula*

British form *melophilus* an abundant resident. Nominate continental form a common passage migrant and winter visitor.

"The north wind doth blow, And we shall have snow, And what will the robin do then, poor thing" Times have changed since the popular rhyme was written in the 16th century. Following the mini ice age of the Maunder Minimum 1645-1715, the hard winters of 1947 and 1963 and the more recent warming of the last three decades Robin Redbreast is doing well. The Lincolnshire BBS index shows a "long term" increase of 79% in the breeding population since 1994 of this most popular of British birds. The *LBC Atlas* estimated the population in the 1980s at 113,000-137,000 pairs and the estimated population in 2016 was around 100,000 an incongruence between the BBS index, the *LBC Atlas* and the adjusted APEP4 estimate that is not easily explained. Autumn falls of the nominate continental race Robin *E. r. rubecula* are an annual Oct feature with the annual peak site count occurring between 8th-17th in the five years to 2018 per LBR reports. Peak county wide annual one day counts (all at Gibraltar Point) ranged from 180 in 2017 to 442 in 2014 averaging at 300.

1968-1972	1980-1989	2008-2011	2016-2019	Incidence	Change
				90%	10%

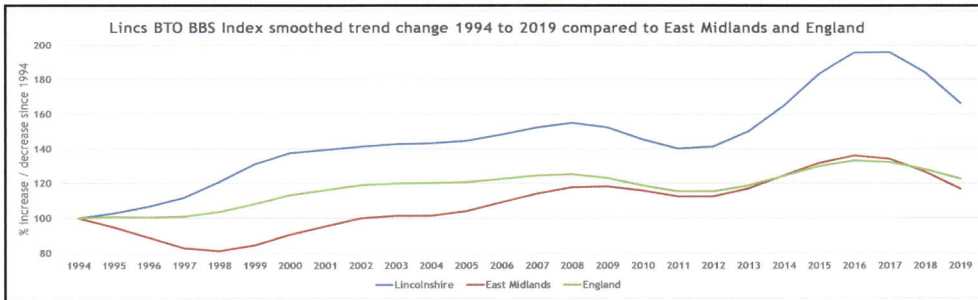

Lincs BTO BBS Index smoothed trend change 1994 to 2019 compared to East Midlands and England

Red Hill, The Wolds

Bluethroat *Luscinia svecica*

Two races. The majority of records are of Red-spotted Bluethroat *L. s. svecica*, rare migrant, northern Europe. White-spotted Bluethroat *L. s. cyanecula*, vagrant, south-western Europe.

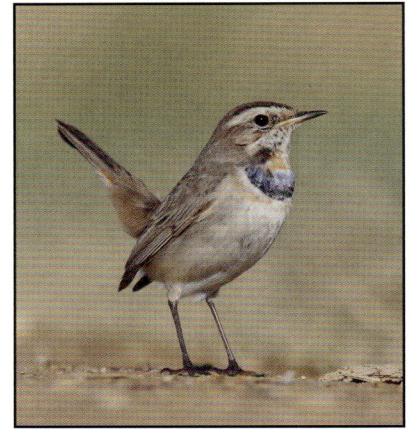

L. s. svecica has contributed most records (though there are quite a few records for which race is not recorded), but it should be noted that distinguishing the subspecies is not wholly straightforward, as some belonging to the white-spotted race *L. s. cyanecula* can show orange spots. Currently, 11 races are recognised, eight of these attributable to red-spotted forms, one to the white-spotted form and two, Iberian Bluethroat *L. s. azuricollis* and Caucasian Bluethroat *L. s. luristanica*, having no coloured spot within the blue throat and breast patch at all (Guzy *et al*, 2020).

Since the first county record at Gibraltar Point in Sep 1949 there have been 42 records involving 65 individuals of *L. s. svecica*, including up to seven at Donna Nook May 12th-16th 1970. In addition, there have been another 40 records of around 40 birds which were racially unassigned, but the majority will also have been *L. s. svecica*.

By contrast confirmed records of *cyanecula* make up only six of the total, five coastal and one exceptional record of a singing male at Whisby NP, Jun 21st-Jul 4th 1987. The white-spotted birds have tended to turn up earlier in the spring and account for all three Apr records, the earliest on Apr 3rd 2018. The majority of *svecica* turn up later in mid-May as befits a northern migrant with most in Sep in the autumn, with the latest record being of one at Horseshoe Point on Nov 4th 2008.

Thrush Nightingale *Luscinia luscinia*

Vagrant. Scandinavia, central and eastern Europe.

This well-known skulker has occurred on nine occasions thanks to coastal ringing groups who have accounted for seven of the records. The bird in May 2001 was not seen but sound-recorded, leaving the Donna Nook bird in May 1994 as the only one which was found and observed in the field. The five spring records turned up between May 15th-22nd and the four in autumn between Sep 2nd-19th. Given the extensive, impenetrable Sea Buckthorn *Hippophae rhamnoides* cover along the coast, it is fair to surmise that more of these birds have been missed than found.

The vast majority of British records have been on the east coast, but with a few inland records too. Thrush Nightingale is a common breeder in southern Sweden and is expanding northwards and westwards further into Europe. The BTO Online Ringing data show that 127 of the 226 birds to 2019 have been trapped. There was just one re-trap, a bird ringed in Norway in Aug 1984 retrapped in Sussex 12 days later.

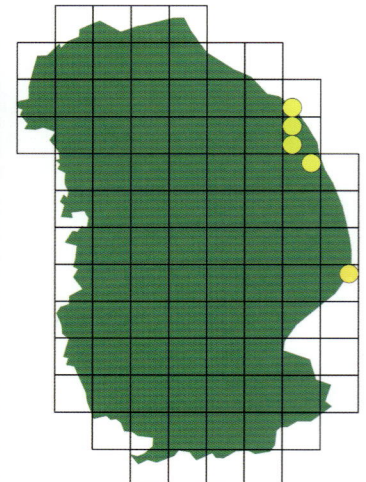

Site Name	First Date	Last Date	Count	Notes
North Somercotes	22/05/1977		1	Trapped
Saltfleetby-Theddlethorpe	03/09/1977	10/09/1977	1	Trapped
Saltfleetby-Theddlethorpe	02/09/1984		1	First calendar year trapped
Donna Nook	22/05/1994		1	
Saltfleet	17/09/1994		1	First winter trapped
Gibraltar Point	15/05/2001		1	Second calendar year +
Gibraltar Point	19/05/2003	26/05/2003	1	First winter trapped
Donna Nook	16/05/2009		1	First summer trapped
Gibraltar Point	19/09/2009	26/09/2009	1	First winter trapped

Nightingale *Luscinia megarhynchos*

Very scarce and declining local summer visitor and very scarce passage migrant.

John Clare the 19th century poet was born at Helpston, Cambridgeshire a stone's throw from West Deeping. Famed for the best phonetic description of any bird song in English literature his "The Progress of Ryme" captures the song of the Nightingale to perfection. For him 200 years ago, Nightingales were commonplace. In 1980, the *LBC Atlas* mentions that a full Nightingale census recorded 103 singing males in Lincolnshire. In 2020 there could have been three. A 1999 sample survey in the county recorded 97 singing males and based on the *LBC Atlas* coverage in the 1980s it was suggested the population could have been 100-150 territories. The *LBC Atlas* concluded by stating "In general, year to year fluctuations probably mask a reasonably stable population". A full Nightingale Survey organised by the BTO in 2012 found 34 singing males in the county of which 15 were at Whisby NP, near Lincoln, the most northerly confirmed breeding outpost of Nightingale in Britain. Their decline has been charted annually in LBR and in 2020 there was one, perhaps two, singing males. It seems unlikely that this northerly toehold can remain for much longer. They still hang on around Helpston, which may soon inherit Whisby's mantle. It is plain to see that Nightingale is a red data species. Why? Deer grazing and general "tidying up" may be having an impact on the scrub that Nightingale rely on for feeding and breeding.

1968-1972	1980-1989	2008-2011	2016-2019	Incidence	Change
				0%	

Red-flanked Bluetail *Tarsiger cyanurus*

Rare. North-eastern Europe.

The first British and county record was in Sep 1903 at North Cotes, apparently an adult male, seen by Caton Haigh and his keeper, but the record wasn't submitted as he was unable to shoot it. The record was eventually published in full in *British Birds* in 1954 (*British Birds*, 47, 28-29). Given the time lapse between the sighting and its appearance in the literature it was prefixed by the editors of *British Birds* as 'probable' but there was no good reason to doubt it and it has been accepted as the first British record.

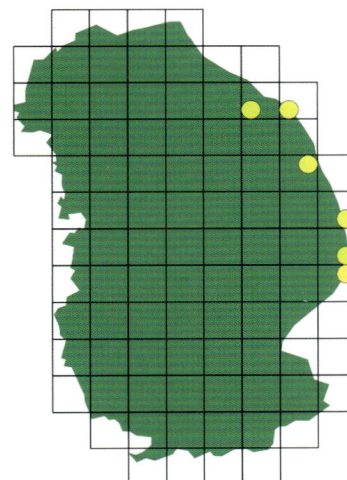

Since then, there have been another 10 records between 1978-2016, all in the autumn between Sep 15th-Nov 15th. This recent increase is concomitant with the species' recent and continuing increase on its nearest breeding grounds in Finland where the record was broken in 2020 when 730 territorial males were counted. From 1990-99 there was an average of one British record per year (including the popular one at Winspit, Dorset in Oct 1993); this increased to four per year 2000-09 and then 14 per year, 2010-18. It became ex-BBRC in 2016. There were three county records in 2016, but despite its increasing frequency of occurrence it remains one of the most sought after 'self-found' rarities.

Site Name	First Date	Last Date	Count	Notes
North Cotes	15/09/1903		1	Adult male
Donna Nook	10/10/1978		1	Immature female
Saltfleetby-Theddlethorpe	12/10/1988		1	First winter female trapped
Skegness	22/10/1999	23/10/1999	1	
Gibraltar Point	15/11/2002	16/11/2002	1	
Chapel St Leonards	06/11/2008		1	First winter male
Saltfleetby-Theddlethorpe	08/11/2008		1	First winter trapped
Gibraltar Point	20/10/2015		1	First winter female trapped
Donna Nook	10/10/2016	16/10/2016	1	First calendar year
Gibraltar Point	12/10/2016	14/10/2016	1	
Chapel St Leonards	22/10/2016	23/10/2016	1	

Red-breasted Flycatcher *Ficedula parva*

Very scarce coastal passage migrant mainly Aug-Oct.

The first Red-breasted Flycatcher in Lincolnshire was a first winter male shot at North Cotes on Sep 16th 1909. Another was found there in Oct 1922 and two more in Sep 1929. In Sep 1949, two were trapped at Gibraltar Point and one seen at Humberston Fitties. From 1950 onwards, the species has been a regular migrant, mainly in autumn, though remaining very scarce. The total number of individuals recorded up to 2015 is 162, and the annual average for the 10 years 2010-19 was between two and three birds. However, totals vary considerably between years. Blank years can still occur, most recently in 2017, but there were eight individuals in both 2013 and 2014.

Autumn accounts for the vast majority of records, which span from Sep 6th to Nov 5th, the last of those remaining to the 7th. There have been just three spring records: May 19th 1999,, May 20th 2013 and May 28th 2008. All have been found on the coast, with the well-watched areas of Donna Nook, Saltfleetby-Theddlethorpe and Gibraltar Point accounting for most.

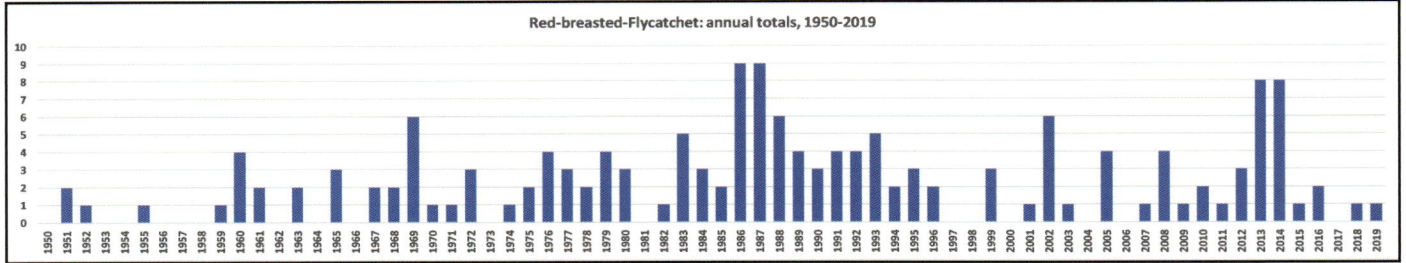

Red-breasted-Flycatchet: annual totals, 1950-2019

Pied Flycatcher *Ficedula hypoleuca*

Scarce passage migrant, mainly coastal; very scarce in spring and scarce in autumn.

Lorand and Atkin (1989) reported that Pied Flycatcher had bred in the western side of the county five times between 1871 and 1901 at locations ranging from Haverholme near Sleaford in the south to Normanby near Scunthorpe in the north. There has been no evidence of breeding since. As a passage migrant Pied Flycatcher is usually very scarce in spring and the four years to 2017 lived up to that with none in 2014 and five birds in 2017. Spring 2018 was exceptional with up to five at Gibraltar Point on Apr 29th (one female, four males) rising to 12 birds (three females, nine males) on May 1st with further singles in late May and four birds elsewhere in the county. Autumn passage is usually busier and based on maximum site counts ranged from three in 2017 to 43 at Gibraltar Point on Aug 24th 2015. During that last week of Aug 2015 there were a further 70 or so birds in the county making it the best autumn passage for many years. The *LBC Atlas* suggests that in the past there had been falls up to 200 birds at a single site but there had been nothing recorded like that since before 1979 at least.

1968-1972	1980-1989	2008-2011	2016-2019	Incidence	Change
■		■	■		

Black Redstart *Phoenicurus ochruros*

Scarce passage migrant, rare and irregular breeder, last in 2012 and very scarce winter visitor.

The European form *gibraltariensis* (Western Black Redstart) was formerly a vagrant but increased from the 1960s and was a regular summer visitor and breeder by the 1980s particularly at industrial sites along the Humber Bank but also occasionally at locations further inland. The *LBC Atlas* estimated an annual population of six pairs during the 1980s. A decline followed in the 1990s and in the 28 years since then two pairs bred in 2003, and a single pair bred in each of the nine years to 2012. There were no records in the other 18 years and none since 2012. From 2014-18 the number of wintering birds ranged from zero to a maximum of four in 2016; spring migration (Mar-May) ranged from six in 2014 to 46 in 2018 and autumn migration (Aug-Nov) ranged from two in 2018 to 19 birds in 2015.

One record of a 1CY male of one of the eastern races *P. o. phoenicuroides/rufiventris/xerophilus* Oct 26th-28th 2016 at Donna Nook. There were seven accepted British records in 2016, doubling the all-time total, in what was a very 'easterly' autumn.

1968-1972	1980-1989	2008-2011	2016-2019	Incidence	Change

Redstart *Phoenicurus phoenicurus*

Very scarce summer visitor and scarce passage migrant in spring, fairly common in autumn.

Redstart is sadly one of the most recently lost breeding birds with the last confirmed breeding taking place at Bulby Wood near Bourne in 2013. The *LBC Atlas* reported that there had probably never been more than 10 pairs a year breeding in the county in the 1980s with confirmed breeding in only five 10km squares compared to 16 in the *BTO Atlas 1968-72*. By the time of the *BTO Atlas 2007-11* it bred in just one 10km square, at Linwood Warren. Overall, Redstart is thought to be holding its own as a breeding bird across England with no obvious explanation for the decline in the East Midlands and Eastern England. It has also become much scarcer in Lincolnshire as an autumn migrant. The *LBC Atlas* reported that "on the coast...parties of up to 30 birds were often recorded in the 1980s. Larger numbers, totalling several hundred, were formerly more frequent but are now irregular". LBR 1996 reported a max combined peak total count across the coast of 708 on Sep 18th. LBR for the five years to 2018 shows that spring passage was very light with peak counts ranging from one in 2015 to four at Gibraltar Point May 7th-10th 2017 with an average peak of five. Spring males sang at two former breeding haunts in 2017. Autumn passage ranged from a peak site count of three in 2017 to 23 at Donna Nook on Sep 18th 2014. The second largest count was 16 at Gibraltar Point on Aug 23rd 2016. The average peak autumn passage 2014-18 count was 10.

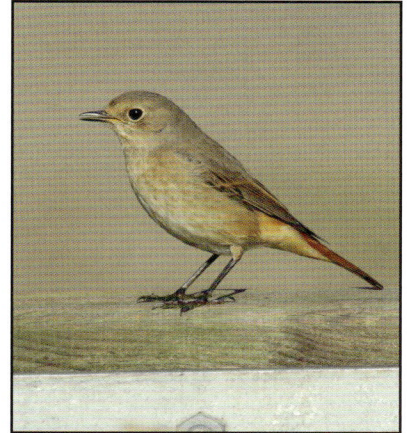

1968-1972	1980-1989	2008-2011	2016-2019	Incidence	Change
				0%	

Whinchat *Saxicola rubetra*

Scarce/fairly common passage migrant. Formerly a common breeder, last bred 1980s.

It's a sad fact that a bird Cordeaux (1872) described as "one of the commonest of our summer visitants" became extinct as a breeder in Lincolnshire with no one noting or lamenting its final passing. While it is now thought of as a bird of rough grasslands and heaths, in the 19th century it thrived in the extensive hay meadows required to power the nation's agriculture. When the horses went, so did the hay meadows and Whinchats with them, along with the Corncrake *Crex crex*. (Newton, 2017). They hung on at Crowle Waste on the Isle of Axholme, but the *LBC Atlas* showed only five confirmed breeding records in the 1980s and LBR does not record the date or location of the last breeding. There were no confirmed breeding records in the 1990s though adult males occasionally held territory and sang, as recorded at Crowle Waste in May 1999. Spring passage in the five years to 2018 as reported in LBR begins from Apr 11th-28th and the total number of birds ranged from eight in 2017 to 22 in 2015, an average of 15 birds per spring passage which finished from May 14th-30th. A notable feature of spring passage is that most birds occur inland. Autumn passage is more coastal, involves many more birds and lasts longer than spring. On average it begins from Jul 3rd-Aug 3rd and the largest peak one day counts ranged from 16 in 2017 and 2018 to 60 at Donna Nook on Sep 4th 2014. Last dates were from Sep 27th-Oct 26th. The UK population is in long term decline having fallen by 57% since 1994. The estimated APEP4 UK population was 50,000 pairs in 2016. Could it return to breed in Lincolnshire? With the right habitat creation, why not?

1968-1972	1980-1989	2008-2011	2016-2019	Incidence	Change
				0%	

Stonechat *Saxicola rubicola*

Scarce/fairly common passage migrant and winter visitor, mainly Sep-Apr. Very scarce and sporadic breeder.

Stonechat was formerly a local breeder in the north-west, the north-east coast and the Fens but died out by the late 1940s. It has been a "nearly" breeder in Lincolnshire for over 50 years. The *LBC Atlas* found one confirmed breeding record in the 1980s at Saltfleet in 1980. It bred in six out of the last 20 years to 2018 with the last record in 2016. It is much better known as a wintering bird and can occur all along the coast during winter from Alkborough to Nene Mouth. The *BTO Atlas 2007-11* showed it wintering in more than two-thirds of all the counties 10km squares. LBR shows that in the five years to 2018 peak numbers usually occur from Oct-Jan with a peak monthly count across the county of 63 birds in Oct 2016. The highest individual day counts for the five-year period come from Gibraltar Point with a peak of 25 in Oct 2017.

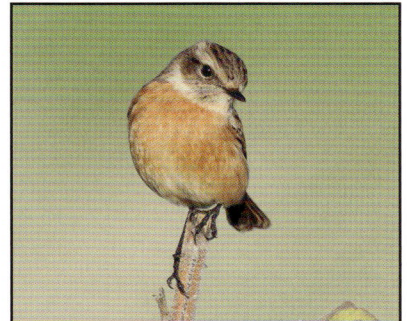

1968-1972	1980-1989	2008-2011	2016-2019	Incidence	Change
				0%	

Siberian Stonechat *Saxicola maurus*

BBRC

Vagrant. Siberia.

The taxonomy of the Stonechats *Saxicola sp.*, in Europe and Asia has been complex and studies on the group continue. Siberian Stonechat was split from what is now African Stonechat, *S. torquatus*, following work by Wittmann *et al* (1995). It is a complex polytypic species which continues to be intensively investigated using morphological and DNA data. IOC taxonomy (10.2) currently recognises five subspecies of which the nominate *maurus* is the one said to have predominantly occurred in Britain. One of these subspecies, *stejnegeri*, was split from Siberian Stonechat by IOC in 2009, becoming Stejneger's Stonechat *S. stejnegeri* and was thus added to the British List following the adoption of IOC taxonomy by BOU and BBRC in 2018; field identification criteria have not yet been fully developed for this new species and BBRC requires confirmation by DNA analysis.

The first county record was an adult male at Donna Nook in May 1978 considered to be *stejnegeri*. Of the 10 records since, the two at Donna Nook in Oct 2016 were also considered to be *stejnegeri*, but no material for DNA analysis was obtained. On current knowledge, the other eight records remain as indeterminate *maurus/stejnegeri*. Apart from the adult male in 1978, those in 2013 and 2016 were also males, in their 1CY; the others were immature/female types. A detailed analysis of the two birds in 2016 can be found in LBR 2016 (Catley and Lorand, undated). All records have been coastal, seven of them at Donna Nook which is clearly the site to watch for further records of this enigmatic group.

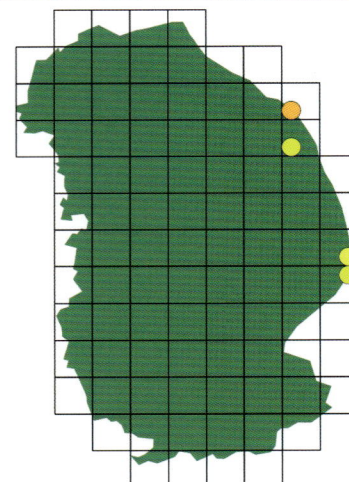

Site Name	First Date	Last Date	Count	Notes
Donna Nook	23/05/1978		1	Adult male
Donna Nook	07/10/1978	08/10/1978	1	Immature male
Donna Nook	09/11/1980		1	
Donna Nook	02/10/1987		1	First winter female
Skegness	21/10/1990	25/10/1990	1	First winter female
Skegness	23/10/1990	25/10/1990	1	First winter female
Donna Nook	22/09/1994		1	First winter female
Saltfleet	01/10/2000	03/10/2000	1	First winter female
Gibraltar Point	05/10/2013	06/10/2013	1	First winter male
Donna Nook	06/10/2016	10/10/2016	1	First calendar year male
Donna Nook	14/10/2016	16/10/2016	1	First calendar year male

Wheatear *Oenanthe oenanthe*

Fairly common passage migrant and former local breeder till end of 19th century. Greenland race (Greenland Wheatear) *leucorhoa* is a scarce migrant in spring/autumn.

Cordeaux (1872) described Wheatear as "nesting sparingly in North Lincolnshire". Lorand and Atkin (1989) could find no confirmed breeding records after 1899. The picture is confused by pairs occasionally holding territory for several weeks in late spring both on the coast and inland but there has been no evidence of confirmed breeding in the last 120 years.

It is a widespread spring migrant inland which accounts for its BBS spring incidence. Information from LBR for the five years to 2018 shows first arrival date from Mar 11th-28th with peak site counts ranging from 19 in 2017 to 42 in 2018 at Gibraltar Point on May 4th. The largest spring count inland was 24 at Risby Warren on Apr 18th.

There was a potential breeding record in 2015 at Greetwell where on Jul 22nd a pair of adults and a single juv. were seen. There had been no reports from this large and complex quarry site earlier in the season. In the autumn peak day counts ranged from 10 in 2018 to 51 in 2017 at Gibraltar Point on Sep 9th.

Greenland Wheatears were reported in four years out of five with numbers ranging from two to 25 per annum.

1968-1972	1980-1989	2008-2011	2016-2019	Incidence	Change
				4%	*

Isabelline Wheatear *Oenanthe isabellina*

BBRC

Vagrant. Eastern Europe to central Asia.

One was present on the beach near the east dunes at Gibraltar Point on Sep 22nd-23rd 2019, representing the first county record. Found feeding along a dune edge and, although there was some discussion over its identity, the photographs clearly showed a large, pale sandy-coloured, pot-bellied Wheatear. In addition, the head markings, wing markings, and especially the shortness of the primary projection establish the identification.

The first British record of modern times was in Norfolk May 28th 1977 and there have been 50 in all to the end of 2019. Most records have come from the Isles of Scilly (eight) and Shetland (seven); there has been just one record inland, in Cambridgeshire (Oct 2016).

The eight records in 2019 equals the previous record total of eight in 2016. This is still a very rare bird for Britain averaging as it does just one to two individuals per year.

Site Name	First Date	Last Date	Count	Notes
Gibraltar Point	22/09/2019	23/09/2019	1	Juvenile

Desert Wheatear *Oenanthe deserti*

Vagrant. North Africa, Central Asia.

An immature male was found at Donna Nook in Sep 1970, the first county record and the 18th for Britain at the time. Its vagrant status remains assured as there have only been a further three records: an immature female at Donna Nook in Oct 1991, a male at Saltfleetby-Theddlethorpe in Nov 1999 and a female at Saltfleet Haven in Nov 2008. These are classically late vagrants and Nov records are frequent. It seems ironic that a bird from hot, arid regions is often the last real rarity of the autumn.

Characteristically encountered on exposed beaches or coastal dunes in inhospitable weather conditions.

Polytypic with three subspecies, all of which are said to have occurred in Britain (Slack 2009). However, the differences are clinal and only the western race *homochroa* and eastern *oreophila* are distinct. No subspecific identity has been attributed to any of the four Lincolnshire records. During 1950-2019 there have been 147 British records, averaging four to five per year, but there have been occasional influxes, the most recent ones in 1997 (17), 2003 (10) and 2011 (15).

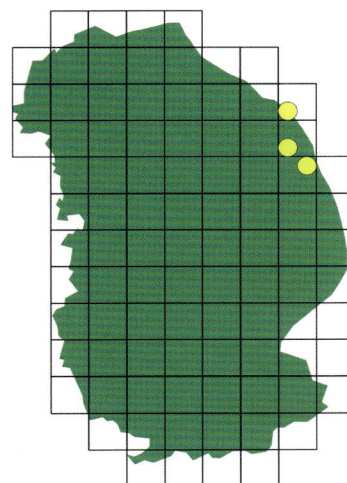

Site Name	First Date	Last Date	Count	Notes
Donna Nook	23/09/1970		1	Immature male
Donna Nook	10/10/1991	13/10/1991	1	First winter female
Saltfleetby-Theddlethorpe	13/11/1999	16/11/1999	1	
Saltfleet	08/11/2008	14/11/2008	1	Female

Western Black-eared Wheatear *Oenanthe hispanica*

Vagrant. Southern Europe.

A very recent addition to the county list, a 2CY female was seen very briefly at Frampton Marsh on Jun 12th 2012 and identified from photographs later. It was identified at the time as the western form of Black-eared Wheatear *O. h. hispanica*, but the Black-eared Wheatear group was recently split into Western Black-eared Wheatear *O. hispanica* and Eastern Black-eared Wheatear *O. melanoleuca*, and adopted by the BOURC in their 51st report (BOURC 2020). This decision has come too soon for a review of previous British records to have been conducted to see how many of each of these two species have occurred, but the total of 'Black-eared Wheatears' before the split stood at 49, 1950-2019, with 11 prior to 1949.

There have been just three British records since the Frampton bird, in 2015, 2016 and 2018, considered to have been *melanoleuca*, indeterminate and *hispanica*, respectively, illustrating the difficulty of identifying the two taxa when confronted by a bird which is not an adult male.

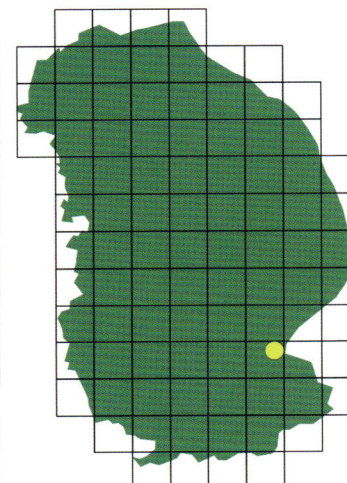

Site Name	First Date	Last Date	Count	Notes
Frampton Marsh	12/06/2012		1	First summer female

Pied Wheatear *Oenanthe pleschanka*

Vagrant Black Sea, Central Asia.

A 1CY female at Gibraltar Point Nov 18th-26th 2000 is the only county record to date. Described at the time as a small, dingy, rather uniform-coloured wheatear with generally cold-toned grey-brown upperparts and a rather bland facial expression, it fitted the description of most Pied Wheatears which have turned up in Britain in late autumn. Slightly later than most previous British records, it arrived during an exceptional period for eastern rarities, which included nine Hume's Warblers *Phylloscopus humei* and two Isabelline Shrikes *Lanius isabellinus*.

There have been 84 British records, 1950-2019, averaging about two to three birds per year. The last record in Nov 2019 came unexpectedly from Worcestershire. Overall, Norfolk has recorded most (12), followed by Shetland (11) and Yorkshire (eight).

Site Name	First Date	Last Date	Count	Notes
Gibraltar Point	18/11/2000	26/11/2000	1	First winter female

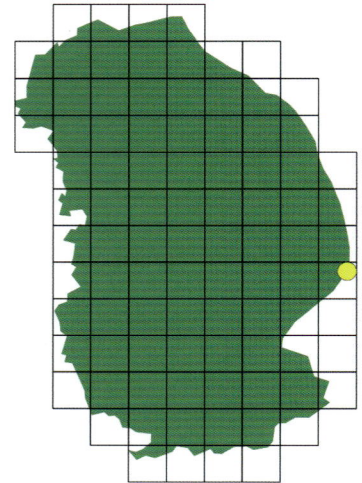

Gibraltar Point

Dipper *Cinclus cinclus*

Vagrant. Britain (*C. c. gularis*; LBRC) and Europe (nominate; BBRC).

Lorand and Atkin (1989) noted that from 1870-1989 there had been about 20 records of *gularis*, mainly Oct-Nov and that others had been found in winter or early spring. One bird wintered on the River Lud at Louth in the five successive winters from 1951-52 to 1955-56; one wintered there 1963-64 and another was in the same location Mar 1967. The nominate race *cinclus* has occurred on at least 15 occasions. Two BBRC-accepted records were in Apr 1967 at Humberston, which was trapped, and the second in 1980 at Donna Nook Nov 21st-25th (*British Birds* 107, 628-629). Prior to 2006, records of *cinclus* were not assessed by the BBRC but their current view is a pragmatic one; birds with little or no chestnut on the belly in eastern Britain, particularly in the Northern Isles and lowland south-east England away from the range of *gularis*, are likely to be nominate *cinclus*. The records in the table have all been accepted by the county records committee in good faith at the time. However, where possible we intend to submit retrospective records and so far, three which have been submitted to BBRC are Lincoln/Heighington 1993/4; Belleau Feb 2008; Gibraltar Point NNR Nov 2008. This subspecies remains a county rarity with a record roughly every 3-4 years since 1963 although there have been none since 2011.

Dipper (British) *Cinclus cinclus gularis*

Site Name	First Date	Last Date	Count	Notes
Louth	08/03/1967		1	

Dipper (Black-bellied) *Cinclus cinclus cinclus*

Site Name	First Date	Last Date	Count	Notes
Tealby	03/02/1963	17/03/1963	1	
Cleethorpes	11/04/1967	13/04/1967	1	Trapped
Louth	15/02/1969		1	
Donna Nook	21/11/1980	25/11/1980	1	
Riseholme	27/11/1980	04/01/1981	1	
Louth	08/12/1980		1	
Little Cawthorpe	05/03/1988	07/03/1988	1	
Stamford	01/12/1990	23/01/1991	1	
Lincoln	14/10/1993	15/10/1993	1	
Branston	03/11/1993		1	Same
Pode Hole	13/11/1993	20/11/1993	1	
Heighington	27/11/1993	05/03/1994	1	Same
Louth	24/01/2002	19/02/2002	1	
Belleau	21/02/2008	24/02/2008	1	
Gibraltar Point	04/11/2008		1	
Saltfleetby Theddlethorpe	13/02/2011		1	

House Sparrow *Passer domesticus*

Abundant resident and partial migrant.

Always a very widespread and abundant bird the humble House Sparrow was estimated in the *LBC Atlas* to have a population somewhere between 70,000 and 130,000 pairs in the late 1980s. In the open countryside it tends to be confined as a breeder to farm buildings. Lincolnshire BBS index data shows that the population has fallen overall by 10% (not significant) during the period 1994-2019, while it has remained more stable across the UK in that time. The big fall that led to its red data designation came during the 1980s. The estimated Lincolnshire population in 2016 was around 130,000 implying that perhaps the *LBC Atlas* estimate should be weighted towards the top end of the range. It generally comes second each year as the most frequently counted bird on Lincolnshire BBS squares after Woodpigeon *Columba palumbus*.

1968-1972	1980-1989	2008-2011	2016-2019	Incidence	Change
				64%	8%

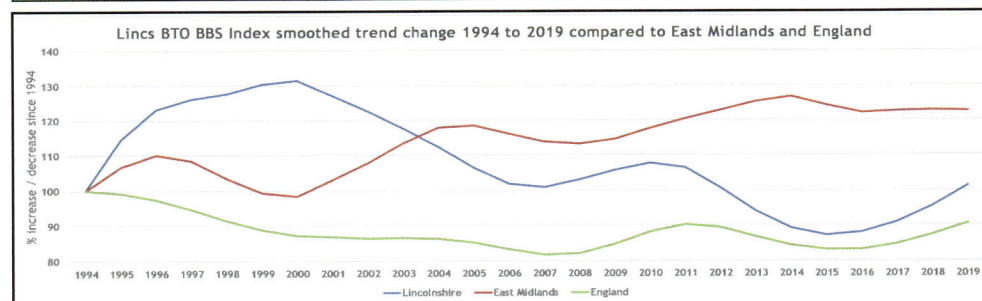

Lincs BTO BBS Index smoothed trend change 1994 to 2019 compared to East Midlands and England

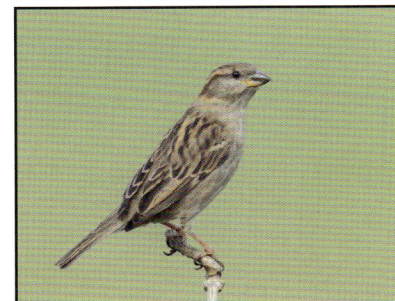

Tree Sparrow *Passer montanus*

Common though localised resident and partial migrant.

Like that of its close relative, the House Sparrow *P. domesticus*, the population crashed through the 1980s but in an even more dramatic fashion. Up to 90% were lost through the *LBC Atlas* period by the end of which, in the late 1980s, it was estimated at 7,000 pairs. Since then, the BBS Index shows that the Lincolnshire population has more than doubled to 2019 from a very low base. The APEP4 estimated population suggests as many as 18,000 pairs in Lincolnshire in 2016 though the numbers have fallen since then.

Tree Sparrows flock up in the autumn and winter and counts of 100 plus are not unusual. The availability of seeds in winter and suitable trees with nesting holes both limit population size. Happily, more farmers are providing suitable stubble and winter seed sources and nest box programmes are often very successful.

1968-1972	1980-1989	2008-2011	2016-2019	Incidence	Change
				28%	-14%

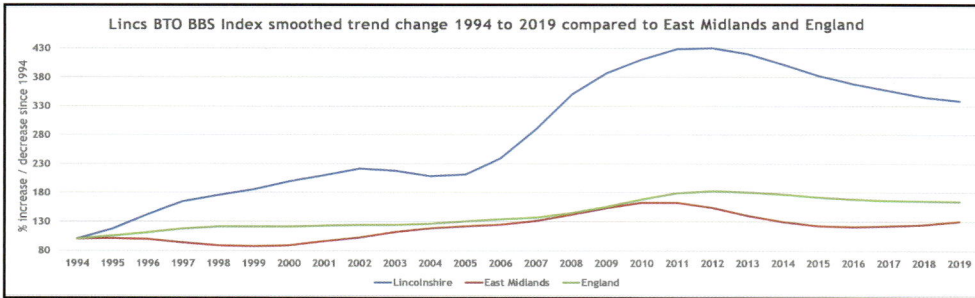

Lincs BTO BBS Index smoothed trend change 1994 to 2019 compared to East Midlands and England

Alpine Accentor *Prunella collaris*

Vagrant. Europe. Two records in 1994 and 2016.

The first Alpine Accentor for the county arrived in mid-Nov and stayed for five days delighting several hundred travelling birders. It frequented the short grassy track, south of Rimac, separating the saltmarsh from the dense dune buckthorn, a far cry from its usual mountain terrain.

The second Alpine Accentor was found at the bird feeders at Sykes Farm, Gibraltar Point on the late afternoon of May 7th and fed at times on the bird table. It disappointed many in disappearing overnight. As with the first record, the bird turned up in coastal habitat never normally associated with the species although vagrants often turn up in the least expected of places.

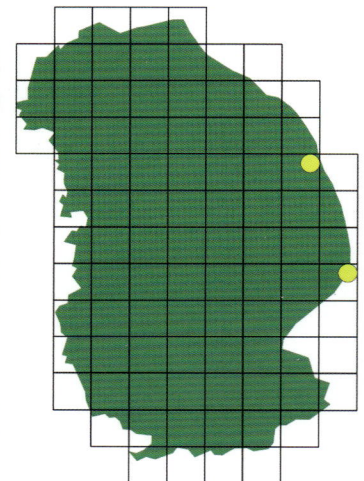

BBRC statistics show that the 1994 record was only the 11th for the UK since their baseline date of 1950 and was the only one of that year. The 2016 record was one of three to arrive in the UK that spring and together these three constituted the 17th-20th records for the UK since 1950. The species remains a true vagrant and averages one record every three years in the UK.

Site Name	First Date	Last Date	Count	Notes
Saltfleetby-Theddlethorpe	14/11/1994	18/11/1994	1	Adult Rimac
Gibraltar Point	07/05/2016		1	Second calendar year

Dunnock *Prunella modularis*

British form *occidentalis* a very common resident and partial migrant. Nominate continental form a scarce passage migrant, mostly in autumn.

This popular and widespread garden resident notable for its breeding system frequently involving two males and one female rather than the conventional pair, is amber listed because the UK population fell by half between the 1970s and 1990s. It has subsequently recovered to some degree. The *LBC Atlas* estimated the Lincolnshire population at around 100,000 pairs in the late 80s. BBS data shows that the Lincolnshire breeding population fell by 4% between 1994 and 2018. The BBS incidence change suggests the species has become a little more widespread over the last 10 years.

1968-1972	1980-1989	2008-2011	2016-2019	Incidence	Change
				88%	14%

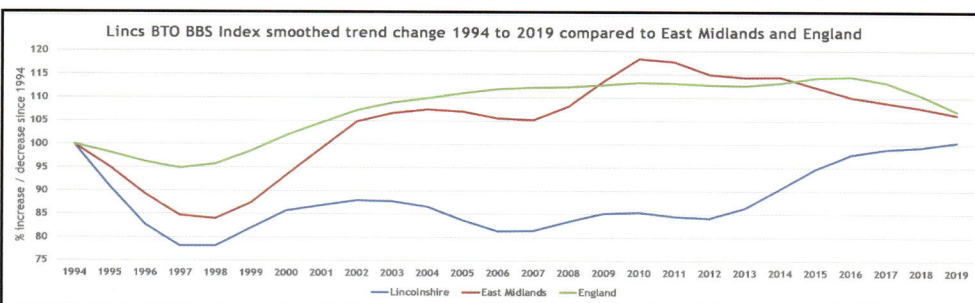

Lincs BTO BBS Index smoothed trend change 1994 to 2019 compared to East Midlands and England

Western Yellow Wagtail *Motacilla flava*

Summer visitor and passage migrant. British race *flavissima* common. Other races listed below are scarce/vagrant.

Cordeaux (1872) described Yellow Wagtail briefly as "The most common and least aquatic of any of the Wagtails". By the late 1980s it was scarcer than Pied Wagtail *M. Alba yarrelli*, when the *LBC Atlas* estimated a population of 2,700 pairs compared to 4,000 for Pied. Since then, in the last three BoCC reviews its UK conservation status has gone from green to amber to red in the space of 20 years. It is a bird of arable land and the decline is thought to be due to agricultural intensification but in Lincolnshire it seems to have become more widespread over the last 10 years. Based on BBS data Lincolnshire holds around 20% of the British population, which is effectively the world population of our endemic *flavissima* race. Perhaps our most important bird of conservation concern? Autumn migration peaks in late Aug- Sep and peak numbers at coastal sites vary considerably from year to year. At Gibraltar Point for instance between 2012 and 2018 the peak varied from 15 in 2014 to 181 in 2013. This may relate to breeding success.

1968-1972	1980-1989	2008-2011	2016-2019	Incidence	Change
				34%	12%

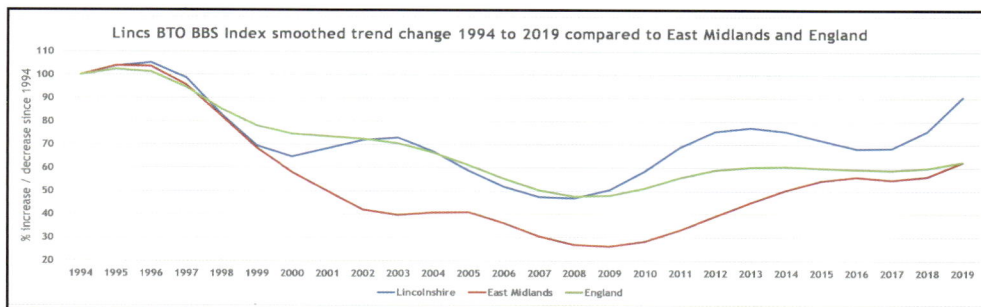

Lincs BTO BBS Index smoothed trend change 1994 to 2019 compared to East Midlands and England

Blue-headed Wagtail *Motacilla flava flava*

Very scarce passage migrant, mainly spring. Has bred sporadically.

This race breeds in north-western Europe. Lorand and Atkin (1989) described it as a sporadic breeder in the 1970s and stated it had bred at Donna Nook, Covenham Reservoir, Cadney and Goxhill. There were no breeding records in the *LBC Atlas* period during the 1980s or the *BTO Atlas 2007-11* or since. In the five years to 2018, LBR reports a total of 21 records all in Apr-May ranging from none in 2018 to 12 in 2015 with one in 2015, three in 2014 and five in 2016. In the same period there were four reports of "Channel Wagtails"; hybrids between this race and the local *flavissima* British race. There were two in 2016 and a single in 2017. Interestingly a Channel Wagtail paired with a Western Yellow Wagtail bred at Marston STW in 2014.

Ashy-headed Wagtail *Motacilla flava cinereocapilla*

Vagrant. Italy

For many years rare wagtail subspecies were not given the attention they deserved and were not formally considered by BBRC. Lorand and Atkin (1989) describe two records of single birds recorded at Covenham Reservoir on May 4th 1983 and May 31st 1984. In the last 30 years there has been a single record of this subspecies, at Covenham Reservoir on Apr 13th 2006. It was submitted late to BBRC and is currently under review. On the face of it, it is an extreme vagrant in Britain with just six accepted records in the period 1950-2019. Many records will have been lost but it is clear this species is a genuine rarity in Britain.

Black-headed Wagtail *Motacilla flava feldegg*

Vagrant. Balkans, Turkey, Caucasus

One at Holbeach Marsh on Apr 6th 2004 is the only county record of this distinctive subspecies. It is an extreme vagrant in Britain with just 21 accepted records in the period 1950-2019.

Grey-headed Wagtail *Motacilla flava thunbergi*

Vagrant. Scandinavia

Lorand and Atkin (1989) reported 18 records between 1969 and 1989 but since then there have been just five, with the last two in 2008, separate birds at Covenham Reservoir and Donna Nook on May 13th-14th and 28th respectively. All have occurred between May 8th-Jun 14th.

Citrine Wagtail *Motacilla citreola*

BBRC

Vagrant. Eastern Europe

The Lincolnshire Citrine Wagtail database is rather slight to say the least with just a single record at Gibraltar Point in Sep 1983. This was one of only two records that year, the other being in Lothian, Scotland in Oct. Whether this dearth of records reflects genuine rarity *per se*, or the difficulty of picking these birds out among the often very large numbers of its commoner congeners is speculative. The advice in the BBRC report for 1983 is still pertinent: "Grey-and-white Yellow Wagtails *M. flava* are still causing problems, but the best pro-Citrine features include a clear-cut and complete pale ear-covert surround, a usually whitish or buffish forehead, and a loud, shrill, almost buzzing call."

Formerly very rare in Britain, since the first on Fair Isle in 1954, it slowly increased through to the 1980s but since the early 1990s the upward trend in British records has been really striking. There have been many years since 2000 with records reaching double figures with 2008 (21), 2011 (20) and 2013 (22) being especially good years. Consequently, it ceased to be considered by BBRC from 2015. Lincolnshire is overdue another one.

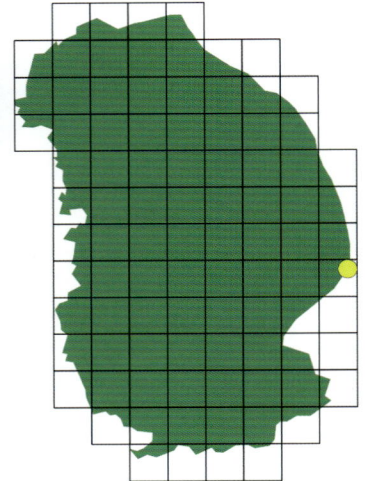

Site Name	First Date	Last Date	Count	Notes
Gibraltar Point	04/09/1983	10/09/1983	1	Juvenile

Grey Wagtail *Motacilla cinerea*

Scarce passage migrant (mainly coastal) and winter visitor (mostly inland). Scarce and fairly local breeder increased in recent years.

Lorand and Atkin (1989) suggest that Grey Wagtail was a sporadic breeder in Lincolnshire from around 1868. It remained rare until the 1980s when it started to spread to the few fast-flowing streams in Lincolnshire, especially on the edge of the Wolds. The *LBC Atlas* reported up to five pairs a year bred across the county in the 1990s and the BTO Atlas 2007-11 showed it had become more widespread in the south-west of the county. It remains fairly local and the BBS does not give a true reflection of its Lincolnshire breeding status. On a typical chalk stream, the river Lud through Louth, there were five breeding pairs reported in 2020. The overall trend of the English population over the last 22 years has been up 7% but that masks ups and downs over that period. It is difficult to accurately estimate the breeding population of Lincolnshire but it has noticeably increased over the last 30 years and probably now lies in the range of 50-100 pairs, a drop in the ocean compared to the 37,000 pairs estimated for Britain in 2016. There is a pronounced coastal autumn passage in Sep-Oct and peak counts across the county from 2013 to 2019 ranged from 49 in 2013 to 77 in 2015. Very few Grey Wagtails are ringed so it is not well understood where these birds are heading for the winter but the ringing evidence such as it is, suggests no further than France.

1968-1972	1980-1989	2008-2011	2016-2019	Incidence	Change
				0.8%	-26%

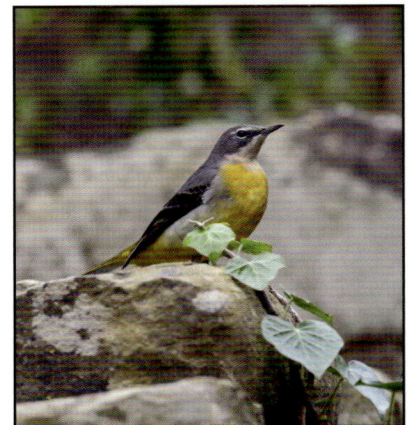

Pied/White Wagtail *Motacilla alba*

Common resident and passage migrant. Nominate continental form White Wagtail *M. a. alba* scarce passage migrant. Members of the species that cannot be assigned to a race are referred to as "Alba Wagtails"

Pied Wagtail has been very widespread in Lincolnshire throughout the last 60 years. The *LBC Atlas* estimated a population of around 4,000 pairs and the Lincolnshire BBS index suggests there has been a (not significant) decline in Lincolnshire of 25% between 1994 and 2018. It is one of our earliest migrants appearing from late Mar. The *LBC Atlas* reported winter roosts of up to 500 birds in Scunthorpe and 1,000 in Lincoln during the 1990s. Sadly, these roosts are not counted regularly but in recent years the largest roosts reported have been at Brayford Pool in Lincoln city centre with 400 there in Jan 2016 and 300 in Feb 2016.

White Wagtail of the nominate race *M. a. alba*, which breeds in nearby mainland Europe, can only be reliably identified in the field and separated from the much more numerous indigenous British *yarrelli* race in adult breeding plumage and this account focuses on their spring passage when identification is most certain and birds are most numerous. It is an early migrant and LBR for the five years to 2018 indicates that first birds arrived from Mar 12th-31st. Peak site counts across the county tend to be in Apr and ranged from eight in 2015 to 84 in 2016. The largest day counts tend to be in a very narrow window from Apr 14th-18th and are usually from Frampton Marsh where there were 40 on Apr 16th 2017 and 30 on Apr 15th 2016

1968-1972	1980-1989	2008-2011	2016-2019	Incidence	Change
				53%	7%

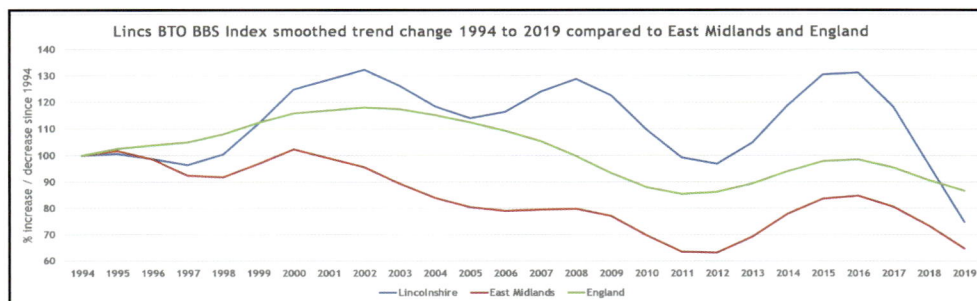

Lincs BTO BBS Index smoothed trend change 1994 to 2019 compared to East Midlands and England

Richard's Pipit *Anthus richardi*

Very scarce passage migrant, mainly Sep-Nov, rarely Dec-Jan and Apr-May.

The first two records were at Tetney, Oct 12th 1887 and shot at Marshchapel on Nov 16th 1912. There was one at Saltfleetby in Oct 1951 and then a gap until Oct 1967 since when it has been all but annual with blank years in only 1996 and 2008. The numbers found per year have been variable, with most in 1968 (12), 1994 (12), 2013 (15) and 2016 (10). Two and three birds together have been recorded in 15 of the years, 1967-2019, but note that with 'flyover' birds and onward migration on the coast it can be difficult in some years to assess how many have actually been present over short periods of time. A conservative estimate of total numbers during this time is of 204 birds; the average per year has increased from 3.3 during 1980-99 to 4.4 during 2000-19, but with nearly six per year in the last decade.

The species is predominantly an autumn migrant with most found between mid-Sep and mid-Nov at coastal locations. Most are seen on single dates, but some stay longer, and a few have wintered. The first to have wintered was at Skidbrooke from Dec 19th 1992 to Jan 10th 1993; another wintered at the same site Jan 12th-Apr 12th 2008, a total of 92 days. Two recent birds, which had most likely over-wintered, were found on the Humber at Goxhill, Feb 26th-Mar 14th 2016 and Alkborough Feb 6th-Apr 24th 2019. The latter bird was present for 78 days, the second longest after that in 2008. Both had a partial or full moult into 2CY/pre-breeding plumage. Interestingly the Apr departure date of the bird in 2019 is very similar to the finding dates of three of the other spring records, *viz* Apr 14th 1995; Apr 16th 1998; Apr 22nd 2015. Two others in spring were found May 9th 1985 and May 16th 2004. One highly unusual occurrence was a bird at Saltfleetby on Aug 15th 1975.

The majority of Richard's Pipits have been found on the coast but a handful have occurred in each of The Wash and the Humber and three further inland, the first being one at Wisbech sewage farm on 7th Oct 1968 and others at Cadney Reservoir on Oct 14th 1988 and Baston-Langtoft gravel pits on Nov 4th 2006.

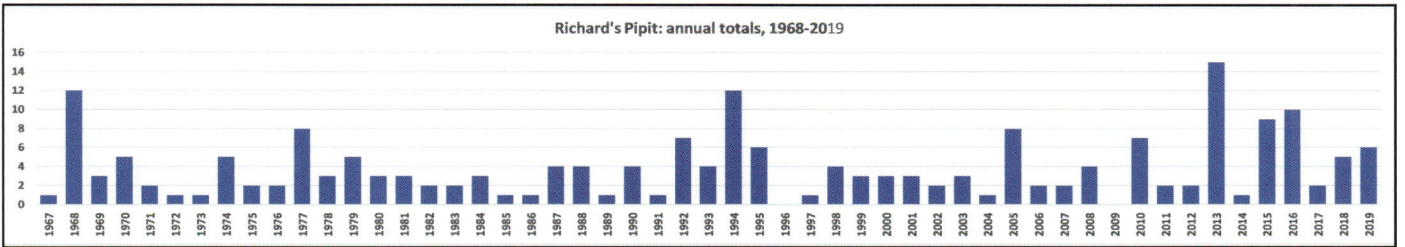

Tawny Pipit *Anthus campestris*

Rare. Southern Europe.

This rare pipit has been recorded on 12 occasions, eight in spring (Apr 25th-Jun 8th) and four in autumn (Sep 20th-Oct 15th). The only inland record was one at the Wisbech Sewage Farm, Sep 1970. Observers of the bird at Tetney outfall sluice in Apr 2011 had the double pleasure of watching it along with a nearby trip of seven Dotterel.

Along with Red-throated Pipit *A. cervinus* the species had a period when it was dropped from BBRC consideration (1983-2014) but it was reinstated from 2015. Overall, there have been more than twice the number of Tawny Pipits in Britain compared to Red-throated, with peaks in the 1980s (56, 1983) and 1990s (57, 1992) but the average 2010-19 was just six to seven birds per year. Although the species is polytypic and has a vast breeding range from Europe across Asia, populations in West and Central Europe have declined markedly for familiar reasons - from afforestation of open habitats, scrub encroachment and intensification of agriculture. In Central Europe there are only isolated breeding sites, mainly at inland sand dunes, open-cast mines and forest clearings. With only four records since 2000 it remains a difficult bird to see in the county.

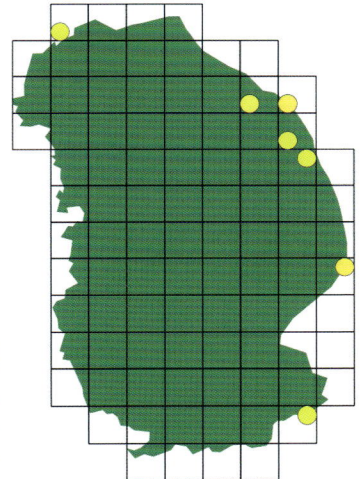

Site Name	First Date	Last Date	Count	Notes
Donna Nook	24/05/1970		1	
Wisbech STW	25/09/1970		1	
Donna Nook	08/05/1975	17/05/1975	1	
Tetney	15/10/1975		1	
Donna Nook	10/05/1980		1	
Gibraltar Point	10/06/1987		1	
Gibraltar Point	13/05/1990		1	
Gibraltar Point	29/05/1993		1	
Saltfleetby-Theddlethorpe	20/09/2000	21/09/2000	1	
Fockerby	01/10/2001		1	
Tetney	25/04/2011	28/04/2011	1	
Saltfleet	08/06/2014		1	

Saltfleet Haven

Meadow Pipit *Anthus pratensis*

Common/very common resident, passage migrant and winter visitor.

Meadow Pipit is primarily an upland bird in Britain but a glance at the breeding Relative Abundance map in *BTO Atlas 2007-11* shows that in lowland England there is an area of distribution on the English east coast stretching from the southern foothills of the North Yorkshire Moors down to the Fens where Meadow Pipit does well compared to the rest of lowland England. Gaps in the *LBC Atlas* of the 1980s, mainly in the Kesteven area, appear to have been filled in by 2007-11. The BBS long term trend for England was down 21% from 1994-2019. There is insufficient data to compute a Lincolnshire BBS index and the estimated population in 2016 was 20,000 pairs. Meadow Pipit is much more numerous on migration especially in autumn and it has a strong wintering presence too. The biggest day numbers each year come from autumn migration counts at Gibraltar Point Bird Observatory. LBR reports for the five years to 2018 show a range in peak day counts from 2,437 in 2015 to 4,800 on Sep 11th 2016 with an average of 3,400. In winter it is found in every 10km square in the county.

These hordes of Meadow Pipits arriving on the coast and passing through the county have given plenty of opportunity to catch and ring them. BTO ringing data showed that to date 17,302 have been ringed in Lincolnshire. There have been six recoveries of birds ringed at Gibraltar Point and one near Donna Nook in Sep-Oct in various years between 1963-2002. Four recoveries were from Portugal, three of them in Oct-Nov the following year, one, two years later in Feb. There were single recoveries from The Netherlands in the following Sep; Morocco in Nov three years later; Italy in Oct the same year; France in Jan two years later; and lastly one was picked up freshly dead in Mar in the North Sea on an Ekofisk oilrig six months after being ringed. Debate over geographical variation has ensued over the years but IOC treat the species as monotypic. Paler birds are noted during migration periods and these are thought to come from Scandinavia and Russia; the variation is thought to be clinal with the palest birds coming from the furthest points east.

1968-1972	1980-1989	2008-2011	2016-2019	Incidence	Change
				27%	4%

Tree Pipit *Anthus trivialis*

Scarce summer visitor and passage migrant. Breeds locally, mainly in the western half of the county.

Smith and Cornwallis (1955) noted the status of the Tree Pipit as follows: "A locally common summer visitor in many northern, western and south-western districts, but very scarce in most eastern parts of the county. It is particularly numerous on the heaths of the north-west and in similar country around Market Rasen, Woodhall Spa and in the district west and south-west of Lincoln." Thirty-four years later, Lorand and Atkin (1989) still regarded the species was "a fairly common summer visitor and passage migrant. It is widespread in the north-west, central and south-west of the county where it breeds in open woodland, young conifer plantations and on heathland." By 1990 the species was confined to less than 10 breeding sites with the bulk of the population in the Laughton Forest-Scotton Common area and Crowle Moors. At Laughton-Scotton the peak in occupied territories was in 2003 when 52 males were recorded but since then numbers have fallen with the last full survey locating 26 males in 2015. The combined county breeding population is now in the region of 40-45 males with studies showing that up to 26% of singing males may be unmated in a typical year. The earliest arrivals appear on breeding sites in the first week of Apr. Birds occur along the coastal strip on passage in Apr-May and Aug-Oct with late records in early Nov. Inland migrants are less obvious in the same periods.

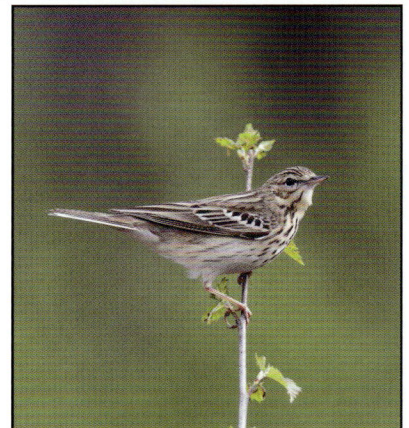

1968-1972	1980-1989	2008-2011	2016-2019	Incidence	Change
				0%	-100%

Olive-backed Pipit *Anthus hodgsoni*

Vagrant. Siberia.

The first county record was in Oct 1980, a typically skulking bird found along the edge of a harvested bean field occasionally perching in adjacent willows *Salix sp.* when it characteristically wagged its tail slowly up and down. There was a 10-year wait for a second when there were two, both from Oct 18th-20th, one of which was trapped. There have been another 12 records, eight of them at Gibraltar Point.

There were 43 British records in 1990, and to echo the BBRC Report of that year, this was an unexpected avalanche with twice as many as in any previous year and more than half the (then) British and Irish total. Since 1990, further 'big years' of double-figure records followed including 1993 (34), 2011 (28) and 2012 (58) and BBRC ceased to consider the species after 2012.

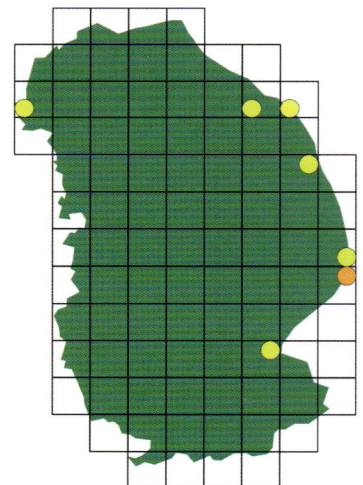

Site Name	First Date	Last Date	Count	Notes
Saltfleetby-Theddlethorpe	19/10/1980		1	
Donna Nook	18/10/1990	20/10/1990	1	
Gibraltar Point	20/10/1990		1	Trapped
Gibraltar Point	12/10/1992		1	
Gibraltar Point	08/11/2000		1	
Gibraltar Point	26/09/2001		1	
Skegness	18/10/2003	20/10/2003	1	
Gibraltar Point	02/11/2008		1	
Wroot	13/10/2011		1	
Donna Nook	12/10/2013		1	
Frampton Marsh	13/10/2013		1	
Gibraltar Point	17/09/2014	20/09/2014	1	
Gibraltar Point	25/10/2017		1	
North Cotes	19/10/2018		1	
Gibraltar Point	08/11/2019		1	

Red-throated Pipit *Anthus cervinus*

BBRC
British Birds Rarities Committee

Vagrant. Scandinavia east across Russia.

This species is inexplicably rare in the county with only three widely-spaced records, one in spring and two in autumn. The first was at Grainthorpe in Sep 1977. The bird at Gibraltar Point, May 26th-28th 1992 was a male, heard in song. That in Oct 2013 was a flyover bird only heard calling by three experienced observers.

The species has been ex-BBRC 2006-14 but was re-listed from 2015 onwards as its annual numbers in Britain declined. Over the last 35 years numbers have fluctuated wildly, between a low of two in 1986 and an extraordinary 47 in 1992. These fluctuations are hard to explain. Red-throated Pipit is a common bird over its vast range and its population is thought to be stable. About 10 per year are being reported in Britain. The vast majority in Britain have been in Shetland (123), Isles of Scilly (93) and Norfolk (68); East and North Yorkshire have recorded 34. It's difficult to believe that this species is not being regularly missed in Lincolnshire amidst the vast numbers of Meadow Pipits *A. pratensis* that crowd our coastline every autumn.

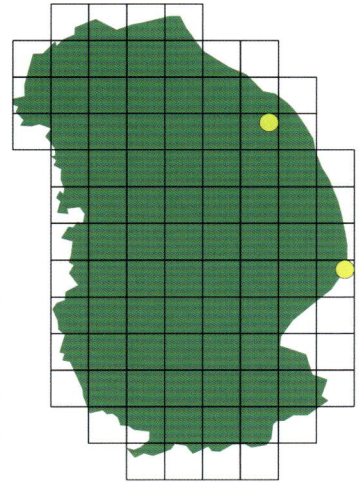

Site Name	First Date	Last Date	Count	Notes
Grainthorpe	21/09/1977	22/09/1977	1	
Gibraltar Point	26/05/1992	28/05/1992	1	Male
Gibraltar Point	12/10/2013		1	

Buff-bellied Pipit *Anthus rubescens*

BBRC
British Birds Rarities Committee

Vagrant. North America, West Greenland

One county record on ploughed fields at Wyberton Marsh on Dec 5th 2005, and then later on wet grassland at the adjacent Frampton Marsh remaining until Jan 29th 2006. It was found associating with a large flock of Meadow Pipits *A. pratensis* which also included a few Water Pipits *A. spinoletta* and Rock Pipits *A. petrosus littoralis*. Initially quite elusive as the flock moved, it often disappeared down runnels in the plough. It was found and seen well during the next two days but there were some tense moments before the landowner agreed access for other birders. The news was finally put out at lunchtime on Dec 13th and some birders saw it late afternoon that day before it disappeared for the rest of 2005. It was re-found by the original finder more than a month later on Jan 24th 2006 along a dyke on the wet grassland at Frampton, undoubtedly the last place anyone would want a rare pipit to frequent. It then disappeared again only to be re-found in the original field where it appeared erratically until the last sighting on Jan 29th.

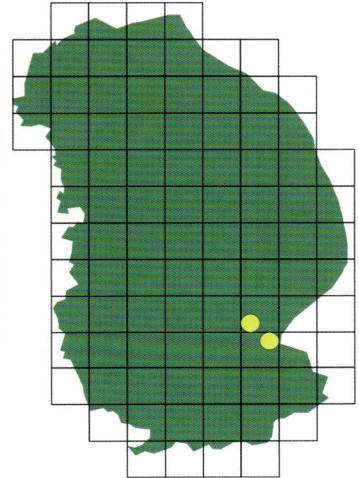

Buff-bellied Pipits are not easy birds to identify with certainty and this was only the third British record at the time and an exceptional find by the observers concerned and a very significant county record. Since then, more has been learned about their distinguishing characteristics with the result that there have been another 34 found between 2007-12 but with only five subsequently. All have been of the nominate race to date. Who will find the first *A. r. japonicus*, the Siberian race?

Site Name	First Date	Last Date	Count	Notes
Wyberton	05/12/2005	13/12/2005	1	
Frampton Marsh	24/01/2006	29/01/2006	1	Same

Water Pipit *Anthus spinoletta*

Scarce passage migrant and winter visitor Oct-Apr. Mainly coastal, possibly increasing but can be elusive, making counting difficult.

Water Pipit can be an elusive bird to get a good view of and when flushed usually flies a good way off. It is thought that the UK wintering population is of the order of 200 birds. The *BTO Atlas 2007-11* suggests most birds winter along the east coast between the Humber and Suffolk. Over the last 10 years to 2018 LBR shows the peak day count in the county has ranged from three in 2011 and 2014 to 23 at Gibraltar Point on Feb 4th 2016. Most birds winter on the coast with Alkborough Flats, Donna Nook and Saltfleetby-Theddlethorpe hosting most birds in different years. The largest count of Water Pipit ever in Lincolnshire was an astonishing flock of 40 at Alkborough Flats seen on Jan 12th 2008. Prior to that the record count was seven. They also winter inland but in fewer numbers with Marston STW and Baston Langtoft GP regularly holding wintering birds.

Rock Pipit *Anthus petrosus*

Scandinavian Rock form *littoralis* is a green listed fairly common passage migrant and winter visitor Sep-Apr; scarce inland. Nominate British and Irish form *petrosus* is amber listed but essentially resident so probably a very scarce winter visitor.

The coastal saltmarshes between Tetney Marshes and Saltfleetby-Theddlethorpe are quite possibly the most important British wintering area for Scandinavian Rock Pipit based on the results reported in the *BTO Atlas 2007-11*. The latter site held 180 birds in Dec 2015. Birds tend to arrive from mid Sep and there is pronounced autumn migration best witnessed at Gibraltar Point where regular visible migration watches are held. During Oct 2017 a total of 1,969 birds were counted flying over heading south. Birds wintering on saltmarshes are hard to count accurately as they don't tend to flock up and are generally scattered in ones and twos over a wide area. The wintering population probably numbers several hundred. Few Rock Pipits have been ringed in the county but Norwegian colour-ringed birds have been spotted at Barton upon Humber in Oct 2013, Pyewipe Marsh, Grimsby in Jan 2016 and Witham Mouth in Oct 2018. Inland records are very scarce with the concrete embankments at Covenham and Toft Newton Reservoirs the most usual places to turn one up.

Chaffinch *Fringilla coelebs*

British form *gengleri* an abundant resident. Nominate continental form a passage migrant and winter visitor.

Our commonest and most widespread finch, the resident local population has been shown to be sedentary from ringing studies. The *LBC Atlas* estimated the population at 146,000-178,000 pairs in the 1980s. The Lincs BBS index shows that the breeding population increased by 37% during the period 1994-2018 and the population estimate in 2016 was 150,000 pairs. The chart indicates that Chaffinch is doing relatively well in Lincolnshire compared to the rest of East Midlands and England. The wider English Chaffinch population appears to be suffering from *Trichomonas gallinae*, a protozoan parasite that has also severely impacted the Greenfinch *Chloris chloris*, since 2006. The BTO Atlas 2007-11 indicates that the Fens was the focal point of a population increase since the previous BTO Atlas 1988-91.

There is a marked spring and autumn passage involving nominate race Scandinavian birds, with LBR reporting peak spring passage at Gibraltar Point in the five years to 2018, ranging from a peak count of 84 on Mar 29th 2016 to 2,526 on Apr 4th 2018. The peak day autumn passage here ranged from 149 in 2016 to 1,511 on Nov 13th 2018. The largest winter flock reported was 300 at Nocton Fen on Jan 1st 2016, with the next largest counts being of 100 from a number of different sites across the county. The largest winter flock reported in the *LBC Atlas* period was 740.

1968-1972	1980-1989	2008-2011	2016-2019	Incidence	Change
				98%	14%

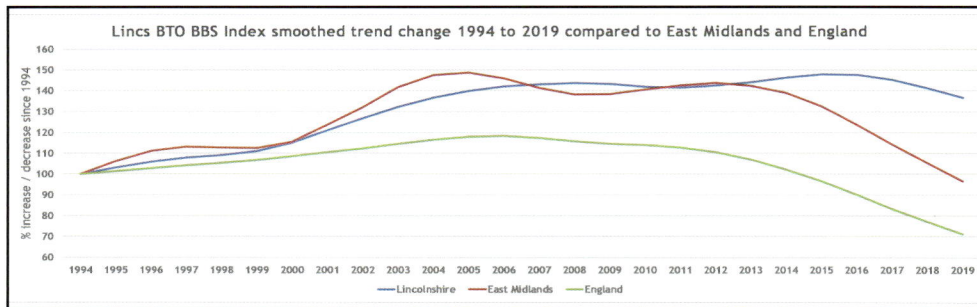

Lincs BTO BBS Index smoothed trend change 1994 to 2019 compared to East Midlands and England

Brambling *Fringilla montifringilla*

Winter visitor and passage migrant. Exceptional in summer. Numbers fluctuate considerably between winters.

Brambling is a fairly common visitor to Lincolnshire from Scandinavia between late Sep-early May and the sight of a spring male in breeding plumage is always a boost. In the 1980s and 1990s The *LBC Atlas* reported that "*in good winters flocks of several hundred are not uncommon*". There has been nothing like that number of birds wintering in recent years and Lincolnshire seems to have done poorly for Brambling compared to neighbouring Yorkshire and Norfolk where such flocks still regularly occur. Wintering numbers tend to be related to beech-mast production, a relatively scarce commodity in Lincolnshire. Wintering flocks of millions are reported from central Europe.

The largest wintering flocks reported in LBR over the five years to 2018 were from Asterby with 60 on Dec 21st 2018 and in the same year 50 at Wroot on Jan 29th. Other than these two exceptions wintering flocks in excess of 25 birds have been rare in recent years.

Migration is a different matter. Several sites on the coast experience good Brambling numbers, with autumn being much better than spring. The best day-count in the five years was 461 birds flying south at Gibraltar Point on Oct 7th 2018. Butterwick Marsh had a flock of 200 on Oct 9th 2018. The only other largish flock reported was of 80 at Gibraltar Point on Oct 13th 2016.

Hawfinch *Coccothraustes coccothraustes*

Rare local resident and rare passage migrant. Occasionally irruptive.

Hawfinch formerly bred in the north-west of the county in the woods around Scunthorpe with the last confirmed breeding taking place in 1999. It had colonised that area in the 1890s having previously been restricted to the south-east of England. Spring flocks of around 20 birds continued to be seen in the Scawby area up until the mid 2000s but there has been no further evidence of breeding. Over the last 10 years it has become increasingly very scarce. LBR for the five years to 2018 reports the maximum number of birds in spring ranged from zero in 2014 and 2017 to four in 2018, with one in 2015 and 2016. Autumn passage was also very light with most birds seen on the coast and numbers ranging from three in 2014 and 2015 and eight in 2016 and 2018.

There was a major exception to this pattern of increased rarity, the largest UK Hawfinch influx in living memory, the Lincolnshire part of which began with two at Gibraltar Point on Oct 14th 2017. The invasion was covered in detail by Hyde (LBR 2018). The autumn peak was 32 birds across 14 sites in Nov with Forest Pines, Broughton having 10 birds in Nov and 21 in Dec. Wintering numbers across the county increased to a peak of 110 in the week Feb 21st-27th 2018, with 60 at Forest Pines. Elsewhere, Scawby Park had 32 on Jan 31st and East Keal churchyard had 24 on Apr 8th. Even at the beginning of Apr there were still 81 birds across the county. By Apr 25th they had all departed, presumably to south-eastern Europe from whence it is thought they came.

In Oct 2020 a colour ringed bird was photographed at Saltfleetby-Theddlethorpe. It had been ringed in the Conway Valley of North Wales in Feb 2018. It was thought by the ringer to be a Scandinavian wintering bird because of its size and weight.

1968-1972	1980-1989	2008-2011	2016-2019	Incidence	Change
				0%	

Bullfinch *Pyrrhula pyrrhula*

British form *pileata* a common resident, but largely absent from the Fens. Nominate Scandinavian form (Northern Bullfinch *P. p. pyrrhula*) a very rare irruptive visitor in autumn and winter.

The Bullfinch is a relatively sedentary, shy and secretive species and these latter traits make assessing its numbers difficult. From the mid-1970s until around 1980 it suffered a 50% decline in population. The decline slowed until it bottomed out around 2000, since when it has gradually started to increase again. No good explanation of the decline is available.

The *LBC Atlas* found it widely distributed in the 1980s but missing from large parts of the Fens, unsurprising for a primarily woodland bird. The population at that time was estimated towards the lower end of 3,700-6,500 pairs. The long-term population change trend shown by BBS in the East Midlands from 1994 to 2018 is an increase of 28% and the estimated population in 2016 was 4,800 pairs.

The peak one-day count recorded in LBR for the five-year period to 2018 ranged from 14 in 2015 to 33 in Jan 2017 at Whisby. Most larger counts come from Whisby near Lincoln. Snipe Dales near Horncastle had 18 in Apr 2018 and 24 in Feb 2020.

1968-1972	1980-1989	2008-2011	2016-2019	Incidence	Change
				15%	14%

Bullfinch (Northern) Pyrrhula pyrrhula pyrrhula

Over the last 25 years there have been records in only two years, indicating the great rarity of the nominate race in Lincolnshire. Four were present at Gibraltar Point (no more than two at once) in Oct-Nov 2003 and two were at Waters Edge Country Park, Barton upon Humber on Nov 27th 2010.

Common Rosefinch *Carpodacus erythrinus*

Rare. North-eastern Europe.

A relative late-comer to the county with the first record being in May 1979 at Donna Nook, there have now been 54 confirmed records in all to 2019. Thirty-four have been in spring with most in May (18) and Jun (15). It is often difficult to decide whether birds in mid-Jul to early Aug should be treated as spring or autumn migrants, but two records, both of singing males, at Donna Nook on Jul 8th 2013, and at Gibraltar Point on Aug 2nd 2019 are considered 'summer' records.

Autumn records (19) span Aug-Oct with the earliest recorded on Aug 26th. The majority have been found in Sep (15) with just two in Oct, on 2nd and 12th, both at Gibraltar Point. There are no records of over-wintering birds. In most years there are one to two birds but three or more have occurred in seven years with most in 2010 (six) and in 2013 (13); there were 201 British records in 2013, the fourth highest ever.

The bulk of the records have come from Gibraltar Point, 38 involving around 43 individuals. The distinctive slowly rising whistle of singing males usually from within deep cover has been heard on at least 17 occasions there late May to early Jun and is clearly the best time to pick one up. The sparrow-like juveniles in the autumn are harder to find and clearly may be overlooked within large autumn flocks of Linnets *Linaria cannabina* and other finches.

Greenfinch *Chloris chloris*

British form *harrisoni* a very common but declining resident, passage migrant and winter visitor. Nominate continental form a winter visitor in unknown numbers. Red list (*harrisoni*) and Green list (nominate).

In the *LBC Atlas* of the 1980s Greenfinch was the second commonest Lincolnshire finch with a population estimated at 17,500 to 29,000 pairs.

With the advent of the *Trichomonas gallinae* epidemic from 2006 onwards, the Lincolnshire population, which has closely followed the English population trend, crashed some 57% over the long term since 1994, with much of that crash in the last 10 years. The estimated population for 2016 was 18,000 pairs and it is now our fourth commonest breeding finch having been overtaken by Goldfinch *Carduelis carduelis* and Linnet *Linaria cannabina*. The impact of *T. gallinae* on Greenfinch is of a similar order to the impact of the "Black Death" on the human population of England in the 14th century. LBR reports show that the largest wintering flocks in the five years to 2018 were of 200 birds at Whisby in Feb 2015 and Dowsby Fen in Nov 2018. These were exceptional, flocks below 60 birds being the norm.

It is salutary to note that the *LBC Atlas* reports flocks in the 1990s of 1,000 in TF26 and 2,000 in TF17. There is a regular autumn passage observed at Gibraltar Point and in the period the highest day count of birds moving south was 312 on Oct 17th 2014.

1968-1972	1980-1989	2008-2011	2016-2019	Incidence	Change
				48%	-28%

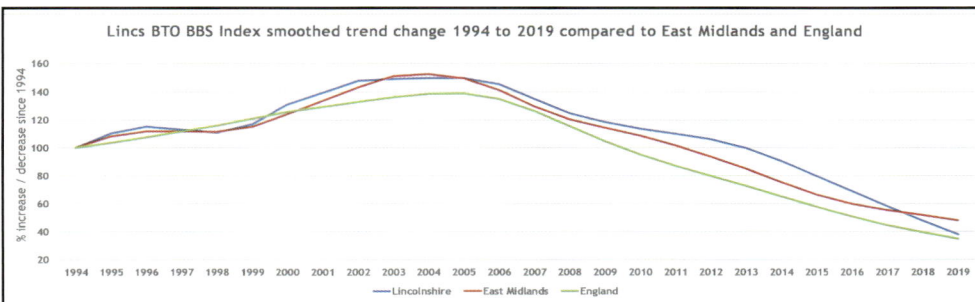
Lincs BTO BBS Index smoothed trend change 1994 to 2019 compared to East Midlands and England

Twite *Linaria flavirostris*

Fairly common, declining to scarce coastal winter visitor in recent years and passage migrant, mainly Sep-Apr. Rare inland.

The *LBC Atlas* reported that "large flocks of several thousand used to be found on The Wash, but in recent years numbers have been drastically reduced." The largest flock reported on The Wash in 2019 was 80 at Gedney Drove End. The endemic race of Twite in Britain *pipilans,* has declined dramatically and BTO BirdFacts reports the population at 7,900 pairs. It is not found sufficiently on BBS squares to produce an index. The epicentre of the decline in the breeding population has been the Pennines, the southern Pennines of Derbyshire particularly and this is probably the origin of many of our wintering birds. Certainly, colour ringed birds from the hills west of Huddersfield, West Yorkshire have been reported at locations ranging from Tetney Marshes to Holbeach Marsh. LBR reports for the six years to 2019 show that Dec-Jan are the peak count months across the coast, and these have fallen steadily from 937 in Dec 2014 to 132 in Dec 2019. The largest flock in recent years has been at Saltfleetby-Theddlethorpe. The peak counts of this flock stood at 451 in Jan 2015 and 127 in Jan 2019. The coast from here to Tetney is now the main centre of the wintering population. It seems that the population decline is attributable to reduced breeding productivity brought about by agricultural intensification in the uplands, fertilising grass fields and heather moorlands for increased dairy potential and a switch from cattle to sheep reducing food availability (Newton 2017). In the six-year period four inland sites had Twite records, Marston STW having six in Mar 2014 and one in Mar 2016. Covenham Reservoir had four in Oct 2014 and two in Nov 2019. Dunsby Fen had three in Nov 2014 and Toft Newton Reservoir also had a single that month.

Linnet *Linaria cannabina*

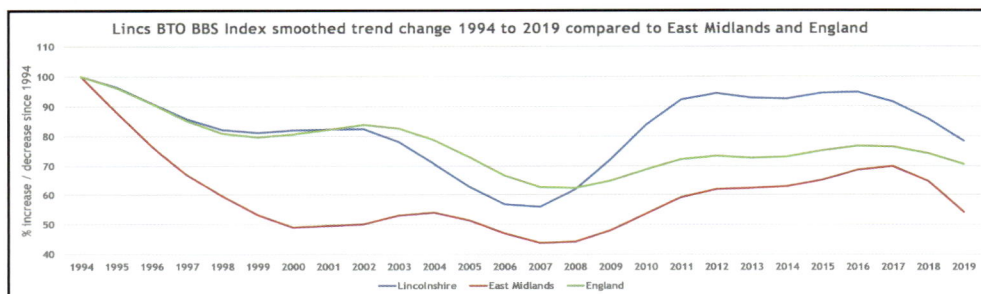

Very common resident, passage migrant and winter visitor.

Linnet was one of the worst-hit birds through agricultural intensification during the 1960s-80s and the English population plummeted from 1966 and hit a low point in the late 1980s. Even after the *LBC Atlas* in the 1980s it was still found breeding in nearly every tetrad and the population was estimated at between 18,000-23,500 pairs. Step forward 30 years to APEP4 and the adjusted population is estimated at 30,000 pairs in 2016. The chart shows Lincolnshire Linnets have done better than the wider English population, but they have shared in a decline over the last three years. LBR reports of the largest wintering flocks reported in the five years to 2018 ranged from 300 in Dec 2014 to 800 at Freiston Shore in Jan 2016 with an average peak of 500. This compares with peak counts from the *BTO Winter Atlas* survey during the 1990s of 750 in TF49 and 800 in TF55. There is a healthy spring and autumn passage along the coast and peak day counts at Gibraltar Point come between Sep 10th-30th and range from 401 on Sep 17th 2015 to 1,827 on Sep 30th 2017.

1968-1972	1980-1989	2008-2011	2016-2019	Incidence	Change
				78%	10%

Lincs BTO BBS Index smoothed trend change 1994 to 2019 compared to East Midlands and England

Common Redpoll *Acanthis flammea*

Nominate Scandinavian and Russian form scarce or very scarce, occasionally fairly common, irruptive winter visitor. Greenland form *rostrata* a very rare vagrant.

Common Redpoll (formerly known as Mealy Redpoll) was split by the BOU from Lesser Redpoll *A. cabaret* in 2001 but confusion about the species still occurs and birds are only accepted in Lincolnshire with supporting evidence, either a ringer's report, photograph or description. It breeds all across the boreal zone of the northern hemisphere, but not in the British Isles. Over the 10 years to 2018 there was only one year, 2011 that could be described as an invasion year. In that year there were 156 birds, of which the largest flock (not only in that year, but the whole decade) was 30 birds at Waters Edge Country Park, Barton Pits on Feb 12th 2011. In the other nine years numbers ranged from one in 2015 to 40 in 2013. The overall average over the decade was 33 per year reducing to 20 omitting the invasion year. Other large flocks reported were 20 at Donna Nook on Oct 12th 2013 and 20 at Laughton Forest on Feb 14th 2017 where five birds remained until Mar 10th.

Lesser Redpoll *Acanthis cabaret*

Fairly common but declining passage migrant and winter visitor. Very scarce breeder.

Lesser Redpoll formerly bred extensively in Lincolnshire in all woodland habitats, gardens and coastal scrub. The *LBC Atlas* estimated the population at 5,000 pairs in the 1980s. It was very widespread and missing from only two 10km squares in the county. It has subsequently suffered one of the worst crashes of any breeding bird in Lincolnshire in the last 50 years. By the *BTO Atlas 2007-11* it was confirmed breeding in only one 10km square in the county. LBR reports indicate that the last confirmed breeding was a nest found near Market Rasen in Jun 2010. Reasons for the decline are not clear but it has been widespread across lowland England and may be related to a decline in birch seed availability. Over the last 10 years there have been numerous ringing controls in the county of birds ringed across the Scottish Highlands, Scottish Borders, Cumbria and Northern Ireland, all locations where it still breeds in good numbers. Peak annual site day counts in the five years to 2018 ranged from 55 at Whisby on Nov 29th 2014 to 300 at Crook Bank, Saltfleetby-Theddlethorpe in Oct 2017 with an average of 120. The fact that it is still relatively numerous during passage and winter gives hope that it may re-establish as a breeding bird at some point in the future.

1968-1972	1980-1989	2008-2011	2016-2019	Incidence	Change
				0.6%	

Arctic Redpoll *Acanthis hornemanni*

BBRC

Vagrant. Scandinavia. First recorded in Oct 1975.

After the first record of two birds seen at Saltfleetby-Theddlethorpe Oct 12th 1975, there have been at least 25 more records involving a minimum of 34 birds. All Lincolnshire records to 2019 have been of the sub-species *A. h. exilipes*, known as Coue's Arctic Redpoll. The taxonomy of the Redpoll complex continues to evolve, and identification of the different taxa needs attention to detail and preferably good photographs.

The northern Redpolls are subject to large irruptive movements, the last in 1995/6 saw huge numbers of Common Redpoll *C. f. flammea* and Coues's Arctic Redpoll with an estimated total of 431 individuals in Britain from Nov 1995 to May 1996 (*British Birds* 93: 59-67).

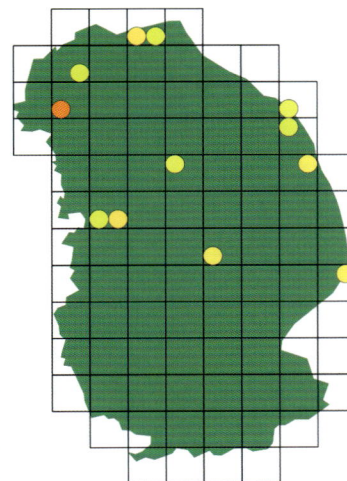

Site Name	First Date	Last Date	Count	Notes
Saltfleetby-Theddlethorpe	12/10/1975		2	
Scunthorpe	28/03/1976		1	Male
Saltfleetby-Theddlethorpe	21/02/1982	13/04/1982	1	Trapped
Gibraltar Point	15/01/1984	25/02/1984	1	
Donna Nook	18/10/1990	21/10/1990	1	
Kirkby Moor	20/01/1991		2	
Kirkby Moor	20/01/1991	27/01/1991	1	
Linwood	21/01/1991	27/01/1991	2	
North Somercotes	26/01/1991	05/02/1991	1	Male
Laughton Forest	20/02/1991	08/03/1991	1	
Donna Nook	12/11/1995		1	First winter
Laughton Forest	27/12/1995		7	
Lincoln	28/12/1995	31/12/1995	1	First winter
Lincoln	28/12/1995	06/01/1996	1	First winter female
Lincoln	31/12/1995	06/01/1996	1	Male
Lincoln	31/12/1995		1	Male
Gibraltar Point	14/01/1996	15/01/1996	1	
Barton upon Humber	19/02/1996		1	
Laughton Forest	05/03/1996		1	
Barton Pits	25/03/2006	28/03/2006	1	
Goxhill	12/02/2009	22/02/2009	1	Juvenile
Laughton Forest	21/03/2010	27/03/2010	1	
Birchwood	21/01/2011	04/03/2011	1	
Barton Pits	26/02/2011	27/02/2011	1	
Gibraltar Point	26/02/2011	27/02/2011	1	
Barton Pits	01/04/2014	06/04/2014	1	First winter

Parrot Crossbill *Loxia pytyopsittacus*

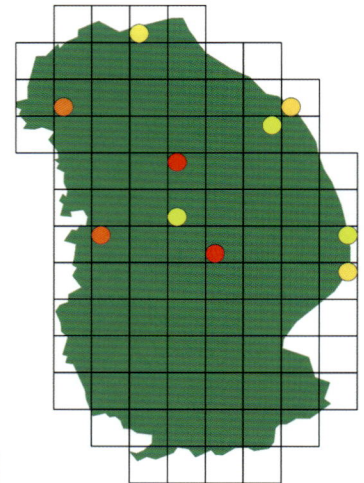

Irruptive vagrant. Scandinavia

Irruptions in the Northern Isles in Sep-Oct 1962 and 1982/83 were described by Catley and Hursthouse (1985) and the first in Lincolnshire was a male found dead at Hartsholme GP on Jan 16th 1963. The largest though, occurred in 1990-91 with at least 86 birds recorded; the biggest in Britain since 1958 with a total of 210 birds. It began on Oct 18th 1990 with five birds at Donna Nook and five at Gibraltar Point from the following day. Barton Pits had one on Oct 20th. Most were recorded at conifer plantations inland where birds stayed through until mid-Mar 1991 with max counts of 35 at Kirkby Moor from Nov 1990-Jan 1st 1991, and 22 remaining until Mar 8th 1991. 19 were at Laughton Forest during Dec 2nd 1990-Mar 13th 1991 and 14 were found at Willingham Woods Dec 2nd 1990 with variable numbers of birds being reported from there through to Jan 21st 1991. Since then, there have been only three records, in 1995, 2013 and 2018.

The immature male at Chambers Farm Wood from Dec 15th-20th 2013 was an early Christmas present for many. The last record, a female at Gibraltar Point on two days from Nov 14th-15th 2018 was the tail end of a large influx that had occurred across Britain in 2017.

The pattern of occurrence of this species in the UK is highly variable as reported by White and Kehoe (2020). During the 1990s there was an average of 27 a year, heavily weighted by the 1990-91 influx, during the 2000s there were none and from 2010-18 an average of 22 per year with at least 93 birds in 2017 and 2013 also having been a good year.

Site Name	First Date	Last Date	Count	Notes
Hartsholme	16/01/1963		1	Male found dead
Hartsholme	19/01/1963	26/01/1963	6	
Hartsholme	19/01/1963	01/02/1963	3	Same
Hartsholme	19/01/1963	31/05/1963	2	Same male and female
Hartsholme	26/01/1963		9	3 same as earlier
Hartsholme	15/03/1963		1	Same female
Hartsholme	17/03/1963		1	Same female
Barton Pits	11/10/1982	14/10/1982	1	Male
Barton Pits	12/10/1982	23/10/1982	1	
Grainthorpe	12/10/1982	13/10/1982	1	Male
Ingoldmells	12/10/1982		1	Immature male
Donna Nook	18/10/1990		2	Immature female
Donna Nook	18/10/1990	20/10/1990	1	Male
Donna Nook	18/10/1990	20/10/1990	1	Female
Donna Nook	18/10/1990		1	Female
Gibraltar Point	19/10/1990	25/10/1990	3	2 Females and 1 male
Gibraltar Point	19/10/1990	26/10/1990	2	Female
Barton Pits	20/10/1990	23/10/1990	1	Female
Kirkby Moor	04/11/1990	01/01/1991	4	
Kirkby Moor	11/11/1990	08/03/1991	22	
Kirkby Moor	11/11/1990	01/01/1991	35	
Kirkby Moor	11/11/1990	30/11/1990	15	
Laughton Forest	02/12/1990	13/03/1991	19	First winter female
Willingham Woods	02/12/1990	10/01/1991	9	
Willingham Woods	02/12/1990	20/12/1990	14	
Willingham Woods	04/12/1990	16/12/1990	2	
Willingham Woods	16/12/1990		1	Same
Kirkby Moor	01/01/1991		20	Same
Willingham Woods	21/01/1991		1	Same
Laughton Forest	08/03/1995		2	Female
Chambers Farm Wood	15/12/2013	20/12/2013	1	Immature male
Gibraltar Point	14/11/2018	15/11/2018	1	Female

Red Crossbill *Loxia curvirostra*

Scarce/fairly common irruptive passage migrant and visitor, mainly Jun-Apr, occasionally all year.

Crossbills are irruptive in Lincolnshire and invasions occur sporadically. It has always been an erratic breeder in Lincolnshire and was not confirmed to breed at all during the *LBC Atlas* years of the 1980s. Lincolnshire does not appear to have a sufficient area of conifers to sustain regular breeding. The *LBC Atlas* mentions breeding took place at Walesby in 1992 and both Walesby and Laughton Forest in 1998. The last year that breeding was confirmed to have taken place was 2012. There were up to eight singing males at Laughton Forest in Mar that year and two broods of chicks were seen in May. Breeding was also confirmed at Snipe Dales in May 2012. By coincidence 2012 was also the last significant irruption year. In Jun 2012 peak counts across the county amounted to 293, a number not seen again this decade. The best day count was 138 over at Gibraltar Point on Jun 5th and 53 at Welton le Marsh on Jul 31st. By contrast in the five years to 2018 the peak flock size ranged from nine at Gibraltar Point on Oct 16th 2016 to 25 at Weelsby Woods, nr Grimsby on Jul 10th 2016. During irruptions birds can turn up anywhere with conifers. Summer migrants on the coast often feed on thistle seeds. A single brood was reported at Laughton Forest in 2019.

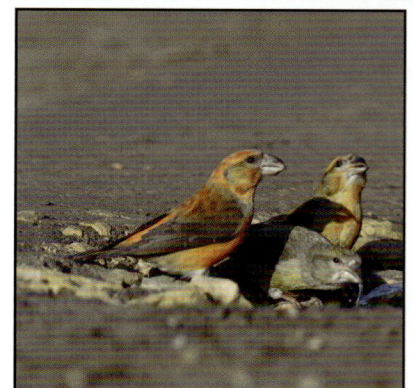

1968-1972	1980-1989	2008-2011	2016-2019	Incidence	Change
				0%	

Two-barred Crossbill *Loxia leucoptera*

BBRC

Irruptive vagrant. Scandinavia and circumpolar in boreal zone

Two-barred Crossbill is an enigmatic bird which has sadly been located only twice in the county. The first in Sep 1889 shot at South Cockerington near Louth was one of only 71 British records prior to 1950. Since then there have been a further 533 in Britain up to 2019 but only one in Lincolnshire. A male was discovered at Forest Pines Golf Course near Broughton on Feb 9th 2014 and stayed with a flock of up to 22 Red Crossbills *L. curvirostra* through to Mar 21st. It was part of an invasion of 101 birds that had arrived during the second half of 2013 and dispersed across the country. Invasions occur during Jun-Aug and begin in the Northern Isles with most birds recorded in Shetland and a few filtering down into mainland Britain and in some cases wintering, like the Forest Pines bird in 2014. Other large invasions to Britain in recent years included 59 in 2008 and a massive 219 in 2019 but sadly there were no reports from Lincolnshire in these invasion years.

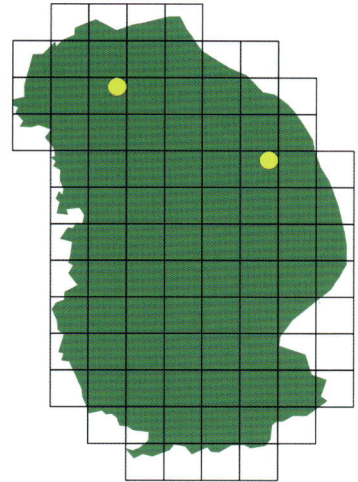

The dearth of extensive conifer forests and associated Crossbills compared to nearby Norfolk and Yorkshire probably accounts for the unattractiveness of the county for this species.

Site Name	First Date	Last Date	Count	Notes
South Cockerington	01/09/1889		1	Adult Male
Broughton	09/02/2014	21/03/2014	1	Adult

Goldfinch *Carduelis carduelis*

British form *britannica* a very common resident and passage migrant. Nominate continental form may also occur.

The Goldfinch is a good news story and an instructive example of how a population can adapt to change. Over 50 years ago it experienced a massive decline typical of many seed-eaters that were starved by agricultural intensification of the "weed" seeds they relied on for survival. Goldfinches switched to feeding on seed provided in gardens and seed crops made available by farmers in response to Environmental Stewardship schemes and that is thought to be the basis of their subsequent rebound and then boom from around 2005 onwards. The pattern of the change to feeding in gardens over the period 2002-12 is shown well by the Lincolnshire Garden Bird Feeding Survey as recounted by Goodall (LBR 2011 p. 200). It is probably the only British finch commoner now than in the 1960s. The *LBC Atlas* put the Lincolnshire population at 8,000-9,000 pairs in the 1980s and the estimated population was 43,000 pairs in 2016, making it Lincolnshire's second commonest finch, up from fourth in the 1980s. Most British Goldfinches winter in Iberia and a heavy autumn passage is witnessed across the county but most obviously on the coast where 3,695 were counted moving south at Gibraltar Point on Oct 18th 2017. It is possible (from ringing recoveries) that many birds wintering in the county come from Scotland. Wintering flocks have rarely exceeded 200 birds and flocks of over 100 are widespread.

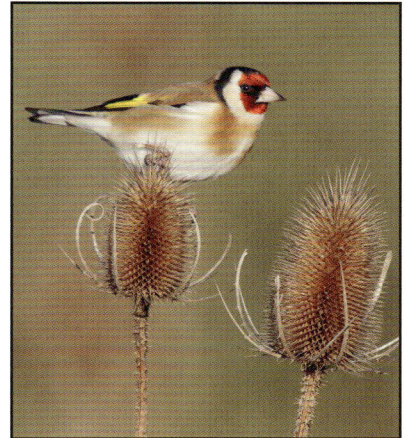

1968-1972	1980-1989	2008-2011	2016-2019	Incidence	Change
				88%	22%

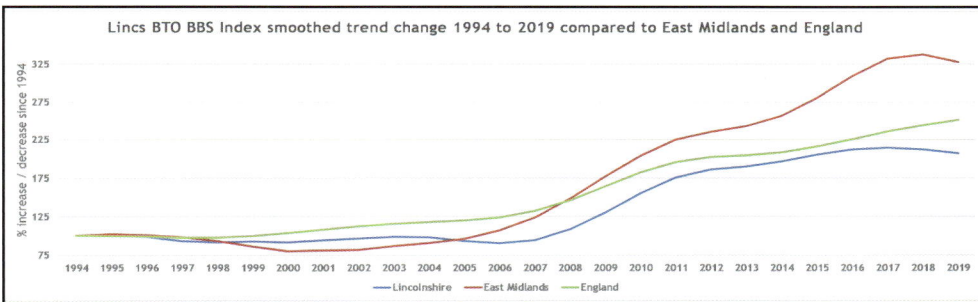

Lincs BTO BBS Index smoothed trend change 1994 to 2019 compared to East Midlands and England

Rape Fields

Serin *Serinus serinus*

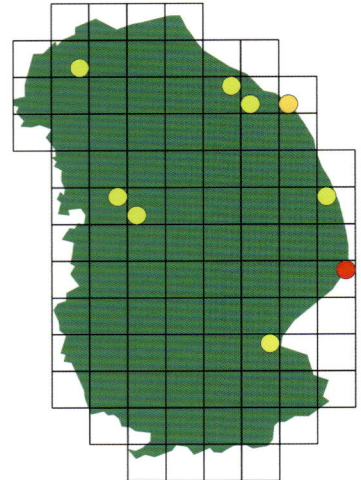

Very scarce migrant, mainly Apr-Jun and Oct-Nov.

The first record was a bird seen at Gibraltar Point on May 16th 1961, since when the total number of records stands at 42 involving 44 individuals in all. Birds have been recorded in all months from Mar-Dec, but the majority has been between Apr-early Jun (27 of the 30 records). Twenty-eight of the 44 birds recorded have been at Gibraltar Point with six others at Donna Nook. Inland records are exceptional and include a bird trapped at Brumby Common near Scunthorpe on Aug 12th 1976 and a singing male at Riseholme near Lincoln present from May 24th-Jun 2nd 1996 which has been the longest staying individual so far. The species remains a surprisingly rare bird in Lincolnshire with almost all recent records involving birds seen briefly flying along the coast on a single date.

European Serin was dropped from the BBRC rarities list in 1982. There were 70 British records prior to 1958, but a massive 2,210 records from 1958-2018. The average in Britain, 2010-18 has been 39 per year. They have bred in Britain, the first record being in Dorset in 1967 and then sporadically since then in Devon, Dorset, Sussex and East Anglia but no more than one or two pairs a year. Just 35 Serins had been ringed in Britain up to 2019 including the one caught at Brumby Common, the sole recovery of a bird ringed in Norfolk in 1993 found dead (hit glass) five days later.

Site Name	First Date	Last Date	Count	Notes
Gibraltar Point	16/05/1961		1	
Donna Nook	04/05/1969		1	
Donna Nook	30/07/1970		1	
Donna Nook	03/11/1971		1	
Donna Nook	05/09/1973		1	
Donna Nook	20/10/1974		1	
Cleethorpes	05/05/1976	07/05/1976	1	Male
Scunthorpe	12/08/1976		1	Male
Gibraltar Point	13/11/1982		1	
Gibraltar Point	28/04/1993		1	Male
Gibraltar Point	06/06/1993		1	
Gibraltar Point	27/05/1994		1	
Gibraltar Point	30/04/1995	03/05/1995	1	Male
Gibraltar Point	08/05/1995		1	Same
Gibraltar Point	26/04/1996		1	Female
Riseholme	24/05/1996	02/06/1996	1	Male
Gibraltar Point	26/05/1996		1	Male
Gibraltar Point	07/04/2011		1	
Anderby	26/04/2013		1	Male
Gibraltar Point	26/04/2013		1	
Gibraltar Point	28/04/2013		1	Adult male trapped
Gibraltar Point	03/06/2013		1	Immature female
Gibraltar Point	07/05/2014	14/05/2014	1	Female
Washingborough	08/03/2015	02/04/2015	1	Male
Gibraltar Point	07/05/2015		1	
Gibraltar Point	23/10/2015		2	
Frampton Marsh	05/05/2016		2	
Gibraltar Point	22/03/2018		1	Female
Gibraltar Point	05/05/2018	06/05/2018	1	
Tetney	28/05/2018		1	
Gibraltar Point	10/06/2018		1	
Gibraltar Point	15/06/2018	16/06/2018	1	
Gibraltar Point	01/07/2018		1	
Gibraltar Point	06/10/2018		1	
Gibraltar Point	29/03/2019		1	Adult male
Gibraltar Point	29/04/2019		1	Adult male
Gibraltar Point	17/05/2019		1	Female
Gibraltar Point	21/05/2019	25/05/2019	1	Male
Gibraltar Point	18/06/2019		1	Female
Gibraltar Point	19/07/2019		1	

View from Mill Hill, Gibraltar Point

Siskin *Spinus spinus*

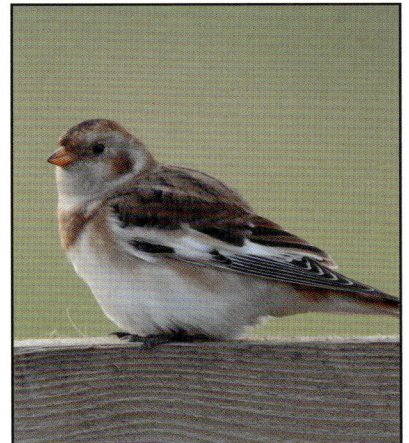

Common passage migrant and winter visitor. Scarce in summer but may breed in some years.

The first breeding record of Siskin in Lincolnshire was in 1973 at Hartsholme Park, Lincoln. During the *LBC Atlas* period of the 1980s the only confirmed breeding site was Snipe Dales to the east of Horncastle in 1985-86. The *BTO Atlas 2007-11* confirmed breeding in five 10km squares. Judging by LBR reports for the five years to 2018 it probably bred in most years in very small numbers. Sites like Laughton Forest, the forests around Market Rasen, Bardney Forest and the Scunthorpe area are likely locations. There is pronounced autumn passage that varies tremendously from year to year. The largest day passage of the period 2009-18 was noted at Gibraltar Point in 2015 when 1,447 birds flew south on Nov 10th. An even bigger passage day there was on Sep 12th 2008, when 3,200 birds flew south. Wintering flocks rarely exceed 100 and winter numbers are thought to have declined to some degree in recent years. In the five years to 2018 the largest flocks reported were from the Lincoln area and included 100 at Bracebridge in Jan 2016 and 100 at Hartsholme Park in Jan 2018.

1968-1972	1980-1989	2008-2011	2016-2019	Incidence	Change
				0.3%	-76%

Lapland Bunting *Calcarius lapponicus*

Nominate continental form a scarce local passage migrant and winter visitor Sep-Apr. Rare inland. The Greenland and Canadian form *subcalcaratus* may also occur.

Lapland Bunting is prone to influxes, and The *LBC Atlas* described two such events in the winters of 1985-86 and 1986-87 which led to 350 birds wintering on Butterwick Marsh in Jan 1986 and 200 in Jan-Feb 1987. In the last 10 years there has been one other event, in Sep-Oct 2010. In that autumn LBR 2010 describes a fast-moving Atlantic depression which caught migrating Greenland birds, causing many thousands to arrive in Iceland and Britain. There was a minimum peak of 229 birds in the county with flocks of 80 at North Cotes, 55 at Donna Nook, 40 at Grainthorpe Marsh and 25 at Tetney Marsh. Birds stayed on to winter and there were 83 at Frampton Marsh in Dec and 85 there in Jan 2011. To put that in context LBR reports for 2010-19 show total peak site counts across the county ranging from 11 in 2018 to 404 in 2010 and 165 in 2011; the peak yearly average being 73 birds. More realistically, the range over the eight years from 2012-19 was 11 in 2018 to 61 in 2017 with an average of 26 birds per year. The best three sites over the eight years to 2019 were Donna Nook which averaged none to eight birds a year and had the best flock of 32 in Feb 2017, Gibraltar Point, averaging five, and Saltfleetby-Theddlethorpe which averaged four to five per year. Birds can occur widely along the North Sea coast from Cleethorpes south and around The Wash to Nene mouth but for some reason are much rarer further up the Humber. Inland records are rare too. In the 10-year period there were just two, both in 2019. One at Middlemarsh Farm, just inland from Skegness in Sep, and another at Scunthorpe in Nov.

Snow Bunting *Plectrophenax nivalis*

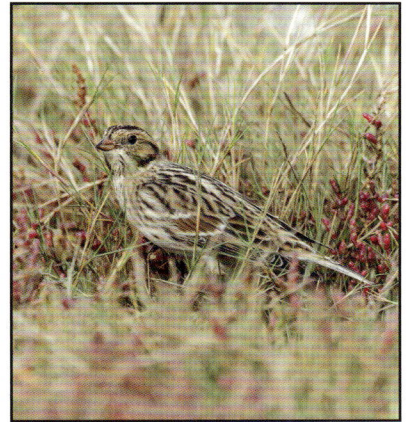

Icelandic form *insulae* a regular fairly common winter visitor and passage migrant on coast, mainly Sep-Apr, exceptional in summer. Scarce but fairly regular inland. Nominate Scandinavian form a scarce winter visitor and passage migrant.

The number of wintering Snow Buntings and the sites they use can change considerably from year to year. The *LBC Atlas* mentions that "exceptional concentrations of up to a 1,000 have occurred in some years in the past" but during the 1980s annual totals in the county rarely reached 1,000 and during the 1990s numbers were much lower, several hundred per annum with a peak of 500. Gregory in LBR (2018 p. 210) provides a useful summary graph of the peak day counts by year of Snow Bunting at Gibraltar Point from 1949-2018. The peak was around 500 in 1963 and since that time numbers have fallen steadily, not exceeding 100 since the late 1990s. LBR reports for the 10 years to 2019 show that the total peak site counts across the county ranged from 94 in 2016 to 667 in 2013 with an average of around 350 per year. The chief wintering area is the north-east coast between Cleethorpes and Mablethorpe, and in the south-east, Gibraltar Point, but birds occur irregularly from Winterton on the Humber round to Gedney Drove End on The Wash. The best three sites which were the only ones to hold birds in each of the 10 years were Cleethorpes which averaged 100 birds a year and had the two best flocks, both of 200 in Dec 2011 and Dec 2013 but had only 15 in 2019, Saltfleetby-Theddlethorpe; averaged 60 per year, and Gibraltar Point averaging 30. Between them these three sites held 55% of the peak counts over the 10 years. There has recently been a run of relatively poor years from 2016-19, average 160 per year but a bad winter may see numbers rise again.

Corn Bunting *Emberiza calandra*

Fairly common but somewhat localised and declining resident.

Corn Bunting is such a sedentary bird that song "dialect" can change within 25 miles. The *LBC Atlas* suggested there is evidence that they move from west Lincolnshire to the coast to winter. In the 1980s the species was still fairly widespread around the county although the population went into free fall from the mid 1970s when agricultural activities intensified. There were thought to be 2,750 pairs in the 1980s but latterly the estimated population in Lincolnshire is just under 700 pairs. Most are found in virtual islands on the Wolds, in the southern and Witham Fens and along The Wash coast. There is a positive association with spring sown barley which is thought to provide an important food source in the absence of "weeds" eliminated by herbicides. Lack of arthropods to feed chicks because of increased pesticide use, lack of food in winter due to a dearth of stubble, previously an important winter food source, may all be significant. Winter flocks are hard to find, probably because they disappear into the agricultural hinterland of the county where few birders venture. The best known and monitored wintering flock is at Gibraltar Point. LBR reports indicate in the five years to 2018 it declined from 131 birds in Dec 2014 to 82 birds in Jan 2019.

1968-1972	1980-1989	2008-2011	2016-2019	Incidence	Change
				11%	-18%

Yellowhammer *Emberiza citrinella*

Very common resident.

Yellowhammer is a very widespread bird in Lincolnshire which is not surprising given that it is a bird of arable farmland. The *LBC Atlas* found it in virtually every tetrad during the 1980s except parts of the Fens. The population at that time was estimated at 42,000-67,000 pairs but it was considered that might be an overestimate as the species was in general decline across the country at that point. The estimated population in 2016 comes out at 39,000 and given that the long term Lincolnshire BBS Index suggests the population has fallen by 15% over the period 1994-2019, the *LBC Atlas* estimate was probably more accurate than the authors thought. Why does this bunting seem more resilient to modern agriculture than its cousin the Corn Bunting *E. calandra*? They are both dependent on winter stubbles to see them through the winter but Newton (2017) points out that Yellowhammer is particularly responsive to the provision of winter seed crops through agri-environment schemes and these may have helped stem its decline from 2005 onwards.

During the 1980s the *LBC Atlas* found winter flocks in the range from single figures to 350. Big winter flocks can be hard to find as they can turn up in stubble and hedges anywhere. LBR for the five years to 2018 shows the largest flock each year ranged from 75 in 2015 to 250 at Fillingham in Dec 2018 with an average peak flock size of 160.

1968-1972	1980-1989	2008-2011	2016-2019	Incidence	Change
				81%	1%

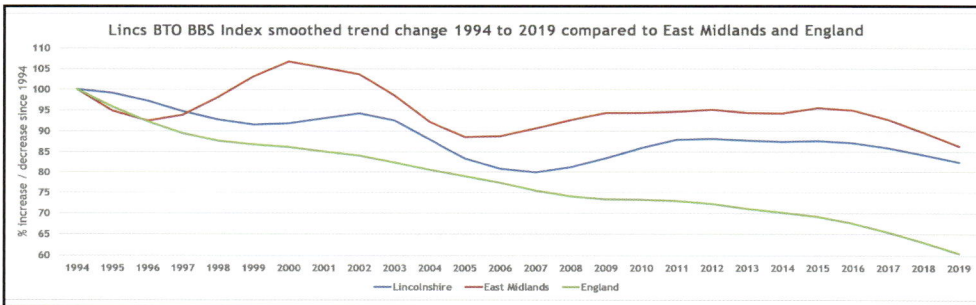

Lincs BTO BBS Index smoothed trend change 1994 to 2019 compared to East Midlands and England

Pine Bunting *Emberiza leucocephalos*

BBRC

Vagrant. Siberia.

One county record at Gibraltar Point Mar 29th-31st 1995 of an elusive 2CY male found in the Plantation feeding station with Chaffinches *Fringilla coelebs* and Yellowhammers *E. citrinella* an exceptional find. The issue of hybridisation with Yellowhammers has been well covered in the general ornithological literature. The bird's entire plumage must be scrutinised for any residual yellow tones with the fringes of the primaries being the best place to look as this is often the last place where yellow is shown in more troublesome hybrids and backcrosses. It has been known for birds with no other suggestion of hybridisation to exhibit conflicting colour only in the underwing coverts or at the concealed bases of the crown feathers.

Three hybrid birds were reported by the BBRC in their 2019 report but there were no issues with 'our' bird and the submission was accepted by BBRC.

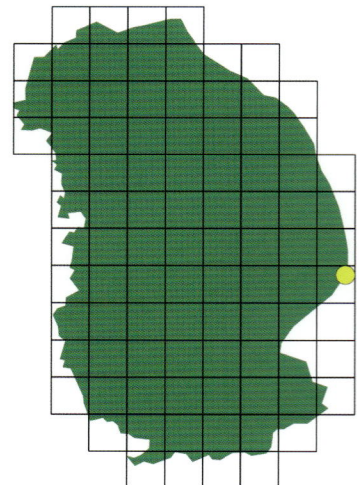

Pine Buntings are widespread breeding birds in Siberia west to the Urals, where they just about get into the European foothills. It migrates south across Asia in the non-breeding season, but in the Western Palearctic it has been a regular if localised winter visitor to northern Israel, and a scarce/rare winter visitor to Italy and rarer still to southern France. Elsewhere in Europe, it is only a rare vagrant, with a tendency to show up during the winter months.

There have been two British records to 1949 and 61 1950-2019 so it remains a much sought-after species averaging only one to two records per year. The exception was nine in a record influx in 2016, an autumn when there were also 13 Siberian Accentors *Prunella montanella* (though none in Lincolnshire).

Site Name	First Date	Last Date	Count	Notes
Gibraltar Point	29/03/1995	31/03/1995	1	First winter male

Ortolan Bunting *Emberiza hortulana*

Vagrant. Continental Europe.

There have been 33 county records involving 36 birds, the first being one at North Coates on May 3rd 1883, with the remainder all since 1963. Donna Nook has accounted for 19 of the records and the concentration along the coastal strip between Tetney and Theddlethorpe is marked. Strangely, Gibraltar Point has had just one record in Aug 1973 while in contrast four birds have been seen inland at Covenham Reservoir. The most recent record at Town's Holt, Grimsby in Aug 2019 was 27 years after the previous at Donna Nook on Sep 19th 1992.

From being a scarce passage migrant in Britain, this species has almost reached vagrant status in recent years and the average per year by decades has fallen from 72 during 1990-99 to 31, 2010-18. This is no doubt linked to the fact the species has undergone the second most pronounced decline of any bird species in temperate western Europe with an estimated population reduction of 82% between 1980 and 2008. It has now been lost as a breeding species from Belgium, the Netherlands and Switzerland in the last decade. Habitat loss and degradation and subsequent reduction in prey items are cited as the main reasons for decline, although illegal captures during migration, particularly in France, cannot be helping. Fortunately, the species has a large range stretching east across into Asia and the species does not reach the threshold for Vulnerable; population estimate across the range is 8,000,000-17,999,999 (BirdLife International 2020 Species factsheet).

Site Name	First Date	Last Date	Count	Notes
North Coates	03/05/1883		1	
Tetney	14/09/1963		1	
Tetney	24/10/1967		1	Male
Donna Nook	18/09/1968		1	
Donna Nook	04/05/1969		1	Female
Donna Nook	17/09/1969		1	First calendar year
Donna Nook	11/05/1970		1	Male
Donna Nook	06/05/1971		1	Female
Saltfleetby-Theddlethorpe	12/05/1971		1	Female
Donna Nook	25/08/1971		2	Immature
Donna Nook	26/08/1971		1	Same immature
Donna Nook	09/05/1973		1	
Donna Nook	02/06/1973		1	
Donna Nook	08/05/1976		1	Female
Saltfleetby-Theddlethorpe	25/08/1976		1	Immature
Donna Nook	26/08/1976		1	
Donna Nook	20/08/1977		1	
Gibraltar Point	06/05/1978		1	Adult male
Donna Nook	08/05/1978		1	Female
Saltfleetby-Theddlethorpe	25/08/1978		1	Immature
Donna Nook	26/08/1978		1	
Donna Nook	16/05/1979		1	Female
Saltfleet	19/05/1979	20/05/1979	1	Female
Donna Nook	30/08/1979		1	Immature
Donna Nook	12/05/1980		1	Male
Saltfleetby-Theddlethorpe	06/09/1981		1	Female
Covenham Reservoir	17/05/1982	21/05/1982	1	Female
Covenham Reservoir	02/05/1983	05/05/1983	1	Female
Covenham Reservoir	06/05/1983		2	One same female
Saltfleetby-Theddlethorpe	05/09/1984		2	Immature
Covenham Reservoir	29/04/1987	08/05/1987	1	Immature male
Tetney	08/05/1988		1	Male
Tetney	10/05/1988		1	Same male
Donna Nook	19/09/1992		1	Immature
Donna Nook	19/09/1993		1	Immature
Grimsby	29/08/2019		1	

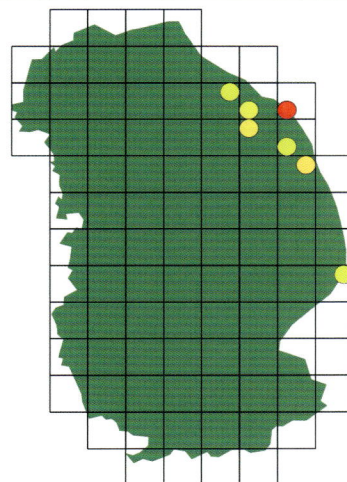

Track from Pyes Hall, Donna Nook

Cirl Bunting *Emberiza cirlus*

Vagrant. Southern England, western Europe.

There are four records of this bunting, none of them recent, with the first two in 1889 followed by the third in 1968 at Donna Nook and fourth at Great Cotes in 1977, an immature or female. Despite some historical speculation there is no reliable evidence that Cirl Buntings ever bred on the heaths in the north-west of the county during the 19th century.

Formerly widespread across southern England and Wales, the species is now restricted to a narrow strip of coast in Devon and south Cornwall where it is being reintroduced. The species is not known for long-distance movements and colour-ringing studies in the small population in Devon reveal that, in general, birds winter within a few kilometres of their breeding sites.

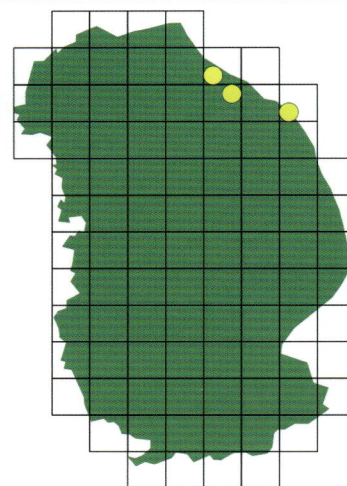

Site Name	First Date	Last Date	Count	Notes
Great Coates	05/01/1889		1	
Grimsby	10/12/1889		1	
Donna Nook	29/09/1968	02/10/1968	1	
Great Coates	24/10/1977		1	

Little Bunting *Emberiza pusilla*

Vagrant. Scandinavia, Russia

With 20 records, the first in 1951, Little Bunting remains a vagrant in Lincolnshire which is hard to understand given that Spurn Point just seven miles across the Humber had 47 birds between 1976 and 2014 (Roadhouse 2016). In Britain as a whole, there were 1,689 records between 1958-2018. Twelve of the Lincolnshire records have been in the last 10 years to 2018 which reflects the recent large increase in the number of annual British records reported by White and Kehoe (2020). From 2000-09 they occurred at a rate of 33 birds a year which shot up to 74 birds per year in the period 2010-18. Of the 19 Lincolnshire records, most (14) have been in autumn between Sep 27th-Oct 31st, three have been in winter and two in spring. Fifteen of the records have been on the coast between Tetney and Gibraltar Point with the latter site and Donna Nook both having had five. There have been four inland records including two recently, in Feb 2014 and Dec 2019, so it clearly pays to check finch and bunting flocks feeding in weedy fields inland as well as on the coast.

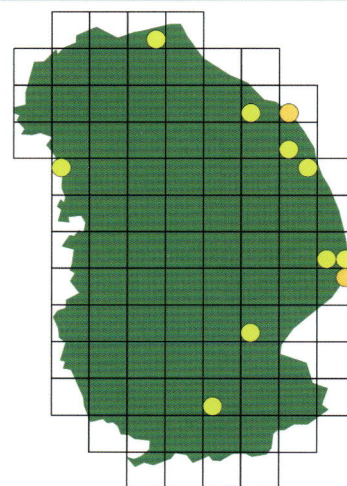

Site Name	First Date	Last Date	Count	Notes
Saltfleet	05/10/1951		1	Immature male
Spalding	02/10/1954		1	Adult male
Boston	04/12/1980		1	
Gibraltar Point	01/10/2000		1	
Donna Nook	25/10/2002		1	
Skegness	21/10/2004	23/10/2004	1	
Gibraltar Point	25/09/2005		1	
Seacroft	31/10/2008		1	
New Holland	21/04/2013		1	
Lea	25/02/2014	10/03/2014	1	
Gibraltar Point	20/04/2015	21/04/2015	1	
Saltfleetby-Theddlethorpe	02/10/2015		1	
Gibraltar Point	10/10/2015		1	Trapped
Donna Nook	11/10/2015		1	
Donna Nook	27/10/2015		1	Same
Donna Nook	04/10/2016		1	
Tetney	15/10/2016		1	
Donna Nook	27/09/2017		1	Trapped
Gibraltar Point	27/09/2017		1	
Gibraltar Point	04/10/2019		1	
Middlemarsh Farm	07/12/2019		1	

Rustic Bunting *Emberiza rustica*

 BBRC

Vagrant. Scandinavia, Russia

A comparatively late addition to the Lincolnshire list, the first was a spring bird at Gibraltar Point in May 1975 the total of seven records, four have been in spring from a relatively early one at Saltfleetby-Theddlethorpe on Mar 22nd 1993 through to May 24th. The three autumn records have been in the narrow window of Oct 1st-10th. All the birds have been on the coast between North Cotes and Gibraltar Point, with five having been one day birds, one for two days and one a three-day bird.

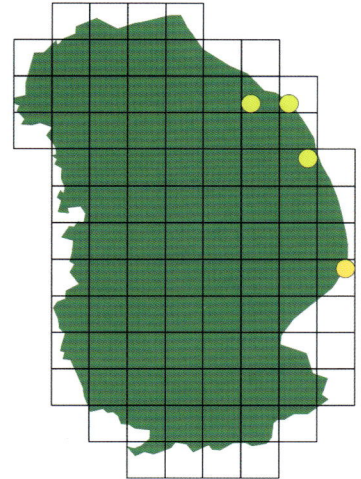

Around 10 birds a year occur in Britain with a total of 567 from 1950-2019. Lincolnshire does not really get its fair share. Numbers in Britain over the last 20 years have been fairly stable with around 8-10 birds per year, but the heyday of this species in Britain was in the 1990s when there were around 20 per year. The last record in the county was in 2015 so we are overdue a few more.

Site Name	First Date	Last Date	Count	Notes
Gibraltar Point	11/05/1975		1	Male
Gibraltar Point	01/10/1978		1	Immature
Saltfleetby-Theddlethorpe	22/03/1992		1	
Gibraltar Point	09/05/1993	11/05/1993	1	Male
North Cotes	23/05/1994	24/05/1994	1	
Donna Nook	03/10/2013		1	
Gibraltar Point	11/10/2015		1	

Yellow-breasted Bunting *Emberiza aureola*

 BBRC

Vagrant. NE Europe to central/south Asia

The first, and only, record for the county concerned a male in full summer plumage at Gibraltar Point on 15th May 1977. At the time this was only the second British spring record. The bird fed on short turf along the saltmarsh edge close to the observatory and often at close range and frequently sang. Anecdotally (R. Lambert *pers. comm.*,) the bird may have been present the previous day as, on seeing the bird on 15th, a visitor with limited experience had said "Oh look, that Yellowhammer is still here". Decades later, this still remains the only county record.

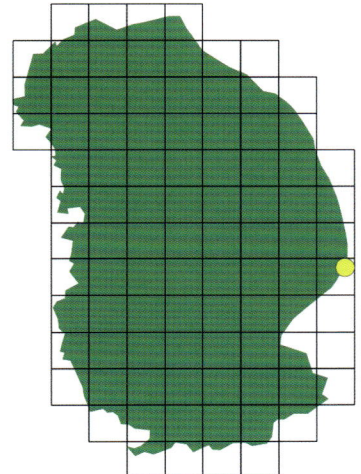

In Britain it is very much a Shetland and Fair Isle speciality, where it used to be a reliable target species for those visiting in mid-Sep but has now become a true vagrant. From the early 1970s and throughout the 1980s and 1990s it was recorded regularly, with an average of around six individuals a year, and that continued into the present century, with six in both 2002 and 2003. Thereafter, a rapid decline set in; there have been just four British records in the past ten years 2010-19, the last in 2018.

Numbers at wintering sites throughout its range have shown rapid declines over the last thirty years driven by excessive trapping at migration and winter roosts. BirdLife International now classifies the species as Critically Endangered (BirdLife International 2017). Although trapping was banned in China in 1997 huge numbers are still trapped and sold illegally, and in 2004 an estimated 10,000 birds were still being sold daily in a single market at Sanshui in China's Guangdong province (Westrip 2017).

Site Name	First Date	Last Date	Count	Notes
Gibraltar Point	15/05/1977		1	Male

Black-headed Bunting *Emberiza melanocephala*

 BBRC

Vagrant. Southern Europe.

Two records, both males, in Sep 1974 at Saltfleetby-Theddlethorpe and in Jun 1980 at Donna Nook. Almost 70% of all British records since 1950 have been males, and almost 60% of those records were in May and Jun. Identifying the eye-catching males is no problem but separating 1CY females from 1CY Red-headed Buntings *E. bruniceps* in autumn remains problematic and may easily escape detection.

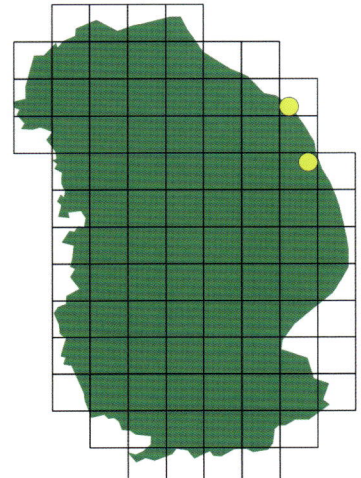

This is another species whose occurrence in Britain has been dogged by the spectre of escapes from captivity but the peak of records in late May and Jun is consistent with overshooting migrants. For example, in 1992 14 were reported, six of which arrived between 4th-14th Jun, five of them on Fair Isle. Likewise, in 1993 eight males turned up between May 24th and Jun 2nd, seven of these along the south coast between Cornwall and Kent.

The BBRC data shows that there have been 220 records, 1950-2018 and the species currently averages about five per year and may be decreasing. This is not an east coast bird with around a quarter of records having occurred in Shetland (58), a pattern shared with other south-eastern vagrants.

Site Name	First Date	Last Date	Count	Notes
Saltfleetby-Theddlethorpe	14/09/1974		1	Male
Donna Nook	16/06/1980	25/06/1980	1	Male

Reed Bunting *Emberiza schoeniclus*

Very common resident and passage migrant.

As its name suggests, Reed Bunting was originally a bird of reedbed and marsh, but it has also exploited arable farmland too. It particularly likes reedy dykes in The Marsh, Fens, Trent valley and Isle of Axholme and along the coastal fringes but it occurs in the Wolds and other higher ground too. Agricultural intensification caused it to decline along with other farmland birds in the 1970s and 1980s and the *LBC Atlas* put the Lincolnshire population at 8,000-10,000 pairs. The chart shows that Reed Bunting has done better in Lincolnshire than other parts of England over the period 1994-2019. The estimated population for Lincolnshire in 2016 is a massive 32,000 pairs. The main reason for this is that Reed Buntings have taken to breeding in oilseed rape and BTO BirdTrends indicates that their density in oilseed rape is four times greater than any other arable crop. Since Lincolnshire has probably the greatest area of this crop in the country it becomes easier to see why they are doing well here. Although British Reed Buntings are thought to winter in Britain a pronounced passage of near continent birds from the Netherlands to Norway occurs along the coast in Sep-Oct. As usual Gibraltar Point Bird Observatory makes the most rigorous counts. In the five years to 2019 LBR reports peak passage days ranged from Sep 30th-Oct 23rd with peak counts ranging from 86 a day in Oct 2015 to 1,086 in Oct 2019. Overall thousands of birds are involved each autumn and passage is often viewed from Donna Nook too. Like other buntings, Reed Bunting flocks up in winter and during the five-year period peak flock counts ranged from 220 in 2018 to 350 at Garthorpe, on the west bank of the mouth of the Trent in Dec 2014 and Jan 2015. This massive flock was feeding on a crop of unharvested wheat next to a reedbed. The average peak winter flock size over the five years was 300. The highest single winter count reported by the *LBC Atlas* during the 1980s was 200.

1968-1972	1980-1989	2008-2011	2016-2019	Incidence	Change
				67%	10%

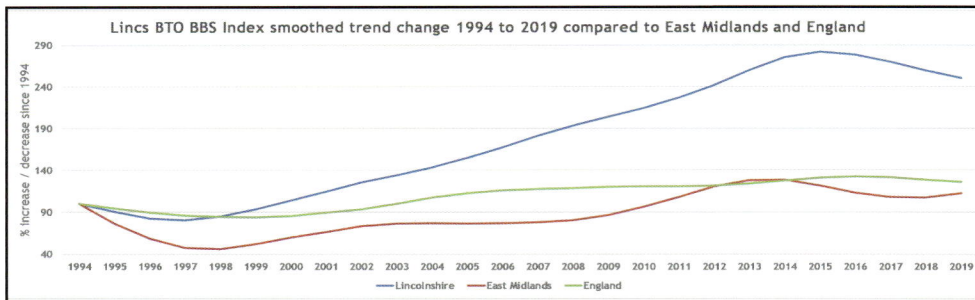

Lincs BTO BBS Index smoothed trend change 1994 to 2019 compared to East Midlands and England

White-throated Sparrow *Zonotrichia albicollis*

Vagrant. North America.

Another species in the line of unexpected North American vagrants with three county records in Dec 1992, Jun 2010 and May 2013. The first of these was caught in Willingham Forest by a group of disbelieving ringers who were trapping in a very weedy field which had attracted a large number of finches. It settled into the area staying until Mar 1993 during which time it was enjoyed by several hundred observers. It is assumed that this bird turned up in the previous autumn and remained unseen until its appearance in a mist net. There are two further records in Jun 2010 (when there were seven in total in Britain) and May 2013, both of them turning up in suburban gardens.

There have been 50 records in Britain to 2018, averaging about one a year. The arrival of North American sparrows has been widely discussed in the literature and, inevitably, ship assistance is suspected as the majority has turned up near shipping routes and White-throated Sparrow is one of the commonest species to be seen on ships at sea. What of the Lincolnshire records? There are port locations not far away to both the north and the south of the county. Lees and Gilroy discussed the issue in Slack (2009) and noted that spring vagrancy in North American passerines is rarer and less predictable than in autumn when sparrows predominate. In addition, spring weather conditions in the Atlantic do not promote vagrancy and granivorous species may stay on deck for longer and survive better. Whether the Lincolnshire birds have arrived in autumn and survived over-winter or have moved onward after a ship-assisted arrival is a moot point, but they are always welcome.

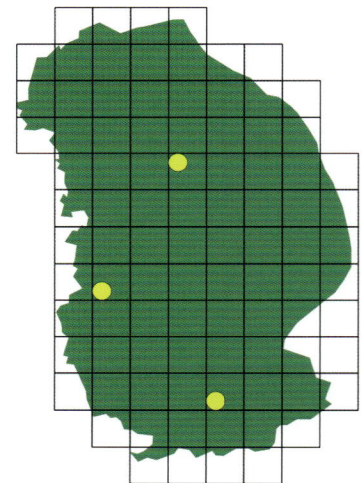

Site Name	First Date	Last Date	Count	Notes
Willingham Woods	05/12/1992	28/03/1993	1	Trapped
Fulbeck	18/06/2010	19/06/2010	1	Adult
Spalding	28/05/2013		1	

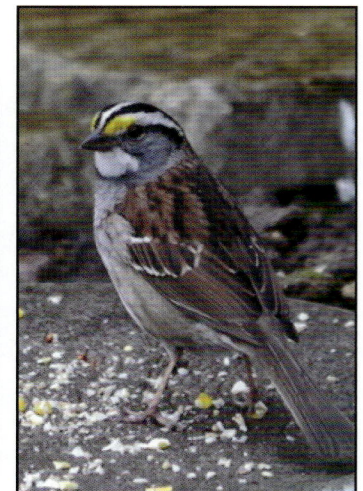

Bobolink *Dolichonyx oryzivorus*

BBRC

Vagrant. North America.

A juvenile Bobolink was found at Crook Bank, Saltfleetby-Theddlethorpe in Oct 2019 and was the first record for Lincolnshire. The morning in question was sunny with strong southerly winds, unlikely weather for a freshly arrived bird which had perhaps arrived earlier and further north in Britain. It frequented dense coastal vegetation along with some other finches and buntings and proved typically elusive. It did stay around long enough for a good number of county birders to see it but was gone by the next day. First recorded in Britain in 1962, on the Isles of Scilly, there has been a total of 33 in all, including this individual. All have occurred between mid-Sep and early Nov. It is a much sought-after species and this is the first in Britain since one on Shetland in 2012.

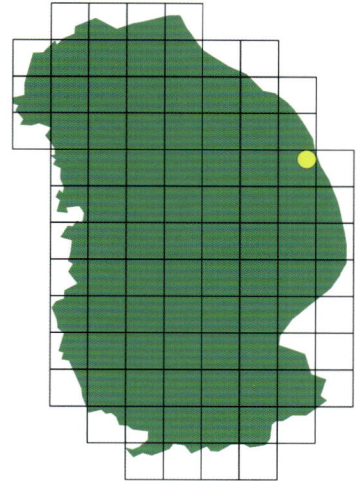

Bobolinks breed in the USA and Canada from British Columbia and Alberta south to Colorado in the west and to Newfoundland and as far south as West Virginia in the east. They are long-distance migrants crossing the Gulf of Mexico to winter in central or southern South America. Their round trips between breeding and over-wintering grounds extend to as much as 20,000 km, one of the longest annual migrations of any New World passerine.

Site Name	First Date	Last Date	Count	Notes
Saltfleetby-Theddlethorpe	18/10/2019		1	Juvenile

Northern Waterthrush *Parkesia noveboracensis*

BBRC

Vagrant. North America.

There is one county record of one at Gibraltar Point on Oct 22nd-23rd 1988 which is by some way the most astonishing of the American landbirds to be found in the county. It was trapped and ringed in the East Dunes on Oct 22nd when it weighed 16.1g. This constituted the fourth British record at the time after records in 1958, 1968 and 1982 (all on the Isles of Scilly) and was the latest by 19 days. Five of the seven records to 2019 have been on the Isles of Scilly, the other was at Portland, Dorset, Oct 1996.

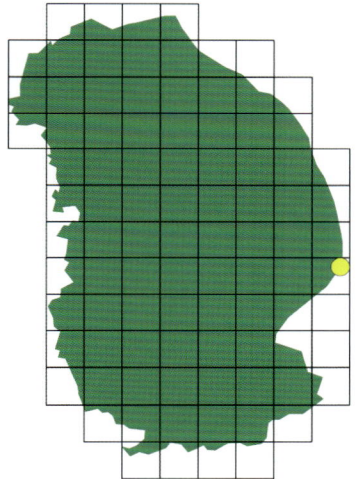

So how did this bird arrive in Lincolnshire? The five records on the Isles of Scilly arrived between Aug 29th-Oct 3rd, that on Portland was later, Oct 14th. The very late date for the Lincolnshire record compared to those on Scilly suggests an earlier arrival somewhere in Britain or even further north in Norway as most recently arrived American landbirds take significantly longer than Siberian birds to feed up and recover. The fact that it was only present a maximum of two days suggests it had most likely made landfall further north and drifted south with Palaearctic migrants. The weather at the time in Lincolnshire caused large falls of thrushes, finches, buntings and Goldcrests *Regulus regulus*. The true mechanism behind the arrival of this absolute gem may never be absolutely known but it was certainly a once-in-a-lifetime experience for those who saw it.

Site Name	First Date	Last Date	Count	Notes
Gibraltar Point	22/10/1988	23/10/1988	1	Trapped

American Redstart *Setophaga ruticilla*

BBRC

Vagrant. North America. One record in 1982.

Arguably one of the least expected vagrants of all time, an American Redstart was found at Gibraltar Point on Nov 7th 1982, remaining until Dec 5th 1982. It was found in a Hawthorn patch along the edge of the Plantation. Weather conditions at the time were typically unsettled with moderate to fresh south-westerlies as a trough of low pressure moved north-east across Britain. It was trapped and ringed on Nov 8th and weighed 10g, at or above the weights seen in birds arriving on their breeding grounds in North America (Smith and Moore 2005) tending to suggest an earlier arrival and recovery elsewhere. Interestingly they normally carry a reduced fat load compared with long-distance migrants such as Blackpoll Warbler *Setophaga striata* which is a major factor in limiting the likelihood of a successful transatlantic flight. It was presumed to be a 1CY male given the orange tinge to the flank patches. Unsurprisingly it remains the only county record. The first record of this species in the UK was in Cornwall in 1967, the Lincolnshire record was narrowly the third arriving as it did three days after one on Islay, Argyll also in 1982. There have been just three further records in the UK: Cornwall 1983, Hampshire 1985 and Outer Hebrides 2017.

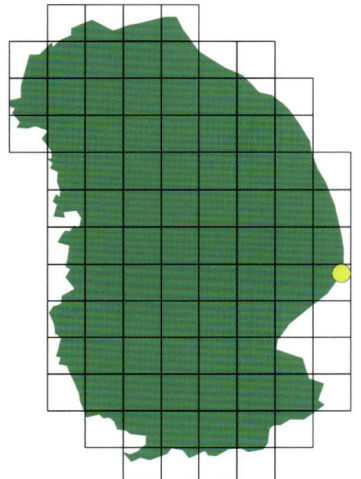

Site Name	First Date	Last Date	Count	Notes
Gibraltar Point	07/11/1982	05/12/1982	1	Trapped and ringed

Birds recorded in Lincolnshire but not in the systematic List

Records relating to species still under assessment by BBRC/BOU, escapes from captivity and birds of uncertain origin.

This somewhat heterogeneous section records the various species which have been recorded in the county but which for various well-known reasons are not included in the county list because they are not adjudged as being Category A, B or C species. The list below is almost certainly incomplete as less attention is often given to "unacceptable" records, but these are the records we have from the LBC Database.

The species in this first section are in Category D and E and it is important to record and submit such species as their status may change over time and local records may inform wider national and international data, including updated taxonomic classifications. For example, Cackling Goose *Branta hutchinsii*, has occurred in the county but the records have never been submitted to BBRC. Our feeling is that it is probably best to submit a record of the occurrence of such a species even if the circumstances appear dubious for BBRC scrutiny. This also avoids the record remaining on the LBC file as "Not proven".

There is one record below still under BBRC/BOU assessment at the time of writing – Bearded Vulture *Gypaetus barbatus*. This has attracted a wide spectrum of views given the proven origin of this bird from a reintroduction scheme in The Alps rather than one of the indigenous populations across southern Europe. The final verdict is awaited with interest by all British birders.

The non-native species reported below include all which are monitored by the RBBP and one or two others which may soon undergo re-assessment. We urge county birders to report them wherever they encounter them. Some of this group may eventually establish breeding, self-sustaining populations and as such need to be monitored.

Some species may of course be attributed to more than one BBRC category such as Lesser White-fronted Goose *Anser erythropus*, for which there is just one accepted record under Category A of an apparently wild bird (see Systematic List). It has also turned up in circumstances suggestive of captive origin, which would be attributed to Category D. It is also conceivable that reintroduced birds to the Scandinavian breeding population, attributable to Category E, could occur.

Helmeted Guineafowl *Numida meleagris*

Native to sub-Saharan Africa. **Cat E**

Not uncommonly seen in and around farmyards as privately kept birds, this species is undoubtedly under-recorded. There is a scatter of records in the LBC database from 2003 onwards from several sites. No breeding records have been received. The species is monitored by the Non-native Breeding Birds Panel which reported two records of breeding in Hertfordshire and Yorkshire in 2014 (Holling *et al* 2017).

Reeves's Pheasant *Syrmaticus reevesii*

Asia, north and south China. Widely kept in captivity, occasionally escape. Reported from Suffolk, West Yorkshire, Norfolk, Gwynedd, Staffordshire and Derbyshire. At least 15 individuals were reported in 2014 from Breckland but there has been no evidence of confirmed breeding anywhere in Britain to date. **Cat E***

This species has been reported from four sites, all escapes from captivity. A female was at Gibraltar Point, Apr-Nov 1994; a second was present Sep 1996-Jul 1997, eventually killed by a car. The others were more recent: a male was photographed at Ruckland in the Wolds in Apr 2011; three were seen at Bonthorpe, Oct 2012; one was at Dunston, Aug 2015.

Golden Pheasant *Chrysolophus pictus*

Native to central China. First released in Britain in 1725, not self-sustaining until 1950s in Breckland. **Cat C1E***

There are only three records of this species, the first a male Jul-Dec 1988 at Swineshead. Another was seen in Sep 1996 at Manton Warren and a more recent record was of another male, Jan-Feb 2017 at Stickford.

There has never been any evidence of breeding birds in the county.

The species is still regularly recorded in several other English counties, but little recent information is available, and the last records of confirmed breeding were from Norfolk and Suffolk in 2011.

It is unknown just how self-sustaining any of the British populations now are (Holling *et al* 2017).

Indian Peafowl *Pavo cristatus*

Native to the Indian subcontinent and Sri Lanka. **Cat E***

During the fieldwork for the *BTO Atlas 2007-11* there were records from 289 10km squares across Britain during that period, although not all of these would have been in potential breeding habitat. This species is regularly seen in various parts of the county, most usually free-ranging birds from private collections and the species is regularly offered for sale by private breeders in the county. A 'feral' population has been surviving at Nocton Wood since the mid-1980s where they have bred ever since. The population has fluctuated and peaked at around 60 birds; more recently, 2015 onwards, there have been around 20 birds. Whether this is a self-sustaining population is unclear.

It is undoubtedly under-reported but may have the capacity to breed in the wild. A similar picture is seen across Britain with suspicions of pockets of breeding birds here and there, but which are going unrecorded or unnoticed (Holling *et al* 2017).

Fulvous Whistling Duck *Chrysolophus pictus*

Native to North and South America, Africa and Asia, it is one of the most widely distributed waterfowl in the world. Usually resident but with limited seasonal movements. **Cat E**

There have been four records, maybe involving just two birds. One was at Boultham Mere in Apr 1996; another was at Frampton Marsh, May-Aug 2010, and presumably the same bird relocated to Alkborough Flats for five days in late Aug 2015, and then moved on to Whisby NR for most of Sep 2015. There have been very few other British records and no evidence that they have ever bred in the wild.

Cackling Goose *Branta hutchinsii*

Native to North America, breeding over a wide range of habitats throughout north temperate, subarctic and arctic regions. **Cat AE**

Four subspecies are recognised, all of which are BBRC taxa. To our knowledge none of the birds which have been recorded in the county have ever been submitted but none have turned up in circumstances suggestive of vagrancy. The first in Jan 2002 at Freiston Shore was in a flock of Canada Geese *B. canadensis*, and recorded as a 'Lesser Canada' in the days before the modern reclassification of the whole group had occurred. Another was at Freiston Shore on and off from Feb 2003 to Jan 2004 and presumed to be nominate *hutchinsii* (so-called Richardson's Canada Goose). The next at Frampton Marsh in Jan 2013 was thought to be of the subspecies *minima* (Ridgway's Goose). It, or another, turned up at Gibraltar Point from late Apr 2013 to mid May 2013 and what may have been the same individual returned to Frampton Marsh on Aug 17th 2013 but was recorded as 'Lesser' Canada Goose without qualification. Submissions of lone birds, birds with feral geese and birds on dates or in locations less indicative of vagrancy will be less likely to gain acceptance.

Bar-headed Goose *Anser indicus*

This species bred intermittently in Norway from the 1950s but has never become established. It has also bred in Belgium, Finland, Germany, Sweden and The Netherlands. It is unclear whether these populations are self-sustaining. **Cat E***

The first record was in Jun 1992 at Gibraltar Point and they have been recorded in most years since. The exact number of individuals involved is hard to assess as they appear to be very mobile. The majority of records refer to single birds often with Greylag Geese *A. anser*, but two together have been recorded in seven years, three in Sep 2000, four in Mar 1996 and Jul 2008, and six in Nov 2001. There have been no reports of breeding attempts. There have been a very few records of breeding elsewhere in Britain but nothing which has led to longer term establishment in any area.

Emperor Goose *Anser canagicus*

Breeds in Arctic regions of western Alaska and northeastern Siberia, winters on ice-free coastal beaches in the Aleutian Islands and the Alaska Peninsula. **Cat E***

At least three records: Feb 14th-17th 1992, Messingham SQ; Jan 13th Kirkby GP; Tattershall Bridge area, Jan 26th 2009. Elsewhere in Britain 15-20 birds have been recorded in Cumbria where a pair first bred on Walney Island in 2001; and one or two pairs nested there subsequently, last in 2010. In the 2009-11 report on non-native breeding birds the population there numbered up to 14 birds (Holling *et al* 2014).

Ross's Goose *Anser rossii*

North America. Breeds northern Canada, winters southern USA. **Cat DE***

One record but not categorised by BBRC, Nov 10th 2002 Read's Island roost, also seen Norfolk, Cumbria, Northumberland (Not categorised on the grounds of likely escape from captivity, but which may, at some future date, merit further consideration; *British Birds* 96: p. 608).

A presumed male escapee was first seen paired with a female Greylag Goose *A. anser*, Sep 2007, at Barton Pits and also seen in ensuing years to Oct 2009. Other records of between one and five individuals, and with no indication of wild vagrancy, have been recorded at Frampton Marsh (2009), Nene Mouth (2013), Donna Nook (2015) and Sutton Bridge (2016).

The origin of most European records of Ross's Geese is uncertain. They are certainly a potential vagrant, but the situation is complicated by escapes from captivity, some of which occasionally breed ferally in the UK. The winter of 2003-04 saw at least three with flocks of Pink-footed Geese *A. brachyrhynchus*, wintering in Norfolk and at least one of these was an escape since it carried a plastic ring. The status of the other two was uncertain.

Lesser White-fronted Goose *Anser erythropus*

Northern Scandinavia to north-east Siberia. Birds from the reintroduction scheme in Swedish Lapland winter in the Netherlands. Others winter in Hungary, Black and Caspian Sea areas, northern Kazakhstan and lower Yangtze valley, China. **Cat AE***

One accepted county record of a wild bird at Holbeach Marsh, Jan 1943 (see Systematic List). Other records seem unlikely to have been of wild origin, but the database is probably incomplete as some records will not have been submitted because of the spectre of escape from captivity. Other records include the following: one May-Oct 1992, Read's Island, eventually being shot; it was unringed but had a pure white outermost primary in right wing. One in Oct 1992 at Saltfleetby but said to be very tame; another at Tattershall Pits Dec-2000 consorting with other ornamental wildfowl; one at Thurlby Sand Pit Dec 2002. Despite the reluctance to submit records because of potential escapes from captivity, BBRC still welcomes submissions, including colour-ringed and satellite-tagged birds from the Scandinavian reintroduction scheme.

Black Swan *Cygnus atratus*

Native to Australia. **Cat E***

The first record of this non-native swan in the county was in 1996, when two were noted at Gibraltar Point in Apr. There were none in 1997 but since then they have occurred in every year 1998-2019 and at more than 60 sites across the county.

These have mostly been of one to three birds but there have been several instances of confirmed breeding with counts of up to eight birds post-breeding. Records supporting breeding in the county have been as follows: pair with two juveniles, Jul 1999, Fillingham; three adults and four juveniles, Jun 2004, St. Peter's Pool, Bourne, presumed relocated to Fillingham, Dec 2004; female on nest, Feb 2004, Fillingham; pair with four downy young, Bourne, Mar 2007; adult with three juveniles, Jul 2010, South Ormsby; three birds, nest-building seen, Mar 2011, Bourne; three adults with four juveniles, Nov 2012, Wellhead, Bourne (derived from adults released in May 2012); adult and two well-grown juveniles (autumn hatched), Bourne, Feb 2013.

Commonly seen in many counties in Britain, Black Swan has been recorded in every month of the year in Lincolnshire and it is the most widely distributed rare non-native breeding bird in Britain with at least probable breeding recorded in 27 counties. By the time of the 2012-14 non-native breeding birds report the five-year average was 19 breeding pairs across Britain (Holling *et al* 2011, 2017).

Wood Duck *Aix sponsa*

Native to North America and Cuba. **Cat DE***

Lorand and Atkin (1989) thought that most of the early records were probably escapes from wildfowl collections and that most were not published. The first on record was a pair at Barton Pits in May 1981, and a pair with young were seen at Tallington Lakes in Jun 1981. Since then there have been records in 10 of the years 1992-2019. These have mostly been of male birds with just one to two birds per year although a pair were seen at Belton Golf Club, Grantham in Apr 2019.

A well-watched male at Frampton Marsh in 2018 sported a pink plastic ring indicative of captive origin. The five-year mean number of breeding pairs in Britain up to 2014 in Holling *et al* (2017) was three.

Marbled Duck *Marmaronetta angustirostris*

Fragmented range from the Mediterranean and North Africa to the Middle East, west and south Asia. **Cat DE**

There is a single county record of one at Huttoft Pit, Aug 30th 1992 (Accepted BBRC, Category D, *British Birds* 89: 529). This species has been in Category D since 1993, was briefly placed in Category E, but returned to Category D in Oct 2017. BBRC statistics, 1950-2017, show that there have been 29 records though none has occurred in circumstances suggesting true vagrancy.

Barrow's Goldeneye *Bucephala islandica*

Resident Western Palearctic birds breed Iceland; migrant Nearctic birds breed Alaska, Canada, south to northern California. **Cat AE**

One undated record of two males at Langtoft Pits in summer 1992. Only three British records to date in 1979, 2005 and 2006. The date and circumstances of the Lincolnshire record together with the small but significant risk of escape from captivity don't suggest wild vagrancy.

Hooded Merganser *Hooded Merganser*

Native to North America. **Cat AE**

An approachable bird was seen on May 4th-13th 1996, Fishtoft, and wearing a white darvic ring. It was considered to be an escape. As with some other wildfowl whose origins have come under a shadow, there have been a substantial number of similar records of this species in Britain and the species didn't make it into Category A of the British List until a candidate on North Uist Oct 23rd- Nov 1st 2000 was admitted by the BOURC. Their extensive breeding range stretches throughout wooded areas of the east and Pacific Northwest of North America. The primary winter range includes the south-eastern United States for eastern birds and the Pacific coast north of California for western birds. There have been records from the Azores and there are three accepted records involving four individuals from Ireland, as well as records of apparent vagrants from Iceland and the Canary Islands. The population of the Hooded Merganser in North America has shown a 1,100% increase over the last 40 years (BirdLife International 2020) and this increase arguably seems to be the reason that the number of British records accepted into Category A 2000-19 stands at 13.

Collared Dove *Streptopelia decaocto*

Europe and Asia. Introduced in North America. Mid-1970s. **Cat A**

This species is in Category A of the British List, and the first accepted breeding record was of a pair in Norfolk in 1955 (Richardson, Seago *et al*). Its story in Lincolnshire is an interesting one. A singing male held territory at Manton for at least six years from 1952 onwards and was the first British record. It was considered at the time that the behaviour of the male at Manton was more typical of an early colonist rather than an escape from captivity, but it was not accepted as such by the BOU because some were known to have been imported by aviculturists from 1947. A pair bred at Manton in 1957, four birds were present there in 1958 and soon after a pair bred in Skegness. The rest is history, as they say, but some in the county still believe the Manton record was unfairly maligned and should have been the first for Britain.

Greater Flamingo *Phoenicopterus roseus*

Widespread in Africa, south-west and south-central Eurasia. **Cat DE**

Two records assumed to refer to the same individual accepted by BBRC as Category D records: Sep 16th 1990, Witham Mouth; then Sep 21st Oct 5th and 7th 1990, Gibraltar Point (BBRC Report *British Birds* 89: p. 529 and 90: p. 518).

A third record of one at Toft Newton Reservoir from Dec 2nd-21st 1991 (and reportedly there since Oct) was not submitted to BBRC as far as is known.

Lesser Flamingo *Phoeniconaias minor*

East and south Africa, rare in west Africa, north-west India and Pakistan. Vagrant north Africa, Iberian peninsula. **Cat E**

One was seen on Aug 7th 1997 at Read's Island on the Humber. Considered an escape from captivity.

White Pelican *Pelecanus onocrotalus*

Widespread and discontinuous distribution from south-east Europe to Kazakhstan, Africa south of the Sahara and in north-west India. **Cat DE**

One at the Nene mouth on Aug 16th 2006, also seen in The Netherlands and Germany, was tracked through various English counties before finally being taken into care in Northumberland in Oct 2006. (Accepted by BBRC, Category D; *British Birds* 101: p. 575). Interestingly it was accepted as the 9th record for The Netherlands by the Commissie Dwaalgasten Nederlandse Avifauna (CDNA) which adjudicates on rare birds there.

Pink-backed Pelican *Pelecanus rufescens*

Native to Africa south of the Sahara. **Cat E**

One at Chapel St. Leonard's Feb 26th 1989* and at Mablethorpe Oct 27th 1990, presumed same both accepted by BBRC (*British Birds* 89: p. 531) as identification proved, but considered an escape from captivity. Also seen in north Norfolk.

African Sacred Ibis *Threskiornis aethiopicus*

Africa south of the Sahara, also south-eastern Iraq. Feral populations in France, Italy. **Cat E**

One was at Scopwick in Feb 2002 and another toured the county Apr-Aug 2012 being seen successively at Bassingham, Louth, Spalding, Alkborough Flats and Torksey.

Yésou and Clergeau (2005) reviewed the species' status in Europe and reported a roost in France of 1,030 birds close to a shopping centre. It has reported that there are now more than 5,000 in France (Centre for Agriculture and Bioscience International, CABI, Datasheet 2020) and these have caused major problems for other waterbird populations through competition and predation.

Bearded Vulture *Gypaetus barbatus*

Widely but sparsely distributed across western and eastern Europe, parts of East Africa, the Middle East and Asia. Reintroduction scheme started in Europe in 1986. **Cat E**

A 2CY female Bearded Vulture toured England Jun-Oct 2020. It spent Jul-Sep at various sites in Yorkshire and Derbyshire before moving on and being seen in late Sep in Leicestershire (one day) and then Norfolk on Sep 28th. There were no further reports until non-birdwatchers videoed the bird at Holbeach St Johns on Oct 7th. It stayed in the Moulton Chapel-Cowbit area until Oct 9th. It then moved on through Cambridgeshire before heading south to Sussex and disappearing out towards continental Europe. Genetic analysis of feathers collected from a roost site in Yorkshire revealed that it fledged from a wild nest in the Haute-Savoie in the French Alps in 2019. The Vulture Conservation Foundation has overseen Bearded Vulture reintroduction in Europe since 1986 and the genetic analysis revealed its parents to be a male of wild origin and a female reared in a Swiss zoo and released into the wild in 2006 (Phipps L., Loercher F., Ball D. et al British Birds 114 (1): 33-37 Jan 2020). Its formal status will remain in limbo until BOURC adjudicates on the record and assigns it to Category A or C (presumably).

Eagle Owl *Bubo bubo*

Central and southern Europe to northern and central Asia. **Not categorised**

The first county record was of a bird shot near Stamford in Apr 1879. It was examined by staff at Durham University Museum who said it appeared wild and in good condition after feeding on Rabbits *Oryctolagus cuniculus*. In the modern era, the first record was of a bird in the Scunthorpe area in Dec 2002, present on and off until Jan 2004. Others were seen in the grounds of Belton House, Grantham Jan 2003; East Ravendale Jan 2010; Melton Ross May 2013; Gunthorpe Dec 2013; Sleaford Feb 2015; and Donna Nook Jun 2015. The 'modern' records are most likely escapes from captivity and for this reason the species was removed from the British List by the BOURC in 1996 (Melling, Dudley *et al* 2008).

Fossil records indicate that Eagle Owls occurred in what is now Britain through most of the Ice Age and possibly just afterwards but why it subsequently disappeared is a mystery. There are certainly birds at large in Britain and two pairs have nested in north-west England 2006-13. The case for reintroduction of this apex predator has recently been made by MacDonald (2019). Whether any past or future records of lone birds on the (mainly) east coast will ever be accepted as being of wild origin is clearly a moot point.

Saker Falcon *Falco cherrug*

Eastern Europe and through Russia and central Asia to central China. **Cat DE**

One record of an adult, Apr 22nd 1995, Kirkby GP; accepted BBRC as a Category D record (*British Birds* 92: 607). A review of captive-bred falcons by Fleming *et al* (2011) reported the following: since registration of birds commenced in 1983, 8,051 Peregrines *F. peregrinus*, 4,273 'other' falcons, and 11,788 hybrid falcons have been registered in captivity; the most abundant hybrid combination at the time was between Gyr and Saker Falcons (1,843 birds); Lanner *F. biarmicus*, and Saker Falcons did not need registering at that time; between 1981-2005 escaped Lanner (71) and Saker (104) Falcons accounted for 97% of escapees. This somewhat gloomy picture is obviously a significant problem for birdwatchers and apart from identifying large falcons in the wild, especially captive-bred hybrids, the numbers escaping from captivity probably precludes there ever being a truly wild and identifiable Saker Falcon in the county.

Alexandrine Parakeet *Psittacula eupatria*

Native to Indian sub-continent, Afghanistan, Sri Lanka and south-east Asia. **Cat E***

A small number of records have been reported between 1990-2016 all of them at sites around The Wash from Frampton Marsh to Gibraltar Point, all assumed to be escapes from captivity, especially as Britain's largest parrot sanctuary at Friskney, Lincolnshire Wildlife Park, has free-flying parrots of several species.

There are no breeding records to date. They bred occasionally in Britain 1996-2008 but not 2009-11 (Holling *et al* 2014).

Siberian Rubythroat *Calliope calliope*

Ural Mountains and Siberia east through Asia to central China.

One found at Donna Nook Oct 14th 1977 was initially accepted by BBRC (*British Birds* 72, p. 533) but was later reviewed by the rarities committee and deemed to be unacceptable (*British Birds* 82, p. 540); Siberian Rubythroat is, therefore, no longer on the county list.

Desert Finch *Rhynchostruthus obsoleta*

South-east Turkey, the Middle East, Arabia, into central and eastern Asia. Mainly resident, also partial migrant. **Cat E**

One record of a female from Nov 11th 1990-Mar 31st 1991, Deeping St Nicholas, regarded as an escape.

Red-headed Bunting *Emberiza bruniceps*

Central Asia. **Cat DE**

Formerly a common cage-bird, there have been several records of adult males in Lincolnshire, the most recent being a singing male at Donna Nook, Jul-Aug 1983, but none have been reported since. There was an upsurge of British records in the 1950s to 1970s but a sharp downturn after 1982 when an export ban was imposed by the Indian government.

In modern times, there have been just three British records from 1999-2005. Before the export ban, this species outnumbered Black-headed Bunting *E. melanocephalus*, by almost 6:1. Since 1998 though Black-headed has outnumbered Red-headed by 10:1. Red-headed Bunting is now rare in captivity and it is rarely bred. It appears to be an abundant and stable species in Kazakhstan where the bulk of its population breeds (Vinicombe 2007).

Exotic species which have escaped from collections

An alphabetic list of exotic species which have clearly come from commercial and/or private collections is tabulated here. This gives a clear indication of the types of bird being kept in captivity, although there is no indication of how many are from collections in the county or from elsewhere in Britain. Some of these birds have survived for lengthy periods of time at the sites where they were first noted whereas others were transient.

Occasionally a species turns up which has the initial feeling of a wild vagrant. The two Azure Tits *Cyanistes cyanus*, trapped at Gibraltar Point on May 11th 2018 was perhaps the most bizarre of these occurrences. One of these birds carried a pink split colour ring on the right tarsus and had damaged primary and tail feathers. The second had minor tail damage and a brood patch. How might the second bird have been categorised had it arrived on its own? Another bird which caused an initial stir was the White-crowned Black Wheatear *Oenanthe leucopyga*, discovered in a suburban area of Scunthorpe in Dec 2017. It proved incredibly tame and was eventually caught by a visiting birder using his hat and returned to the owner who had an aviary nearby. The owner, a caged bird enthusiast commented that "You can get most birds if you pay", despite the EU-wide wild bird import ban in place since Jul 1st 2017 which is thought to have reduced the global trade in wild bird species by up to 90%.

Most of the exotic pheasants reported have come from private collections and may survive in the wild for some time, although the female Reeves's Pheasant *Syrmaticus reevesii* at Gibraltar Point which apparently survived from Apr 1994-Jul 1997 eventually went the way of many of its congeners when it was killed by collision with a car.

Some of the wildfowl present problems other than identifying them when they occur, namely hybridisation. Various hybrids have been recorded involving some of the species tabulated below, in particular Chiloe Wigeon *Mareca sibilatrix*, and the shelduck species, *Tadorna*. The remaining, motley collection of exotic wildfowl, finches, parrots and raptors just serve as a reminder of the array of birds that have been imported to Britain over the years, although many are also bred in captivity. The huge number of captive falcons and their various derivative hybrids (see Saker Falcon *Falco cherrug*, above account) many of which could never occur naturally, look set to pose problems for birders in the field for years to come.

It is most likely the case that this is a conservative list as escapes carry little interest for birders who usually do not bother to report them. For the species listed in the previous section, we ask birders to report their records as the Rare Birds Breeding Panel monitor the fortunes of non-native species which have been proven to be breeding in the wild. The presence of Indian Peafowl *Pavo cristatus* in Nocton Woods for several decades now, breeding on and off, (see account) is a good example of a non-native bird which may have quietly established itself there without the notice of local birders. Please keep an eye out.

Common Name	Scientific Name	Common Name	Scientific Name
Australian Shelduck	*Tadorna tadornoides*	Maned Duck	*Chenonetta jubata*
Azure Tit	*Cyanistes cyanus*	Monk Parakeet	*Myiopsitta monachus*
Baikal Teal	*Sibirionetta formosa*	Nene	*Branta sandvicensis*
Blue-winged Goose	*Cyanochen cyanoptera*	New Zealand Scaup	*Aythya novaeseelandiae*
Budgerigar	*Melopsittacus undulatus*	Northern Bobwhite	*Colinus virginianus*
Cape Teal	*Anas capensis*	Plum-headed Parakeet	*Psittacula cyanocephala*
Chilean Flamingo	*Phoenicopterus chilensis*	Red Siskin	*Spinus cucullatus*
Chiloe Wigeon	*Mareca sibilatrix*	Red-billed Quelea	*Quelea quelea*
Cockatiel	*Nymphicus hollandicus*	Red-headed Lovebird	*Agapornis pullarius*
Crimson Rosella	*Platycercus elegans*	Red-tailed Hawk	*Buteo jamaicensis*
Eastern Rosella	*Platycercus eximius*	Ringed Teal	*Callonetta leucophrys*
Ferruginous Hawk	*Buteo regalis*	South African Shelduck	*Tadorna cana*
Golden-backed Weaver	*Ploceus jacksoni*	Stripe-throated Yuhina	*Yuhina gularis*
Greater Rhea	*Rhea americana*	Sulphur-crested Cockatoo	*Cacatua galerita*
Grey Parrot	*Psittacus erithacus*	Turkey Vulture	*Cathartes aura*
Harris's Hawk	*Parabuteo unicinctus*	White-cheeked Pintail	*Anas bahamensis*
Hottentot Teal	*Spatula hottentota*	White-crowned Black Wheatear	*Oenanthe leucopyga*
Lady Amherst's Pheasant	*Chrysolophus amherstiae*	White-faced Whistling Duck	*Dendrocygna viduata*
Lanner Falcon	*Falco biarmicus*	Yellow-billed Pintail	*Anas georgica*
Long-tailed Paradise Whydah	*Vidua paradisaea*	Yellow-billed Teal	*Anas flavirostris*
Long-tailed Rosefinch	*Carpodacus sibiricus*	Yellow-fronted Canary	*Crithagra mozambica*

Birds new to Lincolnshire since Lorand and Atkin (1989)

The list below gives full details of all the species new to Lincolnshire since the publication of *The Birds of Lincolnshire & South Humberside* (Lorand and Atkin 1989). The details of the species, site, date(s) and finders are correct at the time of going to press.

Common Name	Scientific Name	Site Name	First Date	Last Date	Age	Observers
Blue-cheeked Bee-eater	Merops persicus	Leverton	12/07/1989		Ad	Mr. and Mrs. R. Humberstone
Snowy Owl	Bubo scandiacus	Thornton Curtis	13/12/1990		1CY M	D.A. Robinson
Blyth's Reed Warbler	Acrocephalus dumetorum	Saltfleetby-Theddlethorpe	03/09/1991		Unk	A. Ashley, A.D. Lowe, M. Thompson et al (trapped)
Penduline Tit	Remiz pendulinus	Wolla Bank, Chapel St Leonard's	14/10/1991	15/10/1991	M	K. Atkin, C.J. Jennings
White-throated Sparrow	Zonotrichia albicollis	Willingham Woods	05/12/1992	28/03/1993	1CY M	N. Bray, S.A. Britton (trapped)
Kumlien's Gull	Larus glaucoides kumlieni	North Hykeham	18/12/1992		Ad	K.D. Durose
Sociable Plover	Vanellus gregarius	Kirkby GP	30/05/1993	12/06/1993	Ad	K.D. Durose et al
Lesser Crested Tern	Thalasseus bengalensis	Saltfleetby-Theddlethorpe	20/06/1993		Ad F	G.P. Catley
Alpine Accentor	Prunella collaris	Saltfleetby-Theddlethorpe	14/11/1994	18/11/1994	Ad	P.M. Troake et al
Lesser Scaup	Aythya affinis	Barton upon Humber	13/02/1995	16/02/1995	2CY M	G.P. Catley et al
Falcated Duck	Mareca falcata	Kirkby GP	19/02/1995	21/02/1995	Ad M	K.D. Durose
Pine Bunting	Emberiza leucephalos	Gibraltar Point	29/03/1995	31/03/1995	2CY + M	N.A. Lound.
Whistling Swan	Cygnus columbianus columbianus	Nocton Fen	22/01/1998		Ad	K.D. Durose
Franklin's Gull	Larus pipixcan	Kirkby GP	13/05/1998		2CY	K.D. Durose et al
River Warbler	Locustella fluviatilis	Gibraltar Point	29/05/1998		2CY+ M	K.D. Durose
Little Swift	Apus affinis	Barton upon Humber	26/06/1998		2CY+	G.P. Catley et al
Pied Wheatear	Oenenathe pleschanak	Gibraltar Point	18/11/2000	26/11/2000	1CY F	S. Pettifer et al
Green Heron	Butorides virescens	Messingham SQ	24/09/2001	02/10/2001	Unk	A. Stanworth, A. Travis et al.
Caspian Gull	Larus cachinnans	Bagmoor Floods	16/12/2001		1CY	G. Taylor
Lesser Sand Plover	Charadrius mongolus atrifrons	Saltfleetby-Theddlethorpe	11/05/2002	15/05/2002	2CY+ F	B.M. Clarkson, M.J. Tarrant et al
Hume's Warbler	Phylloscopus humei	Anderby Creek	18/10/2003		Unk	N.P. Senior
American Robin	Turdus migratorius	Pyewipe Industrial Estate	01/01/2004	08/03/2004	2CY F	S. Smith, T. Moyer
Black-headed Wagtail	Motacilla flava feldegg	Holbeach Marsh	04/06/2004		2CY+ M	J.J. Gilroy
Pallid Swift	Aspus pallidus	Skegness	23/10/2004		Unk	K.D. Durose, D.M. Jenkins, J.Wright
Terek Sandpiper	Xenus cinereus	Gibraltar Point	11/07/2005		Unk	G. Garner et al
Steppe Grey Shrike	Lanius excubitor pallidirostris	Sutton Bridge	16/11/2005		1CY	K. Fisher
Buff-bellied Pipit	Anthus rubescens	Wyberton Marsh	05/12/2005	13/12/2005	Unk	P.R. French et al
Sora Rail	Porzana carolina	Gibraltar Point	05/03/2006	19/03/2006	2CY	T. Bagshaw, A. Dobson, P.M. Troake et al
Ashy-headed Wagtail*	Motacilla flava cinereocapilla	Covenham Reservoir	12/04/2006	13/04/2006	2CY+ M	G. Langan et al
Snow Goose	Anser caerulescens	Saltfleetby-Theddlethorpe	13/10/2006		Ad	J.R. Walker
Greater Yellowlegs	Tringa melanoleuca	Freiston Shore	09/04/2007		2CY	J. Badley, S. Keightley, P. Sullivan et al
Atlantic Yellow-nosed Albatross	Thalassarche chlororhynchos	Manton, Messingham	02/07/2007	03/07/2007	2CY+	P. Condon
Audouin's Gull	Larus audouinii	Huttoft	15/08/2008	23/08/2008	Near Ad	K. Atkin, P. Haywood
King Eider	Somateria spectabilis	Witham Mouth	05/09/2009	09/10/2009	2CY M	P.R. French
Oriental Pratincole	Glareola maldivarum	Frampton Marsh	09/05/2010	18/05/2010	Ad	J. Badley, P.A. Hyde, W. Lawrance, P. Sullivan et al
Calandra Lark	Melanocorypha calandra	Gibraltar Point	11/05/2011		2CY+	T. Bagworth, K.M. Wilson
American Black Tern	Chlidonias niger surinamensis	Covenham Reservoir	17/09/2011	07/10/2011	1CY	G.P. Catley et al
Thayer's Gull	Larus glaucoides thayeri	Elsham Wold	03/04/2012	18/04/2012	2CY	T.C. Lowe et al
Bufflehead	Bucephala albeola	Covenham Reservoir	27/04/2012		2CY+	G.P. Langan et al
Pallid Harrier	Circus macrourus	Gibraltar Point	08/05/2012		2CY	K.M. Wilson
Western Black-eared Wheatear	Oenanthe hispanica	Frampton Marsh	12/06/2012		2CY F	P. Sullivan
Pacific Swift	Apus pacificus	Saltfleetby-Theddlethorpe	12/06/2013		2CY+	B.M. Clarkson
Black-winged Pratincole	Glareola maldivarum	Gibraltar Point	14/07/2014		Ad	J.P. Shaughnessy
Azorean Yellow-legged Gull	Larus michahellis atlantis	Marston STW	26/10/2015		10CY+	D. Roberts, B. Ward
Western Bonelli's Warbler	Phylloscopus bonelli	Gibraltar Point	08/05/2016		2CY+ M	K.M. Wilson
Baltic Gull	Larus fuscus fuscus	Norton Disney Quarry	22/07/2016		7CY	D. Nicholson, B. Ward.
Western Swamphen	Porphyrio porphyrio	Alkborough Flats	30/08/2016	23/11/2016	2CY+	P. Clelford et al (Also seen 04/01/2017)
Mandt's Black Guillemot	Cepphus grylle mandtii	Witham Mouth	07/12/2017	11/12/2017	2CY+	D. Roberts et al
Iberian Chiffchaff	Phylloscopus ibericus	Gibraltar Point	07/05/2019	10/05/2019	Ad M	B. Ward, K.M. Wilson et al
Isabelline Wheatear	Oenanthe isabellina	Gibraltar Point	22/09/2019	23/09/2019	1CY	J.R. Clarkson et al
Bobolink	Dolichonyx oryzivorus	Saltfleetby-Theddlethorpe	18/10/2019		Unk	T. Hibbert et al
Black-throated Thrush*	Turdus atrogularis	Grimsby	30/01/2020	02/04/2020		J. Forrester
Bearded Vulture*	Gypaetus barbatus	Holbeach St. Johns	07/10/2020	09/10/2020	Imm F	L. Oliver, W Bowel, M. Weedon, H. Lewis-wright
*Under consideration by BBRC						
Category D species						
Greater Flamingo	Phoenicopterus roseus	Gibraltar Point	21/09/1990		Unk	P.R. Davey, K.M. Wilson et al
Saker Falcon	Falco cherrug	Kirkby GP	22/04/1995		2CY+	P.A. Hyde
Ross's Goose	Anser rossii	Read's Island	10/11/2002		Ad	G.P. Catley et al

Gallery of images

The county is blessed with some excellent photographers, many who live here and others who visit. These superb images all came our way during the compilation of this book, some in response to specific requests and some held in archive by LBC. Needless to say, we could not fit all of them into the species accounts in the Systematic List. We therefore decided to include them for added interest and also of course as a tribute to the photographers concerned, whose initials are included with the photographs. Some photographers capture images of birds they find themselves but many go in pursuit of birds found by others and these images are also a tribute to the finders of the birds, without whose hard work and generous sharing of information, many of these images would not have been captured. There is also a superb set of sketches by Steph Thorpe of the Purple Swamphen which graced Alkborough Flats in 2016-17, which were too good to leave out. We hope that you will enjoy perusing what we think is a high quality gallery.

Black Stork, Dunsby, 2017 (SK)

Slavonian Grebe, Cleethorpes, 2016 (GPC)

Lesser Sandplover, Rimac, 2002 (GPC)

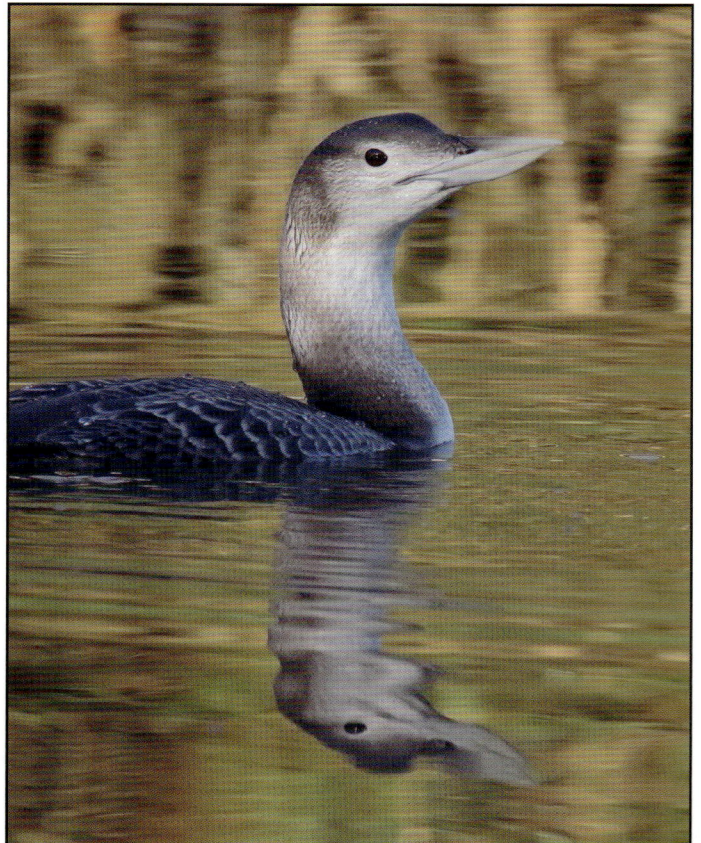

White-billed Diver, Kirkstead Bridge, 2017 (NS)

Barred Warbler, Donna Nook, 2010 (GPC)

Bitten, Humber, 2007 (GPC)

Black Tern, Barton Pits, 2019 (GPC)

Booted Warbler, Donna Nook, 2003 (GPC)

Arctic Redpoll, Barton Pits, 2006 (GPC)

Blue-winged Teal, Donna Nook, 2015 (GPC)

Blue-winged Teal, Messingham, 2009 (JTH)

Black-eared Wheatear, Frampton, 2012 (PS)

White-crowned Black Wheatear Escape 2017 (SG)

Terek Sandpiper, Gibraltar Point, 2009 (NS)

Common Redpoll (Mealy), Barton Pits, 2008 (GPC)

Curlew Sandpiper, Saltfleet, 2010 (GPC)

Barn Owl, Whaplode, 2006 (NS)

Barn Owl (NS)

Bittern, Far ings (GPC)

Bittern, Far Ings (GPC)

Bittern, Far Ings 2019 (PC)

Bittern, Far Ings, 2019 (PC)

Bittern, Far Ings (GPC)

Jack Snipe, Gibraltar Point, 2013 (GPC)

Desert Wheatear, Salftleet, 2008, (GPC)

Ferruginous Duck, Barton Pits, 2001 (GPC)

Great Skua, Barton, 2011, (GPC)

Dusky Warbler, Wolla Bank Huttoft, 2019 (RDH)

Glassy Ibis, Frampton Marsh, 2014 (NS)

Arctic Skua, Saltfleetby 2004 (MJT)

Arctic Tern, Barton Pits 2007 (GPC)

Arctic Warbler, Donna Nook 2014 (MDJ)

Baird's Sandpiper, Kirkby Pits, 2005 (RDH)

Barred Warbler, Covenham, 2004 (NS)

Barred Warbler, Donna Nook, 2010 (SKS)

Bearded Tit (GPC)

Bar-tailed Godwit, Donna Nook (MDJ)

Western (Purple) Swamphen   Alkborough Flats   1st September 2016

"Well upholstered" bird

Massive bill + shield

Paler towards tip

Very blue tinged - indigo? in sunshine really BLUE!

quite awkward jizz

Paler electric blue area on face + breast

Primaries extend beyond tertials which are drooped with wings

Comical bird

flicks tail continually exposing white UTCov's

Very large water bird cp Mhen

Very triangular at times

bulky - strong looking

Long red legs slightly more orangy tt bill

Missing feather

Mad "dread Run" wings flailing high stepping Run!

Long trailing legs in flight   rather 'manic' in flight!

tugging at reeds -

Could look very dark

esp. in shade

Paddling around in front of reeds

Very deliberate movements.

Big thighs.

Jauntly miniature Mhn!

Ministry of Silly Walks

Jogging out into deeper water! Rolling gait!

Jogged out for a bath!

uses feet like hands to hold reed tubers.

Western Swamphen notebook details, Alkborough, 2016 (ST)

Black-necked Grebe, Toft Newton (BMC)

Black-throated Diver, Cleethorpes (NS)

Black-winged Stilt, Grainthorpe 2015 (MDJ)

Black-necked Grebe, Covenham, 2017 (MDJ)

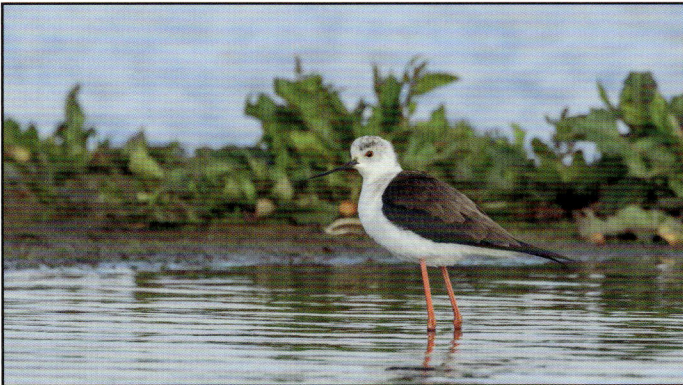

Black-winged Stilt, Grainthorpe 2015 (MDJ)

Bluethroat, Willow Tree Fen, 2017 (MDJ)

Bluethroat, Willow Tree Fen, 2017 (NS)

Bluethroat, Willow Tree Fen, 2017 (MDJ)

Blyth's Reed Warbler, Far Ings, 2020 (PC)

Blyth's Reed Warbler, Far Ings, 2020 (PC)

Blyth's Reed Warbler, Far Ings, 2020 (MDJ)

Blyth's Reed Warbler, Far Ings, 2020 (MDJ)

Brambling, Gibraltar Point (NS)

Brambling, North Somercotes, 2018 (MDJ)

Brent Goose (Black Brant), Freiston Shore, 2008 (SK)

Broad-billed Sandpiper, Frampton Marsh, 2016 (SK)

Buff-breasted Sandpiper, Alkborough (GPC)

Buzzard (GPC)

Caspian Gull, Barton Pits 2006 (GPC)

Cetti's Warbler, Frampton Marsh (SK)

Collared Pratincole, Killingholme, 2011 (GPC)

Dunlin, Donna Nook, 2017, (MDJ)

Dusky Warbler, Wolla Bank, 2019 (GPC)

Fieldfare, Little Grimsby, 2018 (MDJ)

Glaucous Gull, Donna Nook, 2019 (MDJ)

Green-winged Teal, Kirkby on Bain, 2007 (RH)

Black-winged Pratincole, Frampton Marsh, 2019 (SK)

Black-winged Pratincole, Frampton Marsh, 2019 (SK)

Black-winged Pratincole, Frampton Marsh, 2019 (SG)

Black-winged Pratincole, Frampton Marsh, 2019 (SK)

Hume's Warbler, Gibraltar Point, 2015 (GPC)

Iberian Chiffchaff, Croft, 2020 (JRC)

Lapland Bunting, Donna Nook, 2017 (MDJ)

Long-billed Dowitcher, Saltfleet, 2017 (MDJ)

Lesser Yellowlegs, Gibraltar Point, 2006 (RH)

Pectoral Sandpiper, Toft Newton, 2012 (MJD)

Pallid Harrier, Gibraltar Point, 2015 (GPC)

Pallid Harrier, Gibraltar Point, 2015 (GPC)

Purple Sandpiper, Covenham, 2006 (NS)

Purple Sandpiper, Cleethorpes, 2020 (JRC)

Red-backed Shrike, Bonby Carrs, 2013 (GPC)

Red-backed Shrike, Saltfleetby, 2019 (JRC)

Long-tailed Skua, Killingholme, 2013 (GPC)

Arctic Skua, The Wash, 2015 (SK)

Long-tailed Skua, Bardney, 1998 (GPC)

Great Skua, Saltfleetby, 2002 (MJT)

Pomarine Skua, Donna Nook, 2019 (MJT)

Pomarine Skua, Pyewipe, 2010 (GPC)

Pomarine Skua, The Wash, 2015 (SK)

Red-footed Falcon, Willow Tree Fen, 2015 (SK)

Red-footed Falcon, Gibraltar Point, 2015 (SK)

Red-flanked Bluetail, Donna Nook, 2016 (JRC)

Red-flanked Bluetail, Gibraltar Point, 2016 (CRC)

Red-necked Phalarope, Covenham, 2017 (PC)

Red-necked Phalarope, Covenham, 2017 (MDJ)

Red-necked Phalarope, Barton, 2009 (GPC)

Red Kite, Saltfleetby, 2014 (MJT)

Northern Bullfinch, Gibraltar Point, 2003 (GPC)

Black Brant, Freiston Shore, 2008 (SK)

Avocet, Grainthorpe, 2012 (MDJ)

Chetti's Warbler, Barton Pits, 2010 (GPC)

Little Stint, Covenham Reservoir, 2019 (JRC)

Little Stint, Frampton Marsh, 2009 (NS)

Eastern Lesser Whitethroat, Donna Nook, 2011 (GPC)

Purple Heron, Manby, 2019 (JRC)

Purple Heron, Huttoft, 2011 (MDJ)

Red-breasted Flycatcher, Donna Nook, 2010 (MDJ)

Pallas's Warbler, Donna Nook, 2016 (RDH)

Pallas's Warbler, Donna Nook, 2008 (GPC)

Red-breasted Flycatcher, Pyes Hall, 2010 (GPC)

Pallas's Warbler, Donna Nook, 2016 (JRC)

Pink-footed Geese, Alkborough, 2017 (GPC)

Marsh Harriers, Far Ings, 2017 (GPC)

Little Ringed Plover (NS)

Little Ringed Plover, Covenham, 2017 (JRC)

Ringed Plover, Frampton Marsh, 2016 (SK)

Ring Ouzel, RAF Cranwell, 2009 (RH)

Rose-coloured Starling, Mablethorpe, 2020 (GPC)

Ring-necked Parakeet, Grimsby, 2008 (JRC)

Tawny Owl, Baumber, 2020 (RT)

Buzzard, Baumber, 2020 (RT)

Red-necked Phalarope, West Ashby, 2018 (RT)

Icterine Warbler, Donna Nook, 2020 (BMC)

Sanderling, Covenham, 2008 (BMC)

Knot, Cleethorpes, 2019 (IGS)

Wheatear, Donna Nook, 2018 (IGS)

Spotted Flycatcher, 2019 (IGS)

Fieldfare, Frithville, 2018 (SK)

Pectoral Sandpiper, Frampton Marsh, 2019 (SK)

House Martin, Frithville, 2019 (SK)

Kingfisher, Far Ings, 2019 (SB)

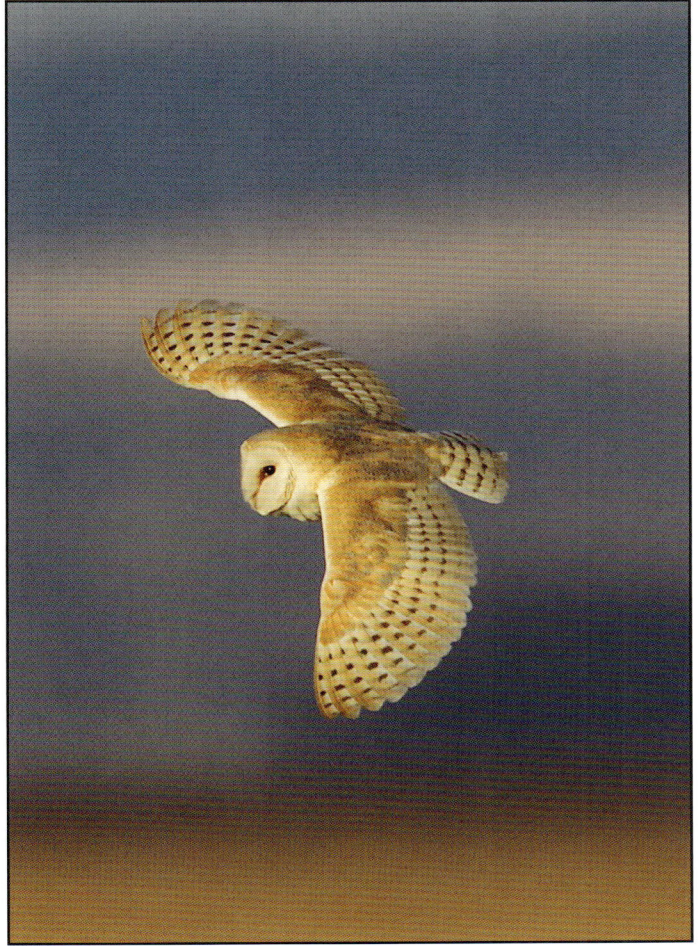

Barn Owl, Alkborough, 2016 (SB)

Merlin, Frampton Marsh, 2020 (SB)

Western Yellow Wagtail, Tetney, 2014 (SB)

Serin, Gibraltar Point, 2013 (KMW)

Short-toed Lark, Gibraltar Point, 2013 (CA)

Sedge Warbler, Frampton Marsh, 2010 (NS)

Female Goldeneye (SB)

Spotted Redshank, Frampton Marsh (SK)

Golden Plover (SK)

Tree Sparrow (SB)

Kingfisher (SB)

Pectoral Sandpiper, Frampton, 2019 (SK)

Waxwing, Boston (SK)

Restart (SK)

Brambling, North Somercotes, 2018 (MDJ)

Fieldfare, Little Grimsby, 2018 (MDJ)

Fieldfare, Thornton Abbey, 2009 (GPC)

Golden Plover (NS)

Honey-Buzzard, Frampton Marsh, 2011 (NS)

Recent Publications

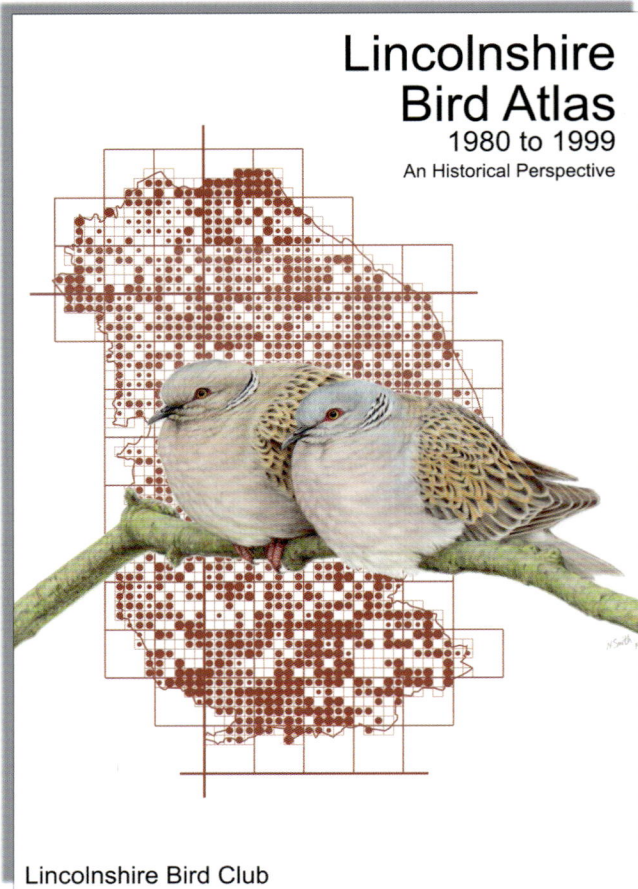

Lincolnshire Bird Atlas
1980 to 1999
An Historical Perspective

Lincolnshire Bird Club

In June 2020 LBC published the *Lincolnshire Bird Atlas 1980 to 1999 An Historical Perspective*, an important source of reference for this book. It contains breeding atlas maps and fascinating narrative details of 121 species breeding in the county in that period, winter atlas maps for those species and a full list with details on status of all species occurring in Lincolnshire up to the year 2000.

The Atlas project was led and overseen by Keith Atkin, and he and Anne Goodall wrote the species accounts. There was a long delay before its publication by essentially the same team producing *Birds of Lincolnshire* (2021).

It is 180 pages in A4 format in black and white with a superb cover drawing of a pair of Turtle Doves by local artist Neil Smith, against a background of their Atlas map from the 1980s. For the modern reader the cover provides a compelling illustration of how a species fortunes can change over 30 years.

The Atlas design matches and complements Birds of Lincolnshire on the shelf. No reviews of the Atlas have been published at the time of writing but David Ballance, author of *Avifaunas, Atlases and Authors A Personal View of Local Ornithology in the United Kingdom from the Earliest Times to 2019* (2020), describes it as being "of exemplary clarity", in a review scheduled for publication in the BOU journal *The Ibis*.

Copies can be obtained from the Lincolnshire Bird Club

LBC publishes an annual Lincolnshire Bird Report. The aim is to publish within 15 months of the year end. LBR 2018 was published in March 2020 and LBR 2019 is on schedule to be published in March 2021.

The report contains a full systematic list detailing all records of rare and scarce species and interesting records of all other species recorded in the year. It also contains finders reports of significant rarities such as firsts for the county and papers on matters relating to the study of birds in Lincolnshire.

Full details of availability and prices of all reports still held can be found at www.lincsbirdclub.co.uk.

Photographers

In this volume of the *Birds of Lincolnshire* we have tried to use the best images available for each species. We owe a huge debt of gratitude to the army of photographers who have freely donated their hard-won images, often after lugging their heavy equipment far and wide in all weathers to the remotest parts of Lincolnshire.

The task of naming all the images in the Systematic List and then crediting each one with the correct photographer's initials would have been huge, and fraught with errors, and unfortunately this was a step too far in the time available. Therefore, the species are only named and credited to the individual photographer in the section called the Gallery. However, the team would like to take this opportunity to say a heartfelt "Thank You" to everyone who supplied us with any images.

Colin Casey, who has been working under serious health constraints, felt he had to make this decision for medical reasons and hopes you will understand. We sincerely apologise if any of your images were not used. A huge quantity of photographs were submitted and we simply did not have room to use all of them. If you see an image that you would like more details on please contact us.

Chris Atkin (CA)	John Clarkson (JRC)	Nige Lound (NL)	Mike Tarrant (MJT)
Keith Atkin (KA)	Paul Coombes (PC)	Ian Mistlebrook (IM)	Alan Tate (AT)
John Badley (JB)	David Cotteridge (DC)	Paul Neale (PN)	Garry Taylor (GT)
Alan Ball (AB)	Jack Dawson (JDa)	Steve Nesbitt (SN)	Russ Telfer (RT)
Nik Barrow (NB)	Neil Drinkall (ND)	Steve Nikols (SNi)	Steph Thorpe (ST)
Steve Bartlett (SB)	Steve Gantlett (SG)	Gerv Orton (GO)	Dave Read (DR)
Owen Beaumont (OB)	George Gregory (GG)	Steve Routledge (SR)	John Walton (JWW)
Derek Bell (DB)	Roy Harvey (RDH)	Peter Roworth (PR)	Ben Ward (BW)
Anthony Bentley (ABe)	Russell Hayes (RH)	Alex Scott (AS)	Rob Watson (RW)
Michael Briggs (MB)	Pete Haywood (PH)	James Siddle (JS)	Adrian Webb (AW)
Dave Bromwich (DRB)	Phil Hyde (PAH)	Andy Sims (ASi)	Jim Welford (JW)
Colin Casey (CRC)	Mark Johnson (MDJ)	Colin Smale (CS)	Barrie Wilkinson (BW)
Graham Catley (GPC)	Steve Keightley (SK)	Neil Smith (NS)	Kevin Wilson (KMW)
Andrew Chick (APC)	Paul Lawrence (PL)	Paul Sullivan (PS)	
Barry Clarkson (BMC)	Jack Levene (JL)	Martin Swannell (MS)	

Acknowledgements

This book has been conceived, written and put together in a record time of under six months. It would have been impossible to write without the foundation of recording and publications of the Lincolnshire Bird Club which have been made possible by the hard work and bird finding of members and supporters. This work especially rests on the shoulders of Lorand and Atkin (1989), the *LBC Atlas* (2020) and the annual LBR 1989-2018. In 2019 LBC reached its 40th anniversary and we hope *Birds of Lincolnshire* is a fitting testament to all those who have contributed, we would like to thank them all and especially those listed below.

Albeit the book is a team effort, the driving inspiration for the work has been Colin Casey (CRC) who has pulled all the data together and produced the tables, maps and charts which are such an informative part of the book. His two years of hard work, structuring and validating the LBC database led to the realisation that the information was there to undertake this enterprise. His leadership and oversight of every aspect of the process, together with his relentless determination, drove the team to ever greater efforts. John Clarkson (JRC), an accomplished bird photographer himself, has, with help from CRC and responses to appeals on Twitter, sourced the photographs from his wide network of photographer contacts, all of which are of birds seen in Lincolnshire. The photographic community of Lincolnshire and beyond, have been generous in sharing their work. Where photographs have not been sourced, CRC has produced illustrations from photographs using Procreate and an Apple Pencil, often of museum specimens of birds taken in the county.

Philip Espin (PE), Chair of LBC and Philip Hyde (PAH), Lincolnshire County Recorder wrote the text. In particular, PAH wrote the rare and scarce species which required extensive checking to validate the records and also wrote the species not on the Lincolnshire list. All BBRC-accepted records of national rarities in Lincolnshire have been checked and included along with county rarities submitted to successive LBC Records Committees. Any omissions or errors are unintentional and observers having unsubmitted records are encouraged to put them in. PE wrote most of the other bird species and much of the other text. PE and PAH mutually checked each other's text, species by species as they were being written and made suggestions to each other for additions and changes. PE is also the BTO Regional Representative for East Lincs and a fanatical BBS surveyor. If that comes across too strongly in the text he makes no apologies and asks all readers to sign up for a BBS square. JRC forensically examined, commented on the text and, with CRC's help, proof-read their output. This process generated an enormous number of emails and phone calls, much re-writing, some misunderstandings and the occasional heated discussion as all four worked to develop and refine the text.

Graham Catley wrote several raptor and heathland species and made suggestions on the texts of many others. LBC is lucky to have a supportive network of experienced birders who have commented on various parts of the book and provided ideas including especially: Robert Carr, Andrew Chick, Kevin Wilson and Steve Keightley. LBC's Hon Sec newshound Jim Wright @Thewryneck commented on the text. Nik Borrow has produced the wonderful colour painting of Pallas's Sandgrouse, a dream county find for most of us, should it ever reappear.

Outside LBC we would like to thank the BTO for the tremendous online resources they make available including BTO BirdFacts, BirdTrack and WeBS Online and many others. Sarah Harris, BTO BBS Co-Ordinator was especially helpful in providing the data from which bird index change charts were prepared. At BBRC Paul French and Chas Holt provided helpful assistance with checking information on rarity occurrences. Charlie Barnes provided data on Lincolnshire habitats and created the hexagram map of predominant habitats across the county. We are especially grateful to David Bromwich, Reserves Manager at LWT and to LWT for the many excellent aerial photographs of their reserves which are a great visual addition to the book. Jack Levene, developer of Wildlife Recorder 4, LBC's bird database software has been as innovative as always, in helping us get the best out of our data.

Last but by no means least we would like to thank our wives: Nicky Clarkson for the final proofreading, Isabel Espin and Trish Hyde, families and friends for their help and support in facilitating the personal sacrifices that producing this book in such a short time have required.

Colin Casey, John Clarkson, Phil Espin, Phil Hyde, Lincolnshire Jan 2021.

Supporters

During the forty plus years that the Lincolnshire Bird Club has been in existence, information and assistance has come from many sources, and here we would like to thank them all.

BBRC British Birds Rarities Committee

Lincolnshire Wildlife Trust

RBBP Rare Breeding Birds Panel

LBRC Lincolnshire Birds Records Committee

GLNP GREATER LINCOLNSHIRE NATURE PARTNERSHIP

Wildlife Recorder (www.wildlife.co.uk)

rspb giving nature a home

Rare Bird Alert Instant and reliable birdnews since 1991

BTO Inspired by birds, informed by science

NATURAL ENGLAND

NL Nature Lister (www.naturelister.co.uk)

Abbreviations used

APEP	Avian Population Estimates Panel
AOT	Annual Occupied Territories
BB	British Birds
BBS	Breeding Bird Survey
BBRC	British Birds Rarities Committee
BoCC	Birds of Conservation Concern
BOU	British Ornithologists' Union
BOURC	British Ornithologists' Union Rarities Committee
BTO	British Trust for Ornithology
BTO Bird Atlas1968-72	see References Sharrock, J.T.R. (1976)
BTO Atlas 1988-91	see References Gibbons D.W.(1993)
BTO Atlas 2007-11	see References Balmer et al (2013)
EA	Environment Agency
GLNP	Greater Lincolnshire Nature Partnership
GP	Gravel Pit
GBFS	Garden Birds Feeding Survey
LBC ATLAS	Lincolnshire Bird Atlas 1980-1999 An Historical Perspective
LBC	Lincolnshire Bird Club
LBRC	Lincolnshire Birds Records Committee.
LBR	Lincolnshire Bird Report
LCC	Lincolnshire County Council
LNU	Lincolnshire Naturalists' Union
Lorand and Atkin (1989)	Birds of Lincolnshire & South Humberside
LWT	Lincolnshire Wildlife Trust
JNCC	Joint Nature Conservation Committee
NNR	National Nature Reserve
NR	Nature Reserve
RBBP	Rare Breeding Birds Panel
RSPB	Royal Society for the Protection of Birds
SQ	Sand Quarry
SMP	Seabird Monitoring Programme
STW	Sewage Treatment Works
WeBS	Wetland Bird Survey
WGBS	Winter Garden Bird Survey

References

Balmer, D.E., Gillings, S., Caffrey, B., Swann, B., Downie, I. & Fuller, R. (2013) Bird Atlas 2007-2011: The breeding and wintering birds of Britain and Ireland. BTO, Thetford.

Beolens, B., and M. Watkins (2003). Whose bird? Men and women commemorated in the common names of birds. Helm, London.

Billerman, S.M., B. K. Keeney, P. G. Rodewald, and T. S. Schulenberg (Editors) (2020). Birds of the World. Cornell Laboratory of Ornithology, Ithaca, NY, USA.

BirdLife International (2020) IUCN Red List for birds. Downloaded from Birdlife International website on 09/10/2020.

Birds of the World (S. M. Billerman, B. K. Keeney, P. G. Rodewald, and T. S. Schulenberg, Editors). Cornell Laboratory of Ornithology, Ithaca, NY, USA. https://birdsoftheworld.org/bow/home.

Blathwayt, F.L. (1915). The Birds of Lincolnshire. Lincolnshire Naturalists Union Transactions, 3: 178-211.

Boyd, H. (1954). The "wreck" of Leach's Petrels in the autumn of 1952. British Birds 47: 137-163.

British Birds Rarities Committee (1997). Rarities Committee News and Announcements. British BIrds 91: 121-123.

British Ornithologists' Union Records Committee (BOURC): 23rd Report (July 1996). Ibis 139: 197-201.

British Ornithologists' Union Records Committee (BOURC): 25th Report (January 1999). Ibis 141: 175-80.

British Ornithologists' Union Records Committee (BOURC): 27th Report (October 2000). Ibis 143: 171-175.

British Ornithologists' Union Records Committee (BOURC): 30th Report (October 2003). Ibis 146: 192-196.

British Ornithologists' Union Records Committee (BOURC): 51st Report (January 2020). Ibis 162: 600-603.

Britton, D. (1980). Identification of Sharp-tailed Sandpipers. British Birds 73: 333-345.

British Ornithologist's Union Records Committee: 27th Report (October 2000). Ibis (2001) 143: 171-175.

CABI (2020). Threskiornis aethiopicus (Sacred Ibis) . In: Invasive Species Compendium. Wallingford, UK: CAB International.

Carboneras, C. and G. M. Kirwan (2020). Falcated Duck (Mareca falcata), version 1.0. In Birds of the World (J. del Hoyo, A. Elliott, J. Sargatal, D. A. Christie, and E. de Juana, Editors). Cornell Lab of Ornithology, Ithaca, NY, USA.

Catley, G.P. (2000). Song and call of a 'Siberian Chiffchaff'. British Birds 93: 456.

Catley, G.P. and S.Lorand (undated). Siberian/Stejneger's Stonechats at Donna Nook, 6th-10th October and 14th-16th October 2016. Lincolnshire Bird Report 2016.

Collinson, J.M. (2012). A genetic analysis of the first British Siberian Stonechat. British Birds 105: 318-331.

Cordeaux, J. (1872) Birds of the Humber District

Cordeaux, J. (1880). Bee-eater, Grey Phalarope and Tengmalm's Owl in Lincolnshire. The Zoologist, Third Series, Vol IV: 511-512.

Cramp, S., Pettet, A., and J.T.R. Sharrock (1960). The irruption of tits in autumn 1957. British Birds 53: p49-77, p99-177, p176-192.

Dean, A.R. and L. Svensson (2005). 'Siberian' Chiffchaff revisited. British Birds 98: 396-410.

Eaton, M.A., Aebischer, N.J., Brown, A.F., Hearn, R.D., Lock, L., Musgrove, A.J., Noble, D.G., Stroud, D.A and Gregory, R.D. (2015). Birds of Conservation Concern 4: the population status of birds in the United Kingdom, Channel Island and Isle of Man. British Birds 108, 708-746

Editors, The (1954). An unrecorded occurence of a probable Red-flanked Bluetail in Lincolnshire. British BIrds 47: 28-29.

European Environment Agency. Breeding population and distribution trends. DAS-209-en. Published 18 Mar 2020.

Fraser, P. (1997). How many rarities are we missing? Weekend bias and length of stay revisited. British BIrds 90: 94-101.

Fraser, P.A. (2013). Report on scarce migrant birds in Britain in 2004-2007 Part 1: non-passerines. British Birds 106: 368-404

Frost, T., Austin, G., Hearn, R., McAvoy, S., Robinson, A., Stroud, D., Woodward, I., and S. Wotton (2019). Population estimates of wintering waterbirds in Great Britain. British Birds 112: 130-145.

Frost, T.M., Calbrade, N.A., Birtles, G.A., Mellan, H.J., Hall, C., Robinson, A.E., Wotton, S.R., Balmer, D.E. and Austin, G.E. (2020). Waterbirds in the UK 2018/2019. The Annual report of the Wetland Bird Survey. BTO, RSPB and JNCC in association with WWT. British Trust for Ornithology, Thetford.

Gregory, G. (2020) Icelandic Redwings (Turdus iliacus coburni) in Lincolnshire. Lincolnshire Bird Report 2018: 197-204. LBC.

Gregory, G. (2020) A Preliminary Phenological Analysis of Ten Species or Subspecies of Wintering Birds at Gibraltar Point National Nature Reserve from the Winter of 1949/50 to the Winter of 2017/18. Lincolnshire Bird Report 2018: 205-210 LBC.

Goodall, A. (2013) Report on the Garden Bird Feeding Survey, 2009-2012. Lincolnshire Bird Report 2011:197-206. LBC.

Guzy, M. J., B. J. McCaffery, and N. Collar (2020). Bluethroat (Luscinia svecica), version 1.0. In Birds of the World (S. M. Billerman, Editor). Cornell Lab of Ornithology, Ithaca, NY, USA.

Harris, S.J., Massimino, D., Balmer, D.E., Eaton, M.A., Noble, D.G., Pearce-Higgins, J.W., Woodcock, P. and Gillings, S. (2020). The Breeding Bird Survey 2019 BTO Research Report 726. British Trust for Ornithology, Thetford.

Henderson. A. (2020) The Hooded Crow in Lincolnshire and Kent. Lincolnshire Bird Report 2018: 189-195. LBC.

Henderson, A. and Wilson, K. Red-throated Divers Gavia stellata wintering off Lincolnshire. Lincolnshire Bird Report 2015: 197-203. LBC.

Holling, M and the Rare Breeding Birds Panel (2010). Rare breeding birds in the United Kingdom in 2007. British Birds 103: 2-52.

Holling, M and the Rare Breeding Birds Panel (2011). Non-native breeding birds in the United Kingdom, 2006, 2007 & 2008. British Birds 104: 114-138.

Holling, M and the Rare Breeding Birds Panel (2013). Rare breeding birds in the United Kingdom in 2011. British Birds 106: 496-554.

Holling, M and the Rare Breeding Birds Panel (2014). Non-native breeding birds in the United Kingdom, 2009-11. British Birds 107: 122-141.

Holling, M and the Rare Breeding Birds Panel (2017). Non-native breeding birds in the United Kingdom, 2012-14. British Birds 110: 92-108.

Holling, M and the Rare Breeding Birds Panel (2019). Rare breeding birds in the United Kingdom in 2017. British Birds 112: 706-758.

Holling, M and the Rare Breeding Birds Panel (2020). Rare breeding birds in the United Kingdom in 2018, in litt.

Hollyer, J.M.(1970). The invasion of Nutcrackers in autumn 1968. British Birds 63: 353-373.

Holt, C.(2013). The changing status of the Great White Egret in Britain. British Birds 106: 246-257.

Hudson, N. and the Rarities Committee (2016). Report on rare birds in Great Britain in 2015. British Birds 109: 566-631.

Hunter, E. (2018). Iberian Chiffchaff: a new breeding species for Great Britain. British Birds 111: 100-108.

Hyde, P. (2020) The invasion of the Hawfinches Coccothraustes coccothraustes in Lincolnshire in 2017-2018. Lincolnshire Bird Report 2018: 183-187. LBC

Hyde, P. (2017) The status of Little Egret Egretta garzetta in Lincolnshire Lincolnshire Bird Report 2015: 191-196. LBC

JNCC (2020) Seabird Monitoring Programme Report 1986-2018

Knox, A.G., Collinson, M., Helbig, A.J., Parkin, D.T., and G. Sangster (2002). Taxonomic recommendations for British birds. Ibis 144: 707-710.

Kampe-Persson, H.(2010). Naturalised Geese in Europe. Ornis Svecica 20: 155-173.

Lincolnshire Bird Club (2020). Lincolnshire Bird Atlas, 1980 to 1999, An Historical Perspective. LBC 2020, ISBN 9781716802638.

Lack, P. (1986) The Atlas of Wintering Birds in Britain and Ireland (Calton)

Lorand, S and Atkin, K (1989). The Birds of Lincolnshire & South Humberside. ISBN 0-948135-11-5.

MacDonald, B (2019). Rebirding: Rewilding Britain and it's Birds. Pelagic Publishing, Exeter. ISBN 978-1-78427-187-9

Madge, S., del Hoyo, J., Christie,D.A., Collar,N., and G. M. Kirwan (2020). Eurasian Nutcracker (Nucifraga caryocatactes), version 1.0. in Birds of the World (S. M. Billerman, B. K. Keeney, P. G. Rodewald, and T. S. Schulenberg, Editors). Cornell Lab of Ornithology, Ithaca, NY, USA.

Marchant, J.H, and A.J. Musgrove (2011). Review of European flyways of the Lesser White-fronted Goose Anser erythropus. BTO Research Report 595.

Massimino, D., Woodward, I.D., Hammond, M.J., Harris, S.J., Leech, D.I., Noble, D.G., Walker, R.H., Barimore, C., Dadam, D., Eglington, S.M., Marchant, J.H., Sullivan, M.J.P., Baillie, S.R. and Robinson, R.A. (2019) BirdTrends 2019: trends in numbers, breeding success and survival for UK Breeding Birds, BTO Research Report 722 BTO Thetford.

Melling, T., Dudley, S. and P. Doherty (2008). The Eagle Owl in Britain. British Birds 101: 478-490.

Neufeldt, I. (1961). Studies of less familiar birds 104. Radde's Bush Warbler. British Birds 53: 117-122.

Newton, I. (2017) Farming and Birds. New Naturalist.

Nisbet, I.C.T. (1961). Dowitchers in Great Britain and Ireland. British Birds 54: 343-356.

Parnaby, D. (2020). Subalpine Warblers on Fair Isle. Fair Isle Observatory website.

Richardson, R.A., Seago, M.J., and A.C. Church (1957). Collared Doves in Norfolk: a bird new to the British List. British Birds 50: 239-246.

Riddington, R., Votier S., and Steele, J. (2000). The influx of redpolls into Western Europe, 1995/96. British Birds 93: 59-67.

Rogers, M.J., and the Rarities Committee. Report on rare birds in Great Britain in 1980. British Birds 74: 453-495.

Rogers, M.J., and the Rarities Committee. Report on Rare Birds in Great Britain in 1997. British Birds 91: 455-517

Roadhouse, A. (2016). The Birds of Spurn. Spurn Bird Observatory Trust, East Yorkshire.

Robinson, K., Clarkson, J., Harrison, M. and Henderson, A. (2016) The Birds of Covenham Reservoir 1970-2014. Lincolnshire Bird Report 2014; 174-188. LBC.

Robinson, R.A. (2005) BirdFacts: profiles of birds occurring in Britain and Ireland. BTO Thetford.

Sangster, G., Collinson, M.J., Helbig, A.J., Knox, A.G., and D.T. Parkin (2002). The specific status of Balearic and Yelkouan Shearwaters. British Birds 95: 636-639.

Sangster, G., Collinson, J.M., Helbig, A.J., Knox, A.G., and D.T. Parkin (2005). Taxonomic recommendations for British birds: third report. (Details of the recognition of Yellow-Legged Gull as a full species).

Sharrock, J.T.R. (1976) The Atlas of Breeding Birds in Britain and Ireland (Tring).

Shipilina, D., Serbyn, M., Ivanitskii, V., Marova, I. and N. Backström (2017). Patterns of genetic, phenotypic, and acoustic variation across a chiffchaff (Phylloscopus collybita abietinus/tristis) hybrid zone. Ecol Evol.7: 2169-2180.

Shirihai, H. and L. Svensson (2018). Handbook of Western Palaearctic Birds. Volume I. Passerines: Larks to Warblers; volume II. Passerines: Flycatchers to Buntings. Helm. London. Oxford. New York. New Delhi. Sydney.

Slack, R. (2009). Rare birds. Where and when. An analysis of status and distribution in Britain and Ireland. Vol 1: Sandgrouse to New World Orioles. Russell Slack, Rare Bird Books, York.

Sims, A. (2017) Observations on breeding Lesser Spotted Woodpeckers at Swanpool, Lincoln, 2015-2016. Lincolnshire Bird Report 2015: 185-189. LBC

Smith, A.E. and R.K. Cornwallis (1955). The Birds of Lincolnshire. Lincolnshire Natural History Brochure, No.2. Lincolnshire Naturalists Union, Lincoln.

Smith, R.J. and F.R. Moore (2005). Fat stores of American Redstarts Setophaga ruticilla arriving at northerly breeding grounds. J. Avian Biol. 36: 117-126.

Smith, M., Bolton, M., Okill D.J., Summers, R.W., Ellis, P., Liechti, F., Wilson, J.D. (2014) Geolocator tagging reveals Pacific migration of Red-necked Phalarope Phalaropus lobatus breeding in Scotland. Ibis 156 (4): 870-873.

Stoddart, A. and the British Birds Rarities Committee (2014). From the Rarities Committee's files: Assessing and recording Subalpine Warblers. British Birds 107: 420-424.

Touzé, H (2019). Lesser Grey Shrike lost as breeding species in France. Bird Guides website, August 8th 2019.

Vinicombe, K, and Cottridge, D. (1996). Rare Birds in Britain and Ireland: a Photographic Record. Collins, London.

Vinicombe, K. (2007). The status of Red-headed Bunting in Britain. British Birds 100: 540-551.

Vincent Fleming, L. (2019) Common Buzzards, Red Kites and Ravens in the South Kesteven uplands: trends in occupied territories and relative abundance from 2003-2017. Lincolnshire Bird Report 2017: 177-182. LBC

White, S and C. Kehoe (2019). Report on scarce migrant birds in Britain in 2017. Part I: non-passerines. British Birds 112: 444-458.

White, S and C. Kehoe (2019). Report on scarce migrant birds in Britain in 2017. Part 2: passerines. British Birds 112: 639-660

White, S and C. Kehoe (2020). Report on scarce migrant birds in Britain in 2018. Part I: non-passerines. British Birds 113: 461-432.

White, S and C. Kehoe (2020). Report on scarce migrant birds in Britain in 2018. Part 2: passerines. British Birds 113: 533-554.

Woodward, I., Aebischer, N., Burnel,l D., Eaton, M., Frost, T., Hall, C., Stroud, S. and Noble, D. (2020). Avian Population Estimates Panel 4 - Population estimates of birds in Great Britain and the United Kingdom British Birds 113, 69-104

Wynn, R.B.and Pierre Yésou (2007). The changing status of Balearic Shearwaters in north-west European waters. British Birds 100: 392-406.

Yésou, P and P. Clergeau (2005). Sacred Ibis: a new invasive species in Europe. Birding World 18: 517-526.

Zuccon, D., Pons, J-M., Boano, G., Chiozzi, G., Gamauf, A., Mengoni, C., Nespoli, D., Olioso, G., Pavia, M., Pellegrino, I., Raković, M., Randi, E., Idrissi, H.R., Touihri, M., Unsold, M., Vitulano, S. and M. Brambilla (2020). Type specimens matter: new insights on the systematics, taxonomy and nomenclature of the subalpine warbler (Sylvia cantillans) complex. Zoo. J. Linn. Soc. 190: 314-341.

Index

Appendix: Checklist of Lincolnshire Birds

The checklist we have included on the following pages is an *aide memoire* for Lincolnshire birders using published information available on Nov 30th 2020. It is based on the full BOU list and birds known to be awaiting acceptance to both it and the Lincolnshire List. The Lincolnshire List stands at 402 species. We have not added each and every subspecies name to the commonly occurring forms, just the less common or rare subspecies where they have occurred in Lincolnshire. In all cases the binomial or trinomial scientific names follow the IOC taxonomy, version 10.2. IOC common names are included as a separate column where they differ from British common usage. The 'List' column combines the species on the Lincolnshire and British Lists. We have omitted those species which are part of a larger species groups which cannot be separated by field observation in most circumstances (e.g. Fea's/Desertas Petrels *Pterodroma feae/deserta*). The key to the 'List' column is as follows:

	On British list but not the Lincolnshire List
	On British and the Lincolnshire List
	Sub-species on the Lincolnshire List
BBRC	BBRC description species
LBRC	LBRC description species

There are several 'Pending' species for the British List: Zino's Petrel *Pterodroma madeira*, Brown Booby *Sula leucogaster*, Bearded Vulture *Gypaetus barbatus*, and one subspecies, Black-eared Kite *Milvus migrans lineatus/formosanus*. The last two would also be added to the Lincolnshire List if accepted by the BOU. There is also one 'Pending' species for the Lincolnshire List alone, Black-throated Thrush *Turdus atrogularis*. We hope that this will stimulate Lincolnshire birders to further their efforts to fill in some of the blanks in this list and offer a brief speculative review of where additions may be made in future.

Cackling Goose *Branta hutchinsii*, has been recorded in Lincolnshire but none has ever been submitted, probably because the automatic designation of 'escape' is always applied. There are a number of other rare wildfowl which could turn up given the increasing acreage of wetland in the county and of the unrecorded marine ducks Surf Scoter *Melanitta perspicillata*, must be top of the wanted list. Pied-billed Grebe *Podilymbus podiceps*, can turn up on any freshwater body and may easily be overlooked amongst large numbers of Little Grebes *Tachybaptus ruficollis*, seen at some of our wetland sites. The county is well overdue another Little Crake *Zapornia parva*, or a first Baillon's Crake *Z. pusilla* – will that be a calling male in spring or an autumn juvenile?

To date the county has been well-served by rare shorebirds with one or two gaps which might reasonably be expected to be filled in the near future – Killdeer *Charadrius vociferus*, Little Whimbrel *Numenius minuta*, Great Knot *Calidris tenuirostris*, Least Sandpiper *C. minuta* and Western Sandpiper *C. mauri*, have all turned up in adjacent coastal counties in recent memory.

It goes without saying that both Ivory Gull *Pagophila eburnea*, and Ross's Gulls *Rhodostethia rosea*, are probably at the top of most county birders wanted lists. Both have occurred in Yorkshire and Norfolk, but not yet in Lincolnshire. Grey Seals *Halichoerus grypus*, coming ashore to pup at Donna Nook provide a huge amount of afterbirth material attractive to gulls and skuas alike and it has long seemed a likely spot for Ivory Gull. The diminution in open landfill sites in the county would seem to offer less opportunity for those species frequently found foraging at such sites – the only British record of Slaty-backed Gull *Larus schistisagus*, was discovered on an Essex landfill site, for example. Gull roosts and open agricultural fields, especially where slurry is spread, are the alternatives and several species on the county list have been added from observation of such sites such as Bonaparte's Gull *Chroicocephalus philadelphia*, Franklin's Gull *Leucophaeus pipixcan* and Thayer's Gull *Larus glaucoides thayeri*. Seawatching offers the best chance of seeing rare petrels, terns and shearwaters and the breeding tern colonies in the county, especially those on or near the coast often attract their rarer cousins too.

Herons have obliged the county with their presence over the years and barring a miracle it doesn't seem likely that further additions will come our way in the near future, although what was the fate of the recent long-staying American Bittern B*otaurus lentiginosus*, in Suffolk? All British birders, let alone those in the county, eagerly await the verdict on the status of Bearded Vulture *Gypaetus barbatus*, in 2020 and a long-awaited taxonomic review of the eastern subspecies of Black Kite *Milvus migrans*, is still pending. Rare owls have given the county a wide berth since the historic records of Eurasian Scop's Owl *Otus scops* and Tengmalm's Owl *Aegolius funereus*, although the latter species afforded the county a near miss when it elected to roost at Spurn Point in 1983, so hope springs eternal.

Another Lesser Grey Shrike *Lanius minor*, would be welcomed by most county birders, as would a resolution of the longstanding record of the 1982 Anderby Creek 'Isabelline' Shrike *L. isabellinus*. The invasion of Nutcrackers *Nucifraga caryocatactes*, in 1968-69 with four Lincolnshire and 339 documented British records has never been repeated and the last accepted British record was 1998; chances of a repeat look remote. Rare larks are equally unpredictable as well as elusive. Crested Lark *Galerida cristatus*, rarely strays much further north than the southern counties and it seems an unlikely addition. A repeat of the 2011 Calandra Lark *Melanocorypha calandra*, is surely not out of the question?

Among the warblers, recent additions are of Hume's Warbler *Phylloscopus humei*, Western Bonelli's Warbler *P. bonelli*, and Iberian Chiffchaff *P. ibericus*. Both Two-barred Warbler *P. plumbeitarsus* and Eastern Crowned Warbler *P. coronatus* also look to be distinct possibilities with several recent British records of both, and Paddyfield Warbler *Acrocephalus agricola*, records have increased of late though they have a distinct northerly bias to them. It still seems incredible that there has not been a single record of Melodious Warbler *Hippolais polyglotta*, the commonest scarce migrant not on the county list. There are also several candidates in the genus *Curruca* (previously in the genus Sylvia) with no county record yet of the newly-minted Moltoni's Warbler *C. subalpina*, perhaps the most likely to fall.

Over the years the county has missed out on rare thrushes, with single records of American Robin *Turdus migratorius* and Black-throated Thrush *T. atrogularis* (BBRC pending). Recent good 'Siberian' autumns have brought plenty to other parts of Britain, mainly the northern areas, but no White's Thrush *Zoothera aurea*, or Siberian Thrush *Geokichla sibirica*, as yet. Chats and flycatchers from the same regions have also avoided the county in the main although the recent upsurge in Red-Flanked Bluetails *Tarsiger cyanurus* has been very welcome following the species eastwards spread and with 730 territorial males reported in Finland in 2020 this looks set to continue. The recent split of Stejneger's Stonechat *Saxicola stejnegeri* from Siberian Stonechat *S. maurus*, has seen an increased awareness of the likely identification features although to date DNA analysis is required to confirm identity; two candidates at Donna Nook in Oct 2016 were well-watched and photographed and may eventually confirm the first county records if descriptions prove adequate.

Wagtails and pipits provide fertile ground for improvement, with just one Citrine Wagtail *Motacilla citreola*, and as yet no Eastern Yellow Wagtail *M. tschutschensis*. The latter has recently come to the fore as an autumn vagrant in Britain, it's characteristic buzzy call and plumage characteristics have been well publicised, and it must be an early candidate for a county first in the very near future. Among the pipits, Olive-backed *Anthus hodgsoni*, has occurred in each of the last three years to 2019 but we seem unable to find Red-throated Pipit *A. cervinus*, of which there have been only six records, surely overlooked?

Parrot Crossbill *Loxia pytyopsittacus*, has been an occasional invader and Two-barred Crossbill *L. leucoptera*, only recently added. Among the buntings, the singing male Yellow-breasted Bunting *Emberiza aureola*, at Gibraltar Point in May 1977 sits alone as a never-to-be-repeated record given its steep decline across it's breeding and wintering range, sadly. Several of the other Asian buntings are possible candidates but, given the topography of the Lincolnshire coastline, may be difficult to find.

It is difficult to anticipate the next North American vagrant. Bobolink *Dolichonyx oryzivorus* at Crook Bank, Theddlethorpe on Oct 2019 was certainly in no-one's mind as a possible vagrant, so what next? Well, given the record of north American vagrants on the east coast, who knows? We are well overdue another American warbler or vireo, eyes peeled.

British Common Name	IOC common name	Scentific name	List	BOU	Notes
Capercaillie	Western Capercaillie	*Tetrax urogallus*		A	
Black Grouse		*Lyrurus tetrix*		AE	
Ptarmigan	Rock Ptarmigan	*Lagopus muta*		A	
Red Grouse	Willow Ptarmigan	*L. lagopus*		A	
Red-legged Partridge		*Alectoris rufa*		C1E*	
Grey Partridge		*Perdix perdix*		AC2E*	
Quail	Common Quail	*Coturnix coturnix*		AE*	
Pheasant	Common Pheasant	*Phasianus colchicus*		C1E*	
Golden Pheasant		*Chrysolophus pictus*		C1E*	
Lady Amherst's Pheasant		*C. amherstiae*		C6E*	
Brent Goose	Brant Goose	*Branta bernicla*		AE	
Brent Goose (Dark-breasted)		*B. b. bernicla*			
Brent Goose (Black Brant)		*B. b. nigricans*	LBRC		
Brent Goose (Pale-bellied)		*B. b. hrota*			
Red-breasted Goose		*B. ruficollis*	BBRC	AE*	
Canada Goose		*B. canadensis*		AC2E*	
Barnacle Goose		*B. leucopsis*		AC2E*	
Cackling Goose		*B. hutchinsii*	BBRC	AE	
Snow Goose		*Anser caerulescens*	LBRC	AC2E*	
Greylag Goose		*A. anser*		AC2C4E*	
Taiga Bean Goose		*A. fabalis*	LBRC	AE*	
Pink-footed Goose		*A. brachyrhynchus*		AE*	
Tundra Bean Goose		*A. serrirostris*		AE	
White-fronted Goose	Greater White-fronted Goose	*A. albifrons*		AE*	
White-fronted Goose (Greenland)		*A. a. flavirostris*	LBRC		
Lesser White-fronted Goose		*A. erythropus*	BBRC	AE*	
Mute Swan		*Cygnus olor*		AC2	
Bewick's Swan	Tundra Swan	*C. columbianus*		AE	
Whistling Swan		*C. c. columbianus*	BBRC	AE	
Whooper Swan		*C. cygnus*		AE*	
Egyptian Goose		*Alopochen aegyptiaca*		C1C5E*	
Shelduck	Common Shelduck	*Tadorna tadorna*		A	
Ruddy Shelduck		*T. ferruginea*	LBRC	BDE*	
Mandarin Duck		*Aix galericulata*		C1E*	
Baikal Teal		*Sibirionetta formosa*	BBRC	AE	
Garganey		*Spatula querquedula*		A	
Blue-winged Teal		*S. discors*	BBRC	AE*	
Shoveler	Northern Shoveler	*S. clypeata*		A	
Gadwall		*Mareca strepera*		AC2E*	
Falcated Duck		*M. falcata*	BBRC	AE	
Wigeon	Eurasian Wigeon	*M. penelope*		AE*	
American Wigeon		*M. americana*	LBRC	AE	
Mallard		*Anas platyrhynchos*		AC2C4E*	
Black Duck	American Black Duck	*A. rubripes*	BBRC	A	
Pintail	Northern Pintail	*A. acuta*		AE	
Teal	Eurasian Teal	*A. crecca*		A	
Green-winged Teal		*A. carolinensis*	LBRC	A	
Red-crested Pochard		*Netta rufina*		AC2E*	
Canvasback		*Aythya valisineria*	BBRC	AE	
Redhead		*A. americana*	BBRC	AE	
Pochard	Common Pochard	*A. ferina*		AE*	
Ferruginous Duck		*A. nyroca*	BBRC	AE	
Ring-necked Duck		*A. collaris*	LBRC	AE	
Tufted Duck		*A. fuligula*		A	
Scaup	Greater Scaup	*A. marila*		A	
Lesser Scaup		*A. affinis*	BBRC	A	
Steller's Eider		*Polysticta stelleri*	BBRC	A	
King Eider		*Somateria spectabilis*	BBRC	A	
Eider	Common Eider	*S. mollissima*		A	
Harlequin Duck		*Histrionicus histrionicus*	BBRC	A	
Surf Scoter		*Melanitta perspicillata*	LBRC	A	
Velvet Scoter		*M. fusca*		A	
White-winged Scoter		*M. deglandi*	BBRC	A	
Common Scoter		*M. nigra*		A	
Black Scoter		*M. americana*	BBRC	A	
Long-tailed Duck		*Clangula hyemalis*		A	

British Common Name	IOC common name	Scentific name	List	BOU	Notes
Bufflehead		*Bucephala albeola*	BBRC	AE	
Goldeneye	Common Goldeneye	*B. clangula*		AE*	
Barrow's Goldeneye		*B. islandica*	BBRC	AE	
Smew		*Mergellus albellus*		A	
Hooded Merganser		*Lophodytes cucullatus*	BBRC	AE	
Goosander	Common Merganser	*Mergus merganser*		A	
Red-breasted Merganser		*M. serrator*		A	
Ruddy Duck		*Oxyura jamaicensis*		C1E*	
Common Nighthawk		*Chordeiles minor*	BBRC	A	
Red-necked Nightjar		*Caprimulgus ruficollis*	BBRC	B	
Nightjar	European Nightjar	*C. europaeus*		A	
Egyptian Nightjar		*C. aegyptius*	BBRC	A	
Needle-tailed Swift	White-throated Needletail	*Hirundapus caudacutus*	BBRC	A	
Chimney Swift		*Chaetura pelagica*	BBRC	A	
Alpine Swift		*Tachymarptis melba*	BBRC	A	
Swift	Common Swift	*Apus apus*		A	
Pallid Swift		*A. pallidus*	BBRC	A	
Pacific Swift		*A. pacificus*	BBRC	A	
Little Swift		*A. affinis*	BBRC	A	
White-rumped Swift		*A. caffer*	BBRC	A	
Great Bustard		*Otis tarda*	BBRC	AE*	
Macqueen's Bustard		*Chlamydotis macqueenii*	BBRC	A	
Little Bustard		*Tetrax tetrax*	BBRC	A	
Great Spotted Cuckoo		*Clamator glandarius*	BBRC	A	
Yellow-billed Cuckoo		*Coccyzus americanus*	BBRC	A	
Black-billed Cuckoo		*C. erythropthalmus*	BBRC	A	
Cuckoo	Common Cuckoo	*Cuculus canorus*		A	
Pallas's Sandgrouse		*Syrrhaptes paradoxus*	BBRC	A	
Rock Dove		*Columba livia*		AC4E*	
Rock Dove (Feral)		*C. l. 'feral'*			
Stock Dove		*C. oenas*		A	
Woodpigeon	Common Wood Pigeon	*C. palumbus*		A	
Turtle Dove	European Turtle Dove	*Streptopelia turtur*		A	
Rufous Turtle Dove	Oriental Turtle Dove	*S. orientalis*	BBRC	A	
Collared Dove	Eurasian Collared Dove	*S. decaocto*		A	
Mourning Dove		*Zenaida macroura*	BBRC	A	
Water Rail		*Rallus aquaticus*		A	
Corncrake	Corn Crake	*Crex crex*	LBRC	AE*	
Sora Rail	Sora	*Porzana carolina*	BBRC	A	
Spotted Crake		*P. porzana*	LBRC	A	
Moorhen	Common Moorhen	*Gallinula chloropus*		A	
Coot	Eurasian Coot	*Fulica atra*		A	
American Coot		*F. americana*	BBRC	A	
Allen's Gallinule		*Porphyrio alleni*	BBRC	A	
Purple Gallinule		*P. martinica*	BBRC	A	
Western Swamphen		*P. porphyrio*	BBRC	AE	
Baillon's Crake		*Zapornia pusilla*	BBRC	A	
Little Crake		*Z. parva*	BBRC	A	
Sandhill Crane		*Antigone canadensis*	BBRC	A	
Crane	Common Crane	*Grus grus*		AE*	
Little Grebe		*Tachybaptus ruficollis*		A	
Pied-billed Grebe		*Podilymbus podiceps*	BBRC	A	
Red-necked Grebe		*Podiceps grisegena*		A	
Great Crested Grebe		*P. cristatus*		A	
Slavonian Grebe	Horned Grebe	*P. auritus*		A	
Black-necked Grebe		*P. nigricollis*		A	
Stone-curlew	Eurasian Stone-curlew	*Burhinus oedicnemus*		A	
Oystercatcher	Eurasian Oystercatcher	*Haematopus ostralegus*		A	
Black-winged Stilt		*Himantopus himantopus*	LBRC	A	
Avocet	Pied Avocet	*Recurvirostra avosetta*		AE	
Lapwing	Northern Lapwing	*Vanellus vanellus*		A	
Sociable Plover	Sociable Lapwing	*V. gregarius*	BBRC	A	
White-tailed Plover	White-tailed Lapwing	*V. leucurus*	BBRC	A	
Golden Plover	European Golden Plover	*Pluvialis apricaria*		A	
Pacific Golden Plover		*P. fulva*	BBRC	A	
American Golden Plover		*P. dominica*	LBRC	A	

British Common Name	IOC common name	Scentific name	List	BOU	Notes
Grey Plover		Pluvialis squatarola		A	
Ringed Plover	Common Ringed Plover	Charadrius hiaticula		A	
Semipalmated Plover		C. semipalmatus	BBRC	A	
Little Ringed Plover		C. dubius		A	
Killdeer		C. vociferus	BBRC	A	
Kentish Plover		C. alexandrinus	BBRC	A	
Lesser Sand Plover		C. mongolus	BBRC	A	
Greater Sand Plover		C. leschenaultii	BBRC	A	
Caspian Plover		C. asiaticus	BBRC	A	
Dotterel	Eurasian Dotterel	C. morinellus		A	
Upland Sandpiper		Bartramia longicauda	BBRC	A	
Whimbrel		Numenius phaeopus		A	
Hudsonian Whimbrel		N. hudsonicus	BBRC	A	
Little Whimbrel	Little Curlew	N. minutus	BBRC	A	
Eskimo Curlew		N. borealis	BBRC	B	
Curlew		N. arquata		A	
Bar-tailed Godwit		Limosa lapponica		A	
Black-tailed Godwit		L. limosa		A	
Black-tailed Godwit (European)		L. l. limosa	LBRC		
Hudsonian Godwit		L. haemastica	BBRC	A	
Turnstone	Ruddy Turnstone	Arenaria interpres		A	
Great Knot		Calidris tenuirostris	BBRC	A	
Knot	Red Knot	C. canutus		A	
Ruff		C. pugnax		A	
Broad-billed Sandpiper		C. falcinellus	BBRC	A	
Sharp-tailed Sandpiper		C. acuminata	BBRC	A	
Stilt Sandpiper		C. himantopus	BBRC	A	
Curlew Sandpiper		C. ferruginea		A	
Temminck's Stint		C. temminckii	LBRC	A	
Long-toed Stint		C. subminuta	BBRC	A	
Red-necked Stint		C. ruficollis	BBRC	A	
Sanderling		C. alba		A	
Dunlin		C. alpina		A	
Purple Sandpiper		C. maritima		A	
Baird's Sandpiper		C. bairdii	BBRC	A	
Little Stint		C. minuta		A	
Least Sandpiper		C. minutilla	BBRC	A	
White-rumped Sandpiper		C. fuscicollis	LBRC	A	
Buff-breasted Sandpiper		C. subruficollis	LBRC	A	
Pectoral Sandpiper		C. melanotos	LBRC	A	
Semipalmated Sandpiper		C. pusilla	BBRC	A	
Western Sandpiper		C. mauri	BBRC	A	
Long-billed Dowitcher		Limnodromus scolopaceus	BBRC	A	
Short-billed Dowitcher		L. griseus	BBRC	A	
Woodcock	Eurasian Woodcock	Scolopax rusticola		A	
Jack Snipe		Lymnocryptes minimus		A	
Great Snipe		G. media	BBRC	A	
Snipe	Common Snipe	G. gallinago		A	
Wilson's Snipe		G. delicata	BBRC	A	
Terek Sandpiper		Xenus cinereus	BBRC	A	
Wilson's Phalarope		Phalaropus tricolor	BBRC	A	
Red-necked Phalarope		P. lobatus	LBRC	A	
Grey Phalarope	Red Phalarope	P. fulicarius	LBRC	A	
Common Sandpiper		Actitis hypoleucos		A	
Spotted Sandpiper		A. macularius	BBRC	A	
Green Sandpiper		Tringa ochropus		A	
Solitary Sandpiper		T. solitaria	BBRC	A	
Grey-tailed Tattler		T. brevipes	BBRC	A	
Lesser Yellowlegs		T. flavipes	LBRC	A	
Redshank	Common Redshank	T. totanus		A	
Marsh Sandpiper		T. stagnatilis	BBRC	A	
Wood Sandpiper		T. glareola		A	
Spotted Redshank		T. erythropus		A	
Greenshank	Common Greenshank	T. nebularia		A	
Greater Yellowlegs		T. melanoleuca	BBRC	A	
Cream-coloured Courser		Cursorius cursor	BBRC	A	

British Common Name	IOC common name	Scientific name	List	BOU	Notes
Collared Pratincole		*Glareola pratincola*	BBRC	A	
Oriental Pratincole		*G. maldivarum*	BBRC	A	
Black-winged Pratincole		*G. nordmanni*	BBRC	A	
Kittiwake	Black-legged Kittiwake	*Rissa tridactyla*		A	
Ivory Gull		*Pagophila eburnea*	BBRC	A	
Sabine's Gull		*Xema sabini*	LBRC	A	
Slender-billed Gull		*Chroicocephalus genei*	BBRC	A	
Bonaparte's Gull		*C. philadelphia*	BBRC	A	
Black-headed Gull		*C. ridibundus*		A	
Little Gull		*Hydrocoloeus minutus*		A	
Ross's Gull		*Rhodostethia rosea*	BBRC	A	
Laughing Gull		*Leucophaeus atricilla*	BBRC	A	
Franklin's Gull		*L. pipixcan*	BBRC	A	
Audouin's Gull		*Ichthyaetus audouinii*	BBRC	A	
Mediterranean Gull		*I. melanocephalus*		A	
Great Black-headed Gull	Pallas's Gull	*I. ichthyaetus*	BBRC	B	
Common Gull	Mew Gull	*Larus canus*		A	
Ring-billed Gull		*L. delawarensis*	LBRC	A	
Great Black-backed Gull		*L. marinus*		A	
Glaucous-winged Gull		*L. glaucescens*	BBRC	A	
Glaucous Gull		*L. hyperboreus*		A	
Iceland Gull		*L. glaucoides*		A	
Iceland Gull (Kumlien's)		*L. g. kumlieni*	LBRC		
Iceland Gull (Thayer's)		*L. g. thayeri*	BBRC		
Herring Gull	European Herring Gull	*L. argentatus*		A	
American Herring Gull		*L. smithsonianus*	BBRC	A	
Caspian Gull		*L. cachinnans*	LBRC	A	
Yellow-legged Gull		*L. michahellis*		A	
Yellow-legged Gull (Azorean)		*L. m. atlantis*	BBRC		
Slaty-backed Gull		*L. schistisagus*	BBRC	A	
Lesser Black-backed Gull		*L. fuscus*		A	
Lesser Black-backed Gull (Baltic)		*L. f. fuscus*	BBRC		
Gull-billed Tern		*Gelochelidon nilotica*	BBRC	A	
Caspian Tern		*Hydroprogne caspia*	BBRC	A	
Royal Tern		*Thalasseus maximus*	BBRC	A	
Lesser Crested Tern		*T. bengalensis*	BBRC	A	
Sandwich Tern		*T. sandvicensis*		A	
Cabot's Tern		*T. acuflavidus*	BBRC	A	
Elegant Tern		*T. elegans*	BBRC	A	
Little Tern		*Sternula albifrons*		A	
Least Tern		*S. antillarum*	BBRC	A	
Aleutian Tern		*Onychoprion aleuticus*	BBRC	A	
Bridled Tern		*O. anaethetus*	BBRC	A	
Sooty Tern		*O. fuscatus*	BBRC	A	
Roseate Tern		*Sterna dougallii*	LBRC	A	
Common Tern		*S. hirundo*		A	
Arctic Tern		*S. paradisaea*		A	
Forster's Tern		*S. forsteri*	BBRC	A	
Whiskered Tern		*Chlidonias hybrida*	BBRC	A	
White-winged Black Tern	White-winged Tern	*C. leucopterus*	LBRC	A	
Black Tern		*C. niger*		A	
Black Tern (American)		*C. n. surinamensis*	BBRC		
Great Skua		*Stercorarius skua*		A	
Pomarine Skua		*S. pomarinus*		A	
Arctic Skua		*S. parasiticus*		A	
Long-tailed Skua		*S. longicaudus*	LBRC	A	
Little Auk		*Alle alle*		A	
Brünnich's Guillemot	Thick-billed Murre	*Uria lomvia*	BBRC	A	
Guillemot	Common Murre	*U. aalge*		A	
Razorbill		*Alca torda*		A	
Great Auk		*Pinguinis impennis*	Extinct	B	
Black Guillemot		*Cepphus grylle*	LBRC	A	
Mandt's Black Guillemot		*C. g. mandtii*	BBRC		
Long-billed Murrelet		*Brachyramphus perdix*	BBRC	A	
Ancient Murrelet		*Synthliboramphus antiquus*	BBRC	A	
Puffin	Atlantic Puffin	*Fratercula arctica*		A	

British Common Name	IOC common name	Scentific name	List	BOU	Notes
Tufted Puffin		Fratercula cirrhata	BBRC	A	
Red-billed Tropicbird		Phaethon aethereus	BBRC	AE	
Red-throated Diver	Red-throated Loon	Gavia stellata		A	
Black-throated Diver	Black-throated Loon	G. arctica		A	
Pacific Diver	Pacific Loon	G. pacifica	BBRC	A	
Great Northern Diver	Common Loon	G. immer		A	
White-billed Diver	Yellow-billed Loon	G. adamsii	LBRC	A	
Wilson's Petrel	Wilson's Storm Petrel	Oceanites oceanicus	LBRC	A	
White-faced Storm Petrel		Pelagodroma marina	BBRC	A	
Black-browed Albatross		Thalassarche melanophris	BBRC	A	
Yellow-nosed Albatross	Atlantic Yellow-nosed Albatross	T. chlororhynchos	BBRC	A	
Storm Petrel	European Storm Petrel	Hydrobates pelagicus	LBRC	A	
Swinhoe's Petrel	Swinhoe's Storm Petrel	Oceanodroma monorhis	BBRC	A	
Leach's Petrel	Leach's Storm Petrel	O. leucorhoa	LBRC	A	
Fulmar	Northern Fulmar	Fulmarus glacialis		A	
Zino's Petrel		Pterodroma madeira		P	Pending (British List)
Capped Petrel	Black-capped Petrel	P. hasitata	BBRC	A	
Scopoli's Shearwater		Calonectris diomedea	BBRC	A	
Cory's Shearwater		C. borealis	LBRC	A	
Sooty Shearwater		Ardenna grisea		A	
Great Shearwater		A. gravis	LBRC	A	
Manx Shearwater		Puffinus puffinus		A	
Yelkouan Shearwater		P. yelkouan	BBRC	A	
Balearic Shearwater		P. mauretanicus	LBRC	A	
Macaronesian Shearwater	Barolo Shearwater	P. baroli	BBRC	A	
Black Stork		Ciconia nigra	BBRC	AE	
White Stork		C. ciconia	LBRC	AE	
Ascension Frigatebird		Fregata aquila	BBRC	A	
Magnificent Frigatebird		F. magnificens	BBRC	A	
Gannet	Northern Gannet	Morus bassanus		A	
Red-footed Booby		Sula sula	BBRC	A	
Brown Booby		S. leucogaster		P	Pending (British List)
Cormorant	Great Cormorant	Phalacrocorax carbo		A	
Shag	European Shag	P. aristotelis		A	
Double-crested Cormorant		P. auritus	BBRC	AE	
Glossy Ibis		Plegadis falcinellus	LBRC	AE	
Spoonbill	Eurasian Spoonbill	Platalea leucorodia		AE	
Bittern	Eurasian Bittern	Botaurus stellaris		A	
American Bittern		B. lentiginosus	BBRC	A	
Little Bittern		Ixobrychus minutus	BBRC	A	
Night-heron	Black-crowned Night Heron	Nycticorax nycticorax	LBRC	AE*	
Green Heron		Butorides virescens	BBRC	A	
Squacco Heron		Ardeola ralloides	BBRC	A	
Chinese Pond Heron		A. bacchus	BBRC	A	
Cattle Egret	Western Cattle Egret	Bubulcus ibis	LBRC	AE	
Grey Heron		Ardea cinerea		A	
Great Blue Heron		A. herodias	BBRC	A	
Purple Heron		A. purpurea	LBRC	A	
Great White Egret	Great Egret	A. alba		A	
Snowy Egret		Egretta thula	BBRC	A	
Little Egret		E. garzetta		A	
Dalmatian Pelican		Pelecanus crispus	BBRC	A	
Osprey	Western Osprey	Pandion haliaetus		AE*	
Bearded Vulture		Gypaetus barbaetus		P	Pending (British list)
Egyptian Vulture		Neophron percnopterus	BBRC	BDE	
Honey-buzzard	European Honey Buzzard	Pernis apivorus	LBRC	A	
Short-toed Eagle	Short-toed Snake Eagle	Circaetus gallicus	BBRC	A	
Spotted Eagle	Greater Spotted Eagle	C. clanga	BBRC	B	
Golden Eagle		Aquila chrysaetos	LBRC	AE	
Sparrowhawk	Eurasian Sparrowhawk	Accipiter nisus		A	
Goshawk	Northern Goshawk	A. gentilis	LBRC	AC3E*	
Marsh Harrier	Western Marsh Harrier	Circus aeruginosus		A	
Hen Harrier		C. cyaneus		A	
Northern Harrier		C. hudsonius	BBRC	A	
Pallid Harrier		C. macrourus	BBRC	A	
Montagu's Harrier		C. pygargus		A	

British Common Name	IOC common name	Scentific name	List	BOU	Notes
Red Kite		*Milvus milvus*		AC3E*	
Black Kite		*M. migrans*	LBRC	AE	
Black Kite (Black-eared)		*M. m. lineatus/formosanus*		P	Pending (British list)
White-tailed Eagle		*Haliaeetus albicilla*	LBRC	AC3E*	
Rough-legged Buzzard		*Buteo lagopus*	LBRC	AE	
Buzzard	Common Buzzard	*B. buteo*		AE*	
Barn Owl	Western Barn Owl	*Tyto alba*		AE*	
Barn Owl (Dark-breasted)		*T. a. guttata*	BBRC		
Scops Owl	Eurasian Scops Owl	*Otus scops*	BBRC	A	
Snowy Owl		*Bubo scandiacus*	BBRC	AE	
Tawny Owl		*Strix aluco*		A	
Hawk Owl	Northern Hawk Owl	*Surnia ulula*	BBRC	A	
Little Owl		*Athene noctua*		C1E*	
Tengmalm's Owl	Boreal Owl	*Aegolius funereus*	BBRC	A	
Long-eared Owl		*Asio otus*		A	
Short-eared Owl		*A. flammeus*		A	
Hoopoe	Eurasian Hoopoe	*Upupa epops*	LBRC	AE	
Roller	European Roller	*Coracias garrulus*	BBRC	A	
Kingfisher	Common Kingfisher	*Alcedo atthis*		A	
Belted Kingfisher		*Megaceryle alcyon*	BBRC	A	
Blue-cheeked Bee-eater		*Merops persicus*	BBRC	A	
Bee-eater	European Bee-eater	*M. apiaster*	LBRC	A	
Wryneck	Eurasian Wryneck	*Jynx torquilla*		A	
Yellow-bellied Sapsucker		*Sphyrapicus varius*	BBRC	A	
Lesser Spotted Woodpecker		*Dryobates minor*		A	
Great Spotted Woodpecker		*Dendrocopos major*		A	
Green Woodpecker	European Green Woodpecker	*Picus viridis*		A	
Lesser Kestrel		*Falco naumanni*	BBRC	A	
Kestrel	Common Kestrel	*F. tinnunculus*		A	
American Kestrel		*F. sparverius*	BBRC	AE	
Red-footed Falcon		*F. vespertinus*	LBRC	A	
Amur Falcon		*F. amurensis*	BBRC	AE	
Eleonora's Falcon		*F. eleonorae*	BBRC	A	
Merlin		*F. columbarius*		A	
Hobby	Eurasian Hobby	*F. subbuteo*		A	
Gyr Falcon	Gyrfalcon	*F. rusticolus*	BBRC	AE	
Peregrine	Peregrine Falcon	*F. peregrinus*		AE	
Ring-necked Parakeet	Rose-ringed Parakeet	*Psittacula krameri*		C1E*	
Eastern Phoebe		*Sayornis phoebe*	BBRC	A	
Acadian Flycatcher		*Empidonax virescens*	BBRC	A	
Alder Flycatcher		*E. alnorum*	BBRC	A	
Eastern Kingbird		*Tyrannus tyrannus*	BBRC	A	
Brown Shrike		*Lanius cristatus*	BBRC	A	
Red-backed Shrike		*L. collurio*		A	
Isabelline Shrike		*L. isabellinus*	BBRC	A	
Isabelline Shrike (Daurian)		*L. i. isabellinus*	BBRC		
Red-tailed Shrike		*L. phoenicuroides*	BBRC	A	
Long-tailed Shrike		*L. schach*	BBRC	A	
Lesser Grey Shrike		*L. minor*	BBRC	A	
Great Grey Shrike		*L. excubitor*		AE	
Great Grey Shrike (Steppe)		*L. e. pallidirostris*	BBRC		
Woodchat Shrike		*L. senator*	LBRC	A	
Masked Shrike		*L. nubicus*	BBRC	A	
Yellow-throated Vireo		*Vireo flavifrons*	BBRC	A	
Philadelphia Vireo		*V. philadelphicus*	BBRC	A	
Red-eyed Vireo		*V. olivaceus*	BBRC	A	
Golden Oriole	Eurasian Golden Oriole	*Oriolus oriolus*	LBRC	A	
Jay	Eurasian Jay	*Garrulus glandarius*		A	
Magpie	Eurasian Magpie	*Pica pica*		A	
Nutcracker	Spotted Nutcracker	*Nucifraga caryocatactes*	BBRC	A	
Chough	Red-billed Chough	*Pyrrhocorax pyrrhocorax*	LBRC	AE*	
Jackdaw	Western Jackdaw	*Coloeus monedula*		A	
Jackdaw (Nordic)		*C. m. monedula*	LBRC		
Rook		*Corvus frugilegus*		A	
Carrion Crow		*C. corone*		A	
Hooded Crow		*C. cornix*		A	

British Common Name	IOC common name	Scentific name	List	BOU	Notes
Raven	Northern Raven	Corvus corax		A	
Waxwing	Bohemian Waxwing	Bombycilla garrulus		AE	
Cedar Waxwing		B. cedrorum	BBRC	AE	
Coal Tit		Periparus ater		A	
Coal Tit (Continental)		P. a. [ater-group]	LBRC		
Crested Tit	European Crested Tit	Lophophanes cristatus	LBRC	A	
Marsh Tit		Poecile palustris		A	
Willow Tit		P. montanus		A	
Blue Tit	Eurasian Blue Tit	Cyanistes caeruleus		A	
Great Tit		Parus major		A	
Penduline Tit	Eurasian Penduline Tit	Remiz pendulinus	LBRC	A	
Bearded Tit	Bearded Reedling	Panurus biarmicus		A	
Woodlark		Lullula arborea		A	
White-winged Lark		Alauda leucoptera	BBRC	A	
Skylark	Eurasian Skylark	A. arvensis		A	
Crested Lark		Galerida cristata	BBRC	AE	
Shore Lark	Horned Lark	Eremophila alpestris		A	
Short-toed Lark	Greater Short-toed Lark	Calandrella brachydactyla	LBRC	A	
Bimaculated Lark		Melanocorypha bimaculata	BBRC	A	
Calandra Lark		M. calandra	BBRC	A	
Black Lark		M. yeltoniensis	BBRC	A	
Lesser Short-toed Lark		Alaudala rufescens	BBRC	A	
Sand Martin		Riparia riparia		A	
Tree Swallow		Tachycineta bicolor	BBRC	A	
Purple Martin		Progne subis	BBRC	A	
Swallow	Barn Swallow	Hirundo rustica		AE	
Crag Martin	Eurasian Crag Martin	Ptyonoprogne rupestris	BBRC	A	
House Martin	Common House Martin	Delichon urbicum		A	
Red-rumped Swallow		Cecropis daurica	LBRC	A	
Cliff Swallow	American Cliff Swallow	Petrochelidon pyrrhonota	BBRC	A	
Cetti's Warbler		Cettia cetti		A	
Long-tailed Tit		Aegithalos caudatus		A	
Wood Warbler		Phylloscopus sibilatrix	LBRC	A	
Western Bonelli's Warbler		P. bonelli	BBRC	A	
Eastern Bonelli's Warbler		P. orientalis	BBRC	A	
Hume's Warbler	Hume's Leaf Warbler	P. humei	BBRC	A	
Yellow-browed Warbler		P. inornatus		A	
Pallas's Warbler	Pallas's Leaf Warbler	P. proregulus	LBRC	A	
Radde's Warbler		P. schwarzi	LBRC	A	
Dusky Warbler		P. fuscatus	LBRC	A	
Willow Warbler		P. trochilus		A	
Chiffchaff	Common Chiffchaff	P. collybita		A	
Chiffchaff (Siberian)		P. c. tristis	LBRC		
Iberian Chiffchaff		P. ibericus	BBRC	A	
Eastern Crowned Warbler		P. coronatus	BBRC	A	
Green Warbler		P. nitidus	BBRC	A	
Two-barred Warbler		P. plumbeitarsus	BBRC	A	
Greenish Warbler		P. trochiloides	LBRC	A	
Pale-legged Leaf Warbler		P. tenellipes	BBRC	A	
Arctic Warbler		P. borealis	LBRC	A	
Great Reed Warbler		Acrocephalus arundinaceus	BBRC	A	
Aquatic Warbler		A. paludicola	BBRC	A	
Sedge Warbler		A. schoenobaenus		A	
Paddyfield Warbler		A. agricola	BBRC	A	
Blyth's Reed Warbler		A. dumetorum	LBRC	A	
Reed Warbler	Eurasian Reed Warbler	A. scirpaceus		A	
Marsh Warbler		A. palustris	LBRC	A	
Thick-billed Warbler		Arundinax aedon	BBRC	A	
Booted Warbler		Iduna caligata	BBRC	A	
Sykes's Warbler		I. rama	BBRC	A	
Eastern Olivaceous Warbler		I. pallida	BBRC	A	
Olive-tree Warbler		Hippolais olivetorum	BBRC	A	
Melodious Warbler		H. polyglotta	LBRC	A	
Icterine Warbler		H. icterina	LBRC	A	
Pallas's Grasshopper Warbler		Helopsaltes certhiola	BBRC	A	
Lanceolated Warbler		Locustella lanceolata	BBRC	A	

British Common Name	IOC common name	Scentific name	List	BOU	Notes
River Warbler		*Locustella fluviatilis*	BBRC	A	
Savi's Warbler		*L. luscinioides*	LBRC	A	
Grasshopper Warbler		*L. naevia*		A	
Zitting Cisticola		*Cisticola juncidis*	BBRC	A	
Blackcap		*Sylvia atricapilla*		A	
Garden Warbler		*S. borin*		A	
Barred Warbler		*Curruca nisoria*		A	
Lesser Whitethroat		*C. curruca*		A	
Western Orphean Warbler		*C. hortensis*	BBRC	A	
Eastern Orphean Warbler		*C. crassirostris*	BBRC	A	
Asian Desert Warbler		*C. nana*	BBRC	A	
Rüppell's Warbler		*C. ruppeli*	BBRC	A	
Sardinian Warbler		*C. melanocephala*	BBRC	A	
Western Subalpine Warbler		*C. iberiae*	BBRC	A	
Moltoni's Subalpine Warbler	Moltoni's Warbler	*C. subalpina*	BBRC	A	
Eastern Subalpine Warbler		*C. cantillans*	BBRC	A	
Eastern Subalpine Warbler (albistriata)		*C. c. albistriata*	BBRC		
Whitethroat	Common Whitethroat	*C. communis*		A	
Spectacled Warbler		*C. conspicillata*	BBRC	A	
Marmora's Warbler		*C. sarda*	BBRC	A	
Dartford Warbler		*C. undata*	LBRC	A	
Firecrest	Common Firecrest	*Regulus ignicapilla*		A	
Goldcrest		*R. regulus*		A	
Wren	Eurasian Wren	*Troglodytes troglodytes*		A	
Red-breasted Nuthatch		*Sitta canadensis*	BBRC	A	
Nuthatch	Eurasian Nuthatch	*S. europaea*		A	
Wallcreeper		*Tichodroma muraria*	BBRC	A	
Treecreeper	Eurasian Treecreeper	*Certhia familiaris*		A	
Treecreeper [familiaris]		*C. f. familiaris*	BBRC		
Short-toed Treecreeper		*C. brachydactyla*	BBRC	A	
Grey Catbird		*Dumetella carolinensis*	BBRC	A	
Northern Mockingbird		*Mimus polyglottos*	BBRC	AE	
Brown Thrasher		*Toxostoma rufum*	BBRC	A	
Rose-coloured Starling	Rosy Starling	*Pastor roseus*	LBRC	AE	
Starling	Common Starling	*Sturnus vulgaris*		A	
Siberian Thrush		*Geokichla sibirica*	BBRC	AE	
White's Thrush		*Zoothera aurea*	BBRC	A	
Varied Thrush		*Ixoreus naevius*	BBRC	A	
Veery		*Catharus fuscescens*	BBRC	A	
Grey-cheeked Thrush		*C. minimus*	BBRC	A	
Swainson's Thrush		*C. ustulatus*	BBRC	A	
Hermit Thrush		*C. guttatus*	BBRC	A	
Wood Thrush		*Hylocichla mustelina*	BBRC	A	
Ring Ouzel		*Turdus torquatus*		A	
Blackbird	Common Blackbird	*T. merula*		A	
Eyebrowed Thrush		*T. obscurus*	BBRC	A	
Black-throated Thrush		*T. atrogularis*	BBRC	A	Pending (Lincs. List)
Red-throated Thrush		*T. ruficollis*	BBRC	A	
Naumann's Thrush		*T. naumanni*	BBRC	A	
Dusky Thrush		*T. eunomus*	BBRC	A	
Fieldfare		*T. pilaris*		A	
Redwing		*T. iliacus*		A	
Song Thrush		*T. philomelos*		A	
Mistle Thrush		*T. viscivorus*		A	
American Robin		*T. migratorius*	BBRC	AE	
Rufous-tailed Scrub Robin		*Cercotrichas galactotes*	BBRC	A	
Spotted Flycatcher		*Muscicapa striata*		A	
Brown Flycatcher	Asian Brown Flycatcher	*M. dauurica*	BBRC	A	
Robin	European Robin	*Erithacus rubecula*		A	
Siberian Blue Robin		*Larvivora cyane*	BBRC	A	
Rufous-tailed Robin		*L. sibilans*	BBRC	A	
Bluethroat (Red-spotted)	Bluethroat	*Luscinia svecica*	LBRC	A	
Bluethroat (Red-spotted)		*L. s. [svecica-group]*	LBRC		
Bluethroat (White-spotted)		*L. s. cyanecula/namnetum*	LBRC		
Thrush Nightingale		*L. luscinia*	BBRC	A	
Nightingale	Common Nightingale	*L. megarhynchos*		A	

British Common Name	IOC common name	Scentific name	List	BOU	Notes
White-throated Robin		*Irania gutturalis*	BBRC	A	
Siberian Rubythroat		*Calliope calliope*	BBRC	A	
Red-flanked Bluetail		*Tarsiger cyanurus*	LBRC	AE	
Taiga Flycatcher		*Ficedula albicilla*	BBRC	A	
Red-breasted Flycatcher		*F. parva*	LBRC	A	
Pied Flycatcher	European Pied Flycatcher	*F. hypoleuca*		A	
Collared Flycatcher		*F. albicollis*	BBRC	A	
Black Redstart		*Phoenicurus ochruros*		A	
Black Redstart (Eastern)		*P. o. [ochruros-group]*	BBRC		
Redstart	Common Redstart	*P. phoenicurus*		A	
Moussier's Redstart		*P. moussieri*	BBRC	A	
Rock Thrush	Common Rock Thrush	*Monticola saxatilis*	BBRC	A	
Blue Rock Thrush		*M. solitarius*	BBRC	AE	
Whinchat		*Saxicola rubetra*		A	
Stonechat	European Stonechat	*S. rubicola*		A	
Siberian Stonechat		*S. maurus*	BBRC	A	
Stejneger's Stonechat		*S. stejnegeri*	BBRC	A	
Wheatear	Northern Wheatear	*Oenanthe oenanthe*		A	
Isabelline Wheatear		*O. isabellina*	BBRC	A	
Desert Wheatear		*O. deserti*	BBRC	A	
Western Black-eared Wheatear		*O. hispanica*	BBRC	A	
Eastern Black-eared Wheatear		*O. melanoleuca*	BBRC	A	
Pied Wheatear		*O. pleschanka*	BBRC	A	
Black Wheatear		*O. leucura*	BBRC	A	
White-crowned Black Wheatear	White-crowned Wheatear	*O. leucopyga*	BBRC	A	
Dipper	White-throated Dipper	*Cinclus cinclus*		A	
Dipper (Black-bellied)		*C. c. cinclus*	BBRC		
House Sparrow		*Passer domesticus*		A	
Spanish Sparrow		*P. hispaniolensis*	BBRC	A	
Tree Sparrow	Eurasian Tree Sparrow	*P. montanus*		A	
Rock Sparrow		*Petronia petronia*	BBRC	A	
Alpine Accentor		*Prunella collaris*	BBRC	A	
Siberian Accentor		*P. montanella*	BBRC	A	
Dunnock		*P. modularis*		A	
Western Yellow Wagtail		*Motacilla flava*		A	
Western Yellow Wagtail (Blue-headed)		*M. f. flava*	LBRC		
Western Yellow Wagtail (Ashy-headed)		*M. f. [cinereocapilla-group]*	BBRC		
Western Yellow Wagtail (Black-headed)		*M. f. feldegg*	BBRC		
Western Yellow Wagtail (Grey-headed)		*M. f. thunbergi*	LBRC		
Eastern Yellow Wagtail		*M. tschutschensis*	BBRC	A	
Citrine Wagtail		*M. citreola*	LBRC	A	
Grey Wagtail		*M. cinerea*		A	
Pied Wagtail	White Wagtail	*M. alba*		A	
Pied Wagtail		*M. a. yarrellii*			
White Wagtail		*M. a. alba*			
Richard's Pipit		*Anthus richardi*	LBRC	A	
Blyth's Pipit		*A. godlewskii*	BBRC	A	
Tawny Pipit		*A. campestris*	BBRC	A	
Meadow Pipit		*A. pratensis*		A	
Tree Pipit		*A. trivialis*		A	
Olive-backed Pipit		*A. hodgsoni*	LBRC	A	
Pechora Pipit		*A. gustavi*	BBRC	A	
Red-throated Pipit		*A. cervinus*	LBRC	A	
Buff-bellied Pipit		*A. rubescens*	BBRC	A	
Water Pipit		*A. spinoletta*		A	
Rock Pipit	Eurasian Rock Pipit	*A. petrosus*		A	
Chaffinch	Common Chaffinch	*Fringilla coelebs*		AE	
Brambling		*F. montifringilla*		A	
Evening Grosbeak		*Hesperiphona vespertina*	BBRC	A	
Hawfinch		*Coccothraustes coccothraustes*	LBRC	A	
Pine Grosbeak		*Pinicola enucleator*	BBRC	AE	
Bullfinch	Eurasian Bullfinch	*Pyrrhula pyrrhula*		A	
Bullfinch (Northern)		*P. p. pyrrhula*	LBRC		
Trumpeter Finch		*Bucanetes githagineus*	BBRC	AE	
Common Rosefinch		*Carpodacus erythrinus*	LBRC	A	
Greenfinch	European Greenfinch	*Chloris chloris*		AE	

British Common Name	IOC common name	Scentific name	List	BOU	Notes
Twite		Linaria flavirostris		A	
Linnet	Common Linnet	L. cannabina		A	
Common Redpoll		Acanthis flammea	LBRC	A	
Lesser Redpoll		A. cabaret		A	
Arctic Redpoll		A. hornemanni	BBRC	A	
Arctic Redpoll (Coues's)		A. hornemanni exilipes	BBRC		
Parrot Crossbill		Loxia pytyopsittacus	LBRC	A	
Scottish Crossbill		L. scottica		A	
Red Crossbill		L. curvirostra		A	
Two-barred Crossbill		L. leucoptera	BBRC	A	
Goldfinch	European Goldfinch	Carduelis carduelis		A	
Citril Finch		C. citrinella	BBRC	A	
Serin		Serinus serinus	LBRC	A	
Siskin	Eurasian Siskin	Spinus spinus		A	
Lapland Bunting	Lapland Longspur	Calcarius lapponicus		A	
Snow Bunting		Plectrophenax nivalis		A	
Corn Bunting		Emberiza calandra		A	
Yellowhammer		E. citrinella		A	
Pine Bunting		E. leucocephalos	BBRC	A	
Rock Bunting		E. cia	BBRC	A	
Ortolan Bunting		E. hortulana	LBRC	AE	
Cretzschmar's Bunting		E. caesia	BBRC	A	
Cirl Bunting		E. cirlus	LBRC	A	
Chestnut-eared Bunting		E. fucata	BBRC	A	
Little Bunting		E. pusilla	LBRC	A	
Yellow-browed Bunting		E. chrysophrys	BBRC	A	
Rustic Bunting		E. rustica	LBRC	A	
Yellow-breasted Bunting		E. aureola	BBRC	AE	
Chestnut Bunting		E. rutila	BBRC	A	
Black-headed Bunting		E. melanocephala	BBRC	AE	
Black-faced Bunting		E. spodocephala	BBRC	AE	
Pallas's Reed Bunting		E. pallasi	BBRC	A	
Reed Bunting	Common Reed Bunting	E. schoeniclus		A	
Lark Sparrow		Chondestes grammacus	BBRC	A	
Dark-eyed Junco		Junco hyemalis	BBRC	AE	
White-crowned Sparrow		Zonotrichia leucophrys	BBRC	AE	
White-throated Sparrow		Z. albicollis	BBRC	AE	
Savannah Sparrow		Passerculus sandwichensis	BBRC	A	
Song Sparrow		Melospiza melodia	BBRC	AE	
Eastern Towhee		Pipilo erythrophthalmus	BBRC	A	
Bobolink		Dolichonyx oryzivorus	BBRC	A	
Baltimore Oriole		Icterus galbula	BBRC	AE	
Red-winged Blackbird		Agelaius phoeniceus	BBRC	A	
Brown-headed Cowbird		Molothrus ater	BBRC	A	
Ovenbird		Seiurus aurocapilla	BBRC	A	
Northern Waterthrush		Parkesia noveboracensis	BBRC	A	
Golden-winged Warbler		Vermivora chrysoptera	BBRC	A	
Black-and-white Warbler		Mniotilta varia	BBRC	A	
Tennessee Warbler		Leiothlypis peregrina	BBRC	A	
Common Yellowthroat		Geothlypis trichas	BBRC	A	
Hooded Warbler		Setophaga citrina	BBRC	A	
American Redstart		S. ruticilla	BBRC	AE	
Cape May Warbler		S. tigrina	BBRC	A	
Northern Parula		S. americana	BBRC	AE	
Magnolia Warbler		S. magnolia	BBRC	AE	
Bay-breasted Warbler		S. castanea	BBRC	A	
Blackburnian Warbler		S. fusca	BBRC	A	
Yellow Warbler	American Yellow Warbler	S. aestiva	BBRC	A	
Chestnut-sided Warbler		S. pensylvanica	BBRC	A	
Blackpoll Warbler		S. striata	BBRC	AE	
Myrtle Warbler		S. coronata	BBRC	A	
Wilson's Warbler		Cardellina pusilla	BBRC	A	
Summer Tanager		Piranga rubra	BBRC	A	
Scarlet Tanager		P. olivacea	BBRC	A	
Rose-breasted Grosbeak		Pheucticus ludovicianus	BBRC	A	
Indigo Bunting		Passerina cyanea	BBRC	AE	